Physics and Chemistry of Finite Systems: From Clusters to Crystals

NATO ASI Series

Advanced Science Institutes Series

A Series presenting the results of activities sponsored by the NATO Science Committee, which aims at the dissemination of advanced scientific and technological knowledge, with a view to strengthening links between scientific communities.

The Series is published by an international board of publishers in conjunction with the NATO Scientific Affairs Division

A Life Sciences **B Physics**	Plenum Publishing Corporation London and New York
C Mathematical and Physical Sciences **D Behavioural and Social Sciences** **E Applied Sciences**	Kluwer Academic Publishers Dordrecht, Boston and London
F Computer and Systems Sciences **G Ecological Sciences** **H Cell Biology** **I Global Environmental Change**	Springer-Verlag Berlin, Heidelberg, New York, London, Paris and Tokyo

NATO-PCO-DATA BASE

The electronic index to the NATO ASI Series provides full bibliographical references (with keywords and/or abstracts) to more than 30000 contributions from international scientists published in all sections of the NATO ASI Series.
Access to the NATO-PCO-DATA BASE is possible in two ways:

– via online FILE 128 (NATO-PCO-DATA BASE) hosted by ESRIN, Via Galileo Galilei, I-00044 Frascati, Italy.

– via CD-ROM "NATO-PCO-DATA BASE" with user-friendly retrieval software in English, French and German (© Springer Science+Business Media Dordrecht 1989).

The CD-ROM can be ordered through any member of the Board of Publishers or through NATO-PCO, Overijse, Belgium.

Series C: Mathematical and Physical Sciences - Vol. 374

Physics and Chemistry of Finite Systems: From Clusters to Crystals

Volume I

edited by

P. Jena
S. N. Khanna

and

B. K. Rao
Department of Physics,
Virginia Commonwealth University,
Richmond, VA, U.S.A.

Springer-Science+Business Media, B.V.

Proceedings of the NATO Advanced Research Workshop on
Physics and Chemistry of Finite Systems: From Clusters to Crystals
Richmond, VA, U.S.A.
October 8–12, 1991

Library of Congress Cataloging-in-Publication Data

Physics and chemistry of finite systems : from clusters to crystals / edited by P. Jena, S.N. Khanna, and B.K. Rao.
p. cm. -- (NATO ASI series. Series C, Mathematical and physical sciences ; vol. 374)
"Published in cooperation with NATO Scientific Affairs Division."
Includes index.
ISBN 978-94-017-2647-4 ISBN 978-94-017-2645-0 (eBook)
DOI 10.1007/978-94-017-2645-0
1. Atomic structure--Congresses. 2. Electronic structure--Congresses. 3. Crystals--Electrical properties--Congresses. 4. Crystals--Optical properties--Congresses. 5. Cluster analysis--Congresses. I. Jena, P. II. Khanna, S.N. III. Rao, B. K. IV. Series: NATO ASI series. Series C, Mathematical and physical sciences ; no. 374.
QC172.P48 1992
530.4'1--dc20 92-16307

ISBN 978-94-017-2647-4

Originally published by Kluwer Academic Publishers in 1992
Softcover reprint of the hardcover 1st edition 1992

This book contains the proceedings of a NATO Advanced Research Workshop held within the programme of activities of the NATO Special Programme on Chaos, Order and Patterns as part of the activities of the NATO Science Committee.

Other books previously published as a result of the activities of the Special Programme are:

ABRAHAM, N.B., ALBANO, A.M., PASSAMANTE, A. and RAPP, P.E. (Eds.) - *Measures of Complexity and Chaos* (B208) 1990 Plenum Publishing Corporation. ISBN 0-306-43387-7.

BUSSE, F.H. and KRAMER, L. (Eds.) - *Nonlinear Evolution of Spatio-Temporal Structures in Dissipative Continuous Systems* (B225) 1990 Plenum Publishing Corporation. ISBN 0-306-43603-5.

CHARMET, J.C., ROUX, S. and GUYON, E. (Eds.) - *Disorder and Fracture* (B235) 1991 Plenum Publishing Corporation. ISBN 0-306-43688-4.

MARESCHAL, M. (Ed.) - *Microscopic Simulations of Complex Flows* (B236) 1991 Plenum Publishing Corporation. ISBN 0-306-43687-6.

MacDONALD, G.J. and SERTORIO, L. (Eds.) - *Global Climate and Ecosystems Change* (B240) 1991 Plenum Publishing Corporation. ISBN 0-306-43715-5.

CHRISTIANSEN, P.L. and SCOTT, A.C. (Eds.) - *Davydov's Soliton Revisted: Self-Trapping of Vibrational Energy in Protein (B243) 1991 Plenum Publishing Corporation. ISBN 0-306-43734-1*

HOLDEN, A.V., MARKUS, M. and OTHMER, H.G. (Eds.) - *Nonlinear Wave Processes in Excitable Media* (B244) 1991 Plenum Publishing Corporation. ISBN 0-306-43800-3.

LING-LIE CHAU, NAHM, W. (Eds.) - *Differential Geometric Methods in Theoretical Physics: Physics and Geometry* (B245) 1991 Plenum Publishing Corporation. ISBN 0-306-43807-0.

ATMANSPACHER, H. and SCHEINGRABER, H. (Eds.) - *Information Dynamics* (B256) 1991 Plenum Publishing Corporation. ISBN 0-306-43912-3.

BABLOYANTZ, A. (Ed.) - *Self-Organisation, Emerging Properties, and Learning* (B260) 1991 Plenum Publishing Corporation. ISBN 0-306-43930-1.

PELITI, L. (Ed.) - *Biologically Inspired Physics* (B263) 1991 Plenum Publishing Corporation. ISBN 0-306-44000-8.

BISHOP, A.R., POKROVSKY, V.L. and TOGNETTI, V. (Eds.) - *Microscopic Aspect of Nonlinearity in Condensed Matter* (B264) 1992 Plenum Publishing Corporation. ISBN 0-306-44001-6.

JIMENEZ, J. (Ed.) - *The Global Geometry of Turbulence: Impact of Nonlinear Dynamics* (B268) 1992 Plenum Publishing Corporation. ISBN 0-306-44014-8.

MOSEKILDE, E. and MOSEKILDE, L. (Eds.) - *Complexity, Chaos and Biological Evolution* (B270) 1992 Plenum Publishing Corporation. ISBN 0-306-44026-1.

ROY, A.E. (Ed.) - *Predictability, Stability and Chaos in N-Body Dynamical Systems* (B272) 1992 Plenum Publishing Corporation. ISBN 0-306-44034-2.

BEN AMAR, M., PELCE, P. and TABELING, P. (Eds.) - *Growth and Form: Nonlinear Aspects* (B276) 1992 Plenum Publishing Corporation. ISBN 0-306-44046-6.

CONTENTS

Preface xv

ATOMIC STRUCTURE

Clusters as a key to the Understanding of Properties as a Function of Size and Dimensionality 1
J. Jortner

Atomic Structure of Clusters through Chemical Reactions 19
S. J. Riley and E. K. Parks

Structure of Decagonal Quasicrystals 29
A. R. Kortan, R. S. Becker, F. A. Thiel, and H. S. Chen

Atomic Structure of Quasicrystals with and without Phason Defects 39
R. Hu, T. Egami, A. -P. Tsai, A. Inoue, T. Masumoto, J. C. Holtzer and K.Kelton

Structure of Quasi-crystals in terms of Atomic Clusters 49
P. Guyot, M. Audier, and M. de Boissieu

STM Studies of Clusters 61
K. Sattler

Structure and Transformation: Clusters as Models for Condensed Phases 71
L. S. Bartell, T. S. Dibble, J. W. Hovick, and S. Xu

Structure Model of the Al-Cu-Co Decagonal Quasicrystal 77
S. E. Burkov

Infrared Spectroscopy of SF_6 attached to Classical and Quantum Clusters (Finite Size Particles attaining Bulk-like Properties) 85
S. Goyal, D. L. Schutt, and G. Scoles

Morphology and Structure of Ultrafine Gold Particles 93
J. Kumar and A. Gupta

Magic Numbers of Bimetallic Cluster Ions Consisting of Group-III Metals: The Icosahedric 13 Atom Clusters 99
A. Nakajima, T. Sugioka, T. Kishi, and K. Kaya

Significant Structure Theory of Molecular Clusters 105
T. Nguyen, K. Cox, and D. D. Shillady

On the Origin of the Dendritic Shapes in Goethite Particles 111
M. Rösler, H. Hofmeister, and E. Held

Structural Fluctuation of Au_{55} and Au_{147} Clusters 119
S. Sawada and S. Sugano

A Face Centered Icosahedral Approximant Phase with Local Icosahedral Symmetry 125
J. E. Shield, M. J. Kramer, R. W. McCallum, and A. I. Goldman

Defect-FCC Structural Models for Argon Clusters Ar_N, $N \geq 500$ 131
B. W. van de Waal

Beyond the Icosahedral Glass Model, The Interpenetrating Pseudo-Icosahedral Clusters (IPIC) Model 139
J. -L. Verger-Gaugry

STABILITY AND EVOLUTION

The Geometric Shell Structure of Metal Clusters 147
T. P. Martin, U. Näher, H. Göhlich, and T. Lange

Electronic Shell Structure in Icosahedral Metal Clusters 157
J. Mansikka-aho, J. Suhonen, S. Valkealahti, E. Hammarén, and M. Manninen

Evolutionary Patterns of Structure and Stability of Clusters 165
U. Landman, R. N. Barnett, C. L. Cleveland, and G. Rajagopal

Chemistry of Stable Quasi-crystals 177
A. -P. Tsai, Y. Yokoyama, A. Inoue, and T. Masumoto

Structure and Evolution of Nano- and Microcrystalline Grain Aggregates 189
N. Rivier

Defects in Quasicrystals 199
M. Kléman

A Computational Quantum Mechanical Study on Weakly Bonded Clusters of NH_3 with SO_2 and HNC 211
S. Chattopadhyay and P. L. M. Plummer

Structural Instability by Phason Softening in Icosahedral Quasicrystals 217
Y. Ishii

Positive and Negative Helium Cluster Ions Formed from Supercritical Expansions 223
T. Jiang, S. Sun, and J. A. Northby

Electronic Structure and Reactivity of Bimetallic Clusters Containing Sodium Atoms 229
K. Kaya, A. Nakajima, T. Naganuma, and K. Hoshino

Metal Clusters: How Many Atoms are Required for Thermionic Emission ? 235
T. Leisner, K. Athanassenas, O. Echt, D. Kreisle, and E. Recknagel

Structural and Electronic Properties of Neutral and Charged Mg_n Clusters ($n \leq 13$ and n = 57, 69) 241
F. Reuse, M. J. Lopez, S. N. Khanna, V. de Coulon, and J. Buttet

Intrinsic Stability of Quasicrystals under the generation of Frenkel Defects 249
J. Roth

Determination of the Spin State for Carbon Clusters: Focussing Effect by Hexapole Magnetic Field 259
H. Shiromaru, M. Mizumachi, Y. Achiba, N. Yanase, N. Kobayashi, and Y. Kaneko

Cluster Relative-Stability Interchanges 265
Z. Slanina

Formation of Clusters in Charged Seeded Expansions of Rare Gas Mixtures 271
S. Sun and J. A. Northby

DYNAMICS

Phase Changes for Clusters and for Bulk Matter 277
R. S. Berry and H. -P. Cheng

Quantum Molecular Dynamics of Clusters 287
J. Bernholc, J-Y. Yi, Q. -M. Zhang, D. J. Sullivan, C. J. Brabec, S. A. Kajihara, E. B. Anderson, and B. N. Davidson

Random Cluster Model for Icosahedral Phase AlMnSi 299
J. L. Robertson

Size-Dependent Thermodynamic and Electronic Properties of Individual Nanometer-size Supported Gold Clusters 309
M. E. Lin, A. Ramachandra, R. P. Andres, and R. Reifenberger

Inelastic Neutron Scattering from Single Grains of Icosahedral and Cubic Al-Li-Cu.................. 323
A. I. Goldman

Structural and Electronic Properties of Pure and Doped Fullerenes 333
W. Andreoni, F. Gygi, and M. Parrinello

Vibrational and Electronic Contributions to the Specific Heat of $Pd_{561}Phen_{36}O_{200}$ 339
J. Baak, H. B. Brom, L. J. de Jongh, and G. Schmid

Vibrational States of Al Adsorbed on Ge Clusters 345
W. S. Bacsa and J. S. Lannin

From Clusters to Liquids: Diffusion, Stokes-Einstein Behavior, and Solvation in Mixed Molecule-Rare Gas Clusters 351
T. L. Beck, J. R. Walker, and T. L. Marchioro II

Proton Spin Relaxation and Thermal History Effects in Organic Molecular Solids 357
P. Beckmann

A Molecular Dynamics Study of the Solid Phase Transition in Sulphur Hexafluoride Clusters .. 363
F. M. Bénière, B. Rousseau, A. H. Fuchs, M. -F. de Feraudy, and G. Torchet

Observation of Electronic Shells in Large Lithium Clusters 369
C. Brechignac, Ph. Cahuzac, M. de Frutos, J. Ph. Roux, and K. Bowen

Structure and Dynamics of Intermediate Size Aluminium Clusters........ 375
S. Debiaggi and A. Caro

Molecular Dynamics Studies of the Melting and Freezing of CCl_4 Clusters 381
J. Chen and L. S. Bartell

Melting and Solidification of Ultrafine Particles Investigated by Optical and Electron Microscopy Techniques 387
P. Cheyssac, R. Kofman, and A. Aouaj

Monte Carlo Studies of Phase Transitions in Clusters of Tellurium Hexafluoride and Benzene. 393
F. J. Dulles, B. J. Chuko, and L. S. Bartell

Vapor Phase Homogeneous Nucleation of Polar Molecules 399
M. Samy El-Shall, D. Wright, and R. Caldwell

Melting of Nickel Clusters 405
I. L. Garzón and J. Jellinek

Phase Changes in Nickel Clusters from an Embedded-Atom Potential 411
Z. B. Guvenç, J. Jellinek, and A. F. Voter

Structure and Dynamics of Argon Cluster Ions 417
T. Ikegami, T. Nagata, and T. Kondow

Molecular Dynamics Simulation of Infrequent Events: Evaporation from Cold Metallic Clusters 423
Z. A. Insepov and E. M. Karatajev

Computer Simulation of Dynamics of Metal Cluster Impact with Crystal Surface 429
Z. A. Insepov and B. Z. Kabdiev

Space/Time Spectroscopic Analysis of Laser Vaporization Process from Si Target under Gas Environment 435
A. Kasuya and Y. Nishina

Structural Transition from Small Clusters to Bulk-like Structures: *Ab Initio* Molecular Dynamics Study of Li, Be, and B Clusters 441
R. Kawai, M. W. Sung, and J. H. Weare

Dynamics of the Cluster-Surface Interaction: Collisional Detachment of Size Selected Carbon Negative Cluster Ions 447
T. Moriwaki, H. Shiromaru, and Y. Achiba

On the Spectra, Structure, and Dynamics of Aniline-Ar_n Clusters: Simulation Studies Compared with the Results of R2P2CI Spectroscopy 453
P. Parneix, P. Hermine, F. G. Amar, and Ph. Bréchignac

Computer Simulation Study of the Evaporation Mechanisms in Lennard-Jones Clusters 459
C. E. Román and I. L. Garzón

Monte-Carlo Geometry Optimization of Sodium Clusters with a Distance-Dependent Monoelectronic Hamiltonian 465
F. Spiegelmann and R. Poteau

Investigation of the Stable Structures of Ni_x, Fe_x, $Ni_{x-y}P_y$, $Ni_{x-y}B_y$, $Fe_{x-y}P_y$, and $Fe_{x-y}B_y$ ($x = 13$, $y < 13$) Microclusters 471
V. S. Stepanyuk, B. L. Grigorenko, A. A. Katsnelson, and A. Szasz

Computer Simulation of Freezing and Melting in Transition Metal Clusters 479
L. T. Wille and H. Dreyssé

ELECTRONIC STRUCTURE

Superconductivity in M_3C_{60} 485
J. L. Martins

Photoelectron Spectroscopy of Alkali Metal Cluster Anions 493
J. G. Eaton, L. H. Kidder, H. W. Sarkas, K. M. McHugh, and K. H. Bowen

Electronic Mechanism for Stability of Quasicrystals 509
T. Fujiwara

Superconductivity in Granular Films 519
B. Weitzel and H. Micklitz

Mössbauer Studies of Quasicrystals 527
R. A. Dunlap, D. W. Lawther, and V. Srinivas

Small Metal Particles Studied by NMR 537
J. J. van der Klink

Shape and Size Dependence of Energy Level Degeneracies 549
V. Subrahmanyam and M. Barma

Measurements of Lithium Clusters Ionization Potentials 555
Ph. Dugourd, D. Rayane, P. Labastie, B. Vezin, J. Chevaleyre, and M. Broyer

Double Zeta Basis Sets in Carbon Cluster Calculations 561
D. W. Ewing and I. Shavitt

Electronic Structure of Deposited Mass-Selected Pt Clusters on Silver Films 569
P. Fayet, F. Patthey, H. -V. Roy, Th. Detzel, and W. -D. Schneider

Homonuclear Metallic Microclusters: Structure and Energetics 575
C. W. Finley, J. Ferrante, and P. Abel

Ionization Energies of Large Sodium Clusters: Direct Evidence for Atomic Shell Structure 581
H. Göhlich, T. Lange, T. Bergmann, U. Näher, and T. P. Martin

Structural and Electronic Properties of Icosahedral Quasicrystals 587
J. Hafner and M. Krajčí

Hartree-Fock Investigations of Pure Antimony and Alkali-Antimony Clusters 593
F. Hagelberg, S. Neeser, N. Sahoo, T. P. Das, and K. G. Weil

Electronic Structures of Metal Clusters on Surface 599
T. Ikeda and S. Ohnishi

A Study of Quantum Monte Carlo Simulation for the Hydrogen Cluster Ions 605
S. Itoh and S. Ihara

A Molecular Orbital Calculation Method Considering Crystal Fields 611
H. Kumahora, K. Tago, K. Kobayashi, and N. Kurita

Cluster Resonances and Bound States and Crystal Band Spectrum 617
P. Kurasov

Calculation of the Electronic Structure and Magnetic Properties of Iron Microclusters 623
L. I. Kurkina, O. V. Farberovich, V. S. Stepanyuk, A. A. Katsnelson, and A. Szasz

Ionization Potentials of Fe Microclusters 629
L. I. Kurkina, O. V. Farberovich, V. S. Stepanyuk, A. A. Katsnelson, and A. Szasz

Measurement of Ionization Potentials for Al_n, and Fe_n Clusters 633
M. Pellarin, J. L. Vialle, J. Lerme, F. Valadier, B. Baguenard, J. Blanc, and M. Broyer

On the Structures and Stabilities of Small Mixed Clusters 639
K. Raghavachari and M. A. Al-Laham

A Hartree-Fock Study of Small Carbon Cluster Anions 645
A. K. Ray and B. K. Rao

Electronic Structure and Adduct Geometries in the Chemisorption of Nitric Oxide in Small Gas-Phase Cobalt Clusters 651
J. Robles, A. Martinez, A. Vela, and M. P. Iñiguez

Stability, Ionization Potential and Electronic Affinity of Small Silver Clusters Calculated in the LSD Approximation 657
V. Russier, C. Mijoule, and J. Langlet

Stability of Hydrogen Clusters near a Transition Metal Surface 663
L. Stauffer, M. Habar, and H. Dreyssé

Theoretical Investigation of Locations of $^{19}F^{*}$ and Nuclear Quadrupole Interactions in Silicon and Diamond 669
S. S. Nunes, S. B. Sulaiman, N. Sahoo, and T. P. Das

Simulation of Ionic Crystals for Electronic Properties using the Hartree-Fock Cluster Procedure 677
D. W. Mitchell, S. B. Sulaiman, N. Sahoo, P. C. Kelires, and T. P. Das

The Electronic Structure of Icosahedral Boron Clusters 683
A. C. Switendick*

Investigation by the Gutzwiller Approach of the ^3He and ^4He Liquid Clusters 691
R. J. Tarento, P. Joyes, and J. van de Walle

A Preliminary Study of Energy Bands of 2D Periodic Quantum Dot Arrays 697
G. Xu, P. Jiang, and Z. Xu

Compound versus Pure Metal Microclusters: How the Photoabsorption Profiles Relate to the Abundance Spectra 703
C. Yannouleas, P. Jena, and S. N. Khanna

MAGNETISM

Magnetism of Finite Systems: From Clusters to Crystals 709
S. N. Khanna and P. Jena

Magnetism of Free Transition Metal and Rare Earth Clusters 721
J. P. Bucher

Band Structure Effects and Magnetsm in Quasicrystals 733
F. Cyrot-Lackmann, C. Berger, A. Gozlan, T. Klein, and D. Mayou

Magnetism in Small Particles 743
S. Gangopadhyay, L. Yiping, G. C. Hadjipanayis, C. M. Sorensen, K. J. Klabunde, V. Papaefthymiou, and A. Kostikas

Size Dependence of the Electronic and Magnetic Properties of ultra-small Chromium and Vanadium Particles Studied Using Synchrotron Radiation 753
C. Binns, C. Norris, H. S. Derbyshire, and S. C. Bayliss

Magnetic Properties of Free Metal Clusters 759
D. C. Douglass, J. P. Bucher, D. B. Haynes, and L. A. Bloomfield

Onset of Magnetism in bcc-like Fe_n Clusters 767
J. Dorantes-Davila, H. Dreyssé, and G. M. Pastor

Theory for the Magnetization of an Ensemble of Transition Metal Clusters 773
P. J. Jensen, S. Mukherjee, and K. H. Bennemann

Diamagnetic Susceptibility and Conductivity of Columnar Structures 779
M. Kuman

Superparamagnetism in Ising Clusters 785
J. Merikoski, M. Manninen, and J. Timonen

Magnetic Nanoparticles in a Metallo-Organic Polymer: Magnetic Properties and Structural and Chemical Characterization 793
F. Palacio, C. Castro, J. Reyes, G. Sturgeon, F. J. Lazaro, and J. González-Calbet

Magnetism in Small Antiferromagnetic Clusters 799
B. V. Reddy and S. N. Khanna

Size Effects in Orbital Magnetism 807
J. M. van Ruitenbeek and D. A. van Leeuwen

Size and Shape Dependence of Small Spin Lattices Described by an Ising Hamiltonian 813
E. E. Vogel, S. Contreras, J. Cartes, and J. Villegas

PREFACE

Recent innovations in experimental techniques such as molecular and cluster beam epitaxy, supersonic jet expansion, matrix isolation and chemical synthesis are increasingly enabling researchers to produce materials by design and with atomic dimension. These materials constrained by size, shape, and symmetry range from clusters containing as few as two atoms to nanoscale materials consisting of thousands of atoms. They possess unique structural, electronic, magnetic and optical properties that depend strongly on their size and geometry. The availability of these materials raises many fundamental questions as well as technological possibilities.

From the academic viewpoint, the most pertinent question concerns the evolution of the atomic and electronic structure of the system as it grows from micro clusters to crystals. At what stage, for example, does the cluster look as if it is a fragment of the corresponding crystal. How do electrons forming bonds in micro-clusters transform to bands in solids? How do the size dependent properties change from discrete quantum conditions, as in clusters, to boundary constrained bulk conditions, as in nanoscale materials, to bulk conditions insensitive to boundaries? How do the criteria of classification have to be changed as one goes from one size domain to another?

Potential for high technological applications also seem to be endless. Clusters of otherwise non-magnetic materials exhibit magnetic behavior when constrained by size, shape, and dimension. Nanoscale metal particles exhibit non-linear optical properties and increased mechanical strength. Similarly, materials made from nanoscale ceramic particles possess plastic behavior. Recent discovery of C_{60} buckyballs and micro-tubules of graphitic carbon and their ease of production show promise in a variety of industrial applications starting from drug delivery to lubricants and superconductivity. The strong dependence of cluster size on their reaction with reagent gases opens new doors for understanding catalysis. Since the properties of clusters can be drastically modified by changing only a few atoms, one can imagine having a three-dimensional periodic table of elements (cluster size providing the third dimension) from which new materials can be custom made. The field, therefore, represents a true interdisciplinary forum that needs coordinated efforts by physicists, chemists, material scientists, and engineers to tap its full potential in aiding our fundamental knowledge and quest for new materials.

The Physics and Chemistry of finite systems consisting of clusters, nano-structures, and quasicrystals was the subject of a recent international symposium held in Richmond, Virginia during the week of October 8-12, 1991. The purpose of the symposium was to bring together researchers from different disciplines including Physics, Chemistry, Mathematics, materials science, and Engineering to discuss the various challenging problems of both fundamental and technological importance. The symposium consisted of fourteen plenary sessions and two poster sessions. There were fifty two invited papers and 211 contributed poster presentations presented by nearly 300 scientists from 28 countries. The symposium concluded with a panel discussion on outstanding problems and future directions of the field. The proceedings of this symposium constitute this book.

The discussions and presentations included a number of topics:

- -Preparation and characterization
- -Atomic and electronic structure
- -Stability and fragmentation
- -Electronic, optical, magnetic, and thermodynamic properties
- -Molecular dynamics simulation
- -Melting
- -Cluster reactions and catalysis
- -Cluster support interaction
- -Cluster assemblies
- -Cluster materials involving C_{60} fullerenes
- -Technological applications

The symposium was in the planning stage for nearly two years, and we owe our gratitude to a large number of colleagues and institutions who have assisted with the organization of the meeting. We are indebted to Professor Linus Pauling for serving as the honorary chairman of this conference and to members of the International Advisory Board for their help in selecting the topics and speakers. We are grateful to the members of the Local Organizing Committee for spending countless numbers of hours during the planning stage. We wish to thank the conferees for the high quality of their participation in the invited and poster presentations, and exchange of ideas. Our special thanks go to Ms. Barbara Martin for her tireless effort in looking after all the arrangements.

This symposium was made possible by generous grants from the North Atlantic Treaty Organization and Philip Morris, U.S.A. The symposium was also supported in part by grants from the National Science Foundation, Office of Naval Research, Oak Ridge Associated Universities, Army Research Office, and Extrel Corporation. We acknowledge, with gratitude, financial assistance from Virginia Commonwealth University, the symposium host.

P. Jena
S. N. Khanna
B. K. Rao

Richmond, Virginia
December 1992

SUPERCONDUCTIVITY IN M_3C_{60}

José Luís Martins
Department of Chemical Engineering and Materials Science
University of Minnesota, Minneapolis, Minnesota 55455, USA

ABSTRACT. The general properties of C_{60} and M_3C_{60} alkali interstitial fullerides will be briefly reviewed. We will argue that the high superconducting transition temperature observed in K_3C_{60} may be the result of the peculiar *s–p* hybridization of the π electrons of C_{60} due to the curvature of its bonding network.

1. Introduction

The discovery that alkali fullerides with stochiometry M_3C_{60} have superconducting transition temperatures as high as 30 K,[1] the highest outside the high-T_c oxide family and two orders of magnitude higher than in similar graphite intercalated compounds, raised the fundamental question of the mechanism of superconductivity in these compounds. Electron correlation in the conduction band of these compounds is expected to be strong and many of the theory for non-phonon superconductivity in correlated electron systems developed for the high-T_c oxides can be adapted to the alkali fullerides. The traditional electron-phonon coupling mechanism is however the simplest explanation for superconductivity in the alkali fullerides.[2] Carbon–carbon bonds are quite strong and carbon is a light element, resulting in high vibrational frequencies, up to a value of $h\nu = 0.2$ eV.[3] We have pointed out[2] that for an average vibrational frequency of 1100 K superconductivity could be explained in K_3C_{60} with a McMillan electron-phonon coupling parameter[4] of $\lambda = 0.5$. This is one third of the value of that parameter for Pb and slightly larger than for Al, therefore phonon mediated superconductivity in M_3C_{60} is quite plausible.

Here the main properties of C_{60} and M_3C_{60} will be review on the basis of density-functional calculations of its properties.[5–9] Analysis of the electron–phonon matrix elements in C_{60} and graphite show that the *s–p* hybridization of the conduction electron wavefunctions of K_xC_{60} allows for new channels for electron-phonon scattering that are not present in graphite.[5] This strengthens the electron–phonon interaction and explains the higher superconducting transition temperature, T_c, in the fullerides.[2,5]

P. Jena et al. (eds.), Physics and Chemistry of Finite Systems: From Clusters to Crystals, Vol. I, 485–491.

2. Properties of M_3C_{60}

2.1. BAND STRUCTURE

The electronic structure of C_{60} and K_3C_{60} has been discussed previously.[6–9] The agreement between the density of states calculated with the first principles local-density aproximation and the shape of the experimental electron photoemission[6] and inverse photoemission[7] shows that the basic theoretical band structure is correct. Because of the weakness of the intermolecular interactions and the high symmetry of C_{60}, the electronic structure of solid C_{60} (fullerite) consists of several "minibands" associated with the molecular states. These molecular states can be characterized as being of σ or π character, and have wavefunctions with a dominant angular momentum component with respect to the center of the molecule.[8,9] The "almost zero-dimensional" molecular character of the solid is responsible for the sharp features observed in both the photoemission and inverse photoemission spectra.

The most important band in M_3C_{60} is the one derived from the lowest empty states of the C_{60} molecule (LUMO-states). Calculations show complete charge transfer of the K valence electron to C_{60},[5,9] and both the photoemission and inverse photoemission spectra show an half filled LUMO-derived band.[2] Analysis of the wavefunctions show that they are of π type with a nodal surface close to the C_{60} spherical surface of radius 3.5 Å, and that the dominant angular momentum is $l = 5$ (we will call them π_5 orbitals). The symetry of the orbital is t_{1u}, meaning that it transforms under the symetry operations like a vector or an atomic p-orbital. However, while the three p-orbitals of a t_{1u} representation have their maximum probability density directed along three mutualy orthogonal axis, the three t_{1u} orbitals of π_5 type have their maxima concentrated around three mutually orthogonal rings.[9] Considered from the tight binding or LCAO point of view, the π_5 orbitals are derived from atomic orbitals that are almost radially oriented and have $\sim$ 15% s-p hybridization.[9,10] A fair aproximation for the coeficients of these hybrid atomic orbitals in the molecular orbital can be obtained from the icosahedral harmonics of angular momentum $l = 5$ that generate the same t_{1u} representation.[9]

Because of computational constraints, most band structures of C_{60} have been calculated for an hypothetical crystal structure with fcc lattice, T_h^3 symmetry and one molecule per unit cell. The observed X-ray diffraction spectra is consistent with the presence of orientational disorder in such a structure (one can choose between two equivalent T_h molecular orientations). An important question is which features of the band structure survive the disorder. For example the necks on the second sheet of the Fermi Surface display the T_h symmetry[11] and may not survive the disorder. However overall the band width depends mainly on the strength of the intermolecular interactions and should be quite independent of orientational disorder. The calculated local-density band width is 0.47 eV.[9] The average density of states in that band is therefore 0.106 states per eV, C atom, and spin. For comparison this is half the value of the density of states at the Fermi level in aluminum.

2.2. PHONONS

There are four kinds of phonons in K_3C_{60}. For each C_{60} molecule there are 3 acoustic modes, 9 "optical modes" associated with the vibration of the cations with respect to the C_{60} molecule, 3 libration modes associated with rigid rotations of the C_{60} molecule, and 174 modes associated with the internal vibrations of the C_{60} molecule. Using a bulk modulus[9] of 18 GPa to calculate the speed of sound, and the size of the Brillouin zone, we estimate that the acoustical phonon branch extends up to $\sim 40\ cm^{-1}$. The modes associated with the rotation should be expected to be strongly anharmonic and have smaller force constants than the acoustic modes. In the absence of solid data about the force constants, one can only say that frequencies of a few cm^{-1} should be expected. The plasma frequency of the potassium ions in K_3C_{60} is 74 cm^{-1}. Due to screening the frequencies of the K "optical" modes should be smaller than this value. The vibrations of the C_{60} molecular cage have been calculated and measured by several authors.[3] They extend from 273 to 1575 cm^{-1}, and have therefore much higher frequencies than all the other modes. Also it is important to notice at this stage that there are 174 high frequency modes compared with 15 low frequency modes.

2.3. PLASMONS

In K_3C_{60} there are 3 electrons in the conduction band and 243 valence electrons. The corresponding (free-electron) plasma frequencies are 2.5 eV and 22 eV. The C_{60} plasmon has been observed[6,7] at 28 eV and at similar energies in K_3C_{60}, indicating that it is a colective mode of all the C valence electrons. The 2.5 eV is in the energy range of the expected interband optical transitions and therefore any plasmon at those energies would be severely damped.

2.4. ELECTRON CORRELATION

Combining the C_{60} ionization potential,[12] $IP = 7.6$ eV, with the electronic affinity,[13] $EA = 2.65$ eV, and the excitation gap of $E_g \sim 1.8$ eV we obtain a molecular value of the electron-electron repulsion parameter $U_{bare} = IP - EA - E_g \sim 3.3$ eV. In solid C_{60} this quantity will be reduced by the dielectric constant of the medium, and would be even smaller for K_3C_{60} crystals. Any reasonable renormalization of the value of U_{bare} will yield a value of the effective U that is of the order of magnitude as the width of the conduction band (a few tenths of an eV). This means that the system may be close to a metal-insulator transition and that a three-band three-dimensional Hubbard model is a possible model for the correlation effects in the conduction band of K_3C_{60}.

2.5. POSSIBLE MECHANISMS FOR SUPERCONDUCTIVITY

We can broadly classify the mechanisms for superconductivity proposed so far into three types: *i*) Non-phonon mechanisms that rely on the strong electron correlation in a relatively narrow band. *ii*) Electron-phonon coupling to the low frequency vibration modes of the alkali atoms or to the acoustic modes. *iii*) Electron-phonon coupling to the high frequency vibrational modes of the C_{60} molecule. The first type of mechanisms are derived from the theoretical efforts to understand the high-T_c cuprates and will not be reviewed here. In the second type of mechanism the ratio of the critical temperature to typical vibration energy, $k_B T_c / \hbar\omega_{vib}$, is of the order of unity, and therefore requires a very strong electron-phonon interaction. One interesting physical idea in this context, is that while in common metals the perturbations in the effective electron potential due to the displacements of the charged ions are screened by the valence electrons, in K_3C_{60} there are no valence electrons around the K^+ ions and therefore its ionic potential is not well screened. The enhanced coupling between the valence electrons and the vibrations of the K ions would be due to this absence of screening electrons around the ion. The third type of mechanism requires "garden-variety" electron-phonon coupling and it will be discussed in the next section.

3. Electron-Phonon Coupling and Superconductivity in K_3C_{60}

The superconducting transition temperature T_c of a material can be estimated from the (logarithmic) average phonon frequency $\omega_{\log}$, the strength λ of the electron-phonon interaction, and the effective coulomb interaction parameter, μ^*, using the Allen and Dynes[14] version of the McMillan[4] equation,

$$T_c = \frac{\hbar\omega_{\log}}{1.20k_B} \exp\left(\frac{-1.04(1+\lambda)}{\lambda - \mu^* - 0.62\lambda\mu^*}\right), \tag{1}$$

which gives good approximate solutions to the Eliashberg equations, for $\lambda < 1.5$. For most metals[14] it has been found that $\mu^* \approx 0.1$, and Equation 1 can be used to estimate λ when T_c and the vibrational spectrum of a material are known.

The vibrational spectrum of the C_{60} molecule has been calculated by several authors using empirical force constants.[15] The logarithmic average of the calculated vibrational frequencies are in the 830 to 870 cm^{-1} range. Considering that those calculations tend to overestimate the observed vibrational frequencies,[2] and that the solid should contain additional low frequency modes associated with the K ions and the translational and rotational degrees of freedom of C_{60}, we estimate that $\hbar\omega_{\log}/k_B \simeq 1100$ K. Using this value, $\mu^* \simeq 0.1$ and the observed $T_c = 18$ K, equation 1 gives $\lambda = 0.5$ for K_3C_{60}. Using reasonable estimates for the range of $\omega_{\log}$ and μ^* we obtain λ values between 0.45 and 0.8. The electron-phonon interaction is therefore reasonably strong in the potassium fulleride crystals.

Carbon has a strong scattering potential, and it should be an effective electron-phonon scatterer, in particular if we take into account the local field effects in the covalent carbon bonds. It should be no surprise that a metallic form of carbon has a strong electron–phonon coupling parameter. However graphite intercalated compounds have small values of T_c, in the 0.12 to 0.15 K range[16] for KC_8, RbC_8, and CsC_8, corresponding to values of $\lambda \approx 0.23$. Even these values can only be explained by assuming scattering between graphite π states and electron state associated with the intercalated atoms.[17] For LiC_6 where there are no Li states near the Fermi level superconductivity has not been observed down to 0.1 K, and λ would be even smaller.

The microscopic expression for the electron-phonon interaction parameter in a crystal is,[18]

$$\lambda = 2N(E_F)\sum_{\mathbf{q}\nu} \frac{1}{\hbar\omega_{\mathbf{q}\nu}} \ll |\, g(m\mathbf{k}, m'\mathbf{k}', \mathbf{q}\nu)\,|^2 \gg_{FS}, \tag{2}$$

where, $N(E_F)$ is the density of states (per spin) at the Fermi level, $\ll \ldots \gg_{FS}$ represents a double average over the Fermi surface (FS), and

$$g(m\mathbf{k}, m'\mathbf{k}', \mathbf{q}\nu) = -\delta_{\mathbf{q}+\mathbf{k}-\mathbf{k}',\mathbf{G}} \sqrt{\frac{\hbar}{2N_c M \omega_{\mathbf{q}\nu}} \frac{1}{n_{at}}} \sum_{\alpha=1}^{n_{at}} \mathbf{e}_\alpha(\mathbf{q}\nu) \cdot \mathbf{I}_\alpha(m\mathbf{k}, m'\mathbf{k}') \tag{3}$$

is the matrix element for electron scattering from the state in band m with wavevector $\mathbf{k}$ to the state in band m' with wavevector $\mathbf{k}'$, by the phonon mode ν with wavevector $\mathbf{q}$. The phonon frequency is $\omega_{\mathbf{q},\nu}$, and its polarization vector for atom α is $\mathbf{e}_\alpha(\mathbf{q}\nu)$. M is the nuclear mass, N_c is the number of unit cells in the crystal and n_{at} is the number of atoms in the unit cell. The Bloch matrix element for the scattering process is

$$I_{\alpha j}(m\mathbf{k}, m'\mathbf{k}') = N_c n_{at} \int \psi^*_{m'\mathbf{k}'}(\mathbf{r}) \frac{\partial V_{sc}(\mathbf{r})}{\partial R_{0\alpha j}} \psi_{m\mathbf{k}}(\mathbf{r}) d^3 r, \tag{4}$$

where the wavefunctions ψ are normalized to the crystal, $V_{sc}(\mathbf{r})$ is the self-consistent potential, and $R_{0\alpha j}$ is the j-th cartesian coordinate of atom α in the unit cell at the origin. Notice that in C_{60} the maximum phonon energy is ≈ 0.2 eV, which is comparable with the 0.47 eV band width,[9] and the Fermi surface averaging of Equation 2 should be replaced by a more appropriate formula.[19]

In an atomic orbital basis set the integral in Equation 4 can be written as[20]

$$\begin{aligned} I_{\alpha j}(m\mathbf{k}, m'\mathbf{k}') = & \sum_{j=1}^{N_c} e^{-i\mathbf{k}.\mathbf{t}_j} \sum_{j'=1}^{N_c} e^{-i\mathbf{k}'.\mathbf{t}_{j'}} \sum_{\beta=1}^{n_{at}} \sum_{\beta'=1}^{n_{at}} \sum_{L} \sum_{L'} n_{at} \bar{f}_{\beta L}(m\mathbf{k}) f_{\beta' L'}(m'\mathbf{k}') \times \\ & \int \phi^*_{L\beta}(\mathbf{r}-\mathbf{t}_j) \frac{\partial V_{sc}(\mathbf{r})}{\partial R_{0\alpha j}} \phi_{L'\beta'}(\mathbf{r}-\mathbf{t}_{j'}) d^3 r, \end{aligned} \tag{5}$$

where $\phi_{L\beta}(\mathbf{r})$ is the atomic orbital of type L at site β, $\mathbf{t}_j$ is a lattice vector, and $f_{\beta Lm}(m\mathbf{k})$ are the coefficients of the wave function expansion. It is out of question to evaluate accurately Equations 2, 3, and 5 from first principles for a complicated crystal such as K_3C_{60}. The fact that T_c depends exponentially on λ (Eq. 1) makes any calculation using shape aproximations or simplified hamiltonians to be qualitative. We will continue therefore our analysis with qualitative arguments. In Equation 5 only one of the atoms in the whole crystal is displaced, therefore only a few of the intgrals

$$\langle L(\beta) \mid V_j(\alpha) \mid L'(\beta') \rangle = \int \phi^*_{L\beta}(\mathbf{r}) \frac{\partial V_{sc}(\mathbf{r})}{\partial R_{0\alpha j}} \phi_{L'\beta'}(\mathbf{r}) d^3, \tag{6}$$

have non negligeable values. The huge sum over all the pairs of atomic orbitals in the crystal (Eq. 5) turns out to contain only a few important terms. The numerical coefficients in that equation are of the order of unity because $\mid f \mid \sim 1/\sqrt{n_{at}}$. In a tight binding approximation where we keep only one center and two center nearest neighbor integrals, we will have three types of non-zero integrals (Eq. 6). They are single site integrals with $\alpha = \beta = \beta'$, electron hopping integrals with $\alpha = \beta$ and β' a nearest neighbor of α, and the intrasite scattering at $\beta = \beta^prime$ due to the atomic displacement at the nearest neighbor site α. The weights of those matrix elements in the sum are of the order of unity because $\mid f \mid \sim 1/\sqrt{n_{at}}$. Examination of Equations 2 and 3 show that λ is proportional to the density of states per spin and per atom, $N(E_F)/N_c n_{at}$, inversely proportional to the phonon force constants, $M\omega^2_{\mathbf{q}\nu}$, and proportional to an average of $\mid \mathbf{I}_\alpha \mid^2$. (We use the normalization $\sum_\alpha \mid \mathbf{e}_\alpha \mid^2 = 1$.)

In intercalated graphite the electron states near the Fermi energy that are derived from the graphitic planes have π-character and are formed from the atomic p_z orbitals. For these orbitals the contribution from single center scattering vanishes by symetry, and only contributions from two center integrals due to the component of the atomic displacement parallel to the bond contribute to the electron-phonon scattering. Notice that atomic displacements perpendicular to the planes do not contribute to the electron scattering. In C_{60} if we form the three *sp* carbon hybrids that point in the direction of the C—C bonds we are left with a "dangling bond" orbital[10] $0.284\phi_s - 0.256\phi_{p_x} + 0.924\phi_{p_z}$, where x is the direction of the bond between hexagons, and xz is a symmetry plane. Angular decomposition of the local density wavefunctions around each atom confirm that this crude approximation is reasonable,[9] although the real wavefunctions are more complicated. This hybridization allows for the single-site scattering of the wavefunction, which have a large matrix element and two site scattering between the two *p* components.

4. Discussion and Conclusions

According to our arguments, superconductivity in K_3C_{60} and Rb_3C_{60} is associated with the vibrations of the carbon cage. Superconductivity should occur for any dopant that can be introduced at concentrations of a few percent (around three dopant electrons per C_{60} molecule) and which turn the fullerite into a metal. This dopants could be interstitials like alkali atoms, atoms trapped inside the cage, or B and N substitutionals. It is crucial that this dopants give their electrons to the carbon cage and do not form localized impurity states. Experimental studies that correlate structural properties and electronic states with the occurrence of superconductivity will be extremely helpful for further theoretical work on the origins of superconductivity in these systems. When superconductivity exists the isotope effect should be observed for C but should be minimal for the dopants (if charge transfer is complete).

Because of its exponential dependence on λ, the dependence of T_c on dopant is difficult to predict accurately. Narrowing of the conduction band by increasing the C_{60} intermolecular distance will increase $N(E_F)$ and consequently λ and T_c. This is the likely explanation of the reduction of T_c with pressure and is consistent with the dependence of T_c in composition. The highest phonon energy in C_{60} is 0.2 eV. This is comparable to the width of the conduction band. If the Fermi level is near the edge of the band, the highest energy phonons will see a reduced number of states available for scattering with detrimental effects on the superconductivity. If the band gets too narrow, superconductivity may disapear because there are no states available for scattering for the high phonon frequencies. Another danger of narrowing the band is that the system may be driven into an insulating phase, destroying again superconductivity. Since the curvature of the bonding network is smaller in C_{70}, the *sp* hybridization of the π system will be smaller and that will decrease T_c. Finally, since the conduction band of C_{60} is narrow, correlation effects are expected to be important, and the system may be near a metal–insulator transition. It would be very interesting to observe the interplay of both phenomena in the fullerides, particularly to see if the superconducting T_c is enhanced in the proximity of the metal–insulator transition.

AKNOWLEDGEMENT This work would not have been possible without the collaboration with Dr. Norm Troullier and Matthias Schabel on the electronic properties of C_{60}, graphite and their compounds.

REFERENCES

[1] A. F. Hebard *et al.*, Nature (London) **350**, 600 (1991). M. J. Rosseinsky, *et al.* Phys. Rev. Lett. **66**, 2830 (1991).
[2] P. J. Benning, J. L. Martins, J. H. Weaver, L. P. F. Chibante, and R. E. Smalley, Science **252**, 1417 (1991).
[3] C. I. Frum, R. Engleman Jr., H. G. Hedderich, P. F. Bernath, L. D. Lamb, and D. R. Huffmann, Chem. Phys. Lett. **176**, 504 (1991). D. D. Bethune, G. Meijer, W. C. Tang, and H. J. Rosen, Chem. Phys. Lett. **174**, 219 (1990).
[4] W. L. McMillan, Phys. Rev. **167**, 331 (1968).
[5] J. L. Martins, N. Troullier, and M. C. Schabel, University of Minnesota preprint.
[6] J. H. Weaver, J. L. Martins, T. Komeda, Y. Chen, T. R. Ohno, G. H. Kroll, N. Troullier, R. E. Haufler, and R. E. Smalley, Phys. Rev. Lett. **66**, 1741 (1991).
[7] P. J. Benning, D. M. Poirier, N. Troullier, J. L. Martins, J. H. Weaver, R. E. Haufler, L. P. F. Chibante, and R. E. Smalley, Phys. Rev. B - Rapid Commun. **44**, 1962–1965 (1991).
[8] J. L. Martins, N. Troullier, and J. H. Weaver, Chem. Phys. Lett. **180**, 457–460 (1991).
[9] N. Troullier and J. L. Martins, submited.
[10] R. C. Haddon, L. E. Brus, and K. Raghavachari, Chem. Phys. Lett. **125**, 459 (1986).
[11] S. C. Erwin and W. E. Pickett, preprint.
[12] J. A. Zimmerman, J. R. Eyler, S. B. H. Bach, and S. W. McElvany, J. Chem. Phys. **94**, 3556 (1991).
[13] L.-S. Wang, J. Conceicao, C. Jin, and R. E. Smalley, Chem. Phys. Lett. **182**, 5 (1991).
[14] P. B. Allen and R. C. Dynes, Phys. Rev. B **12**, 905 (1975).
[15] E. Brensdal, B. N. Cyvin, J. Brunvoll, and S. J. Cyvin, Spectroscopy Lett. **21**, 313 (1988). D. E. Weeks and W. G. Harter, J. Chem. Phys. **90** 4744 (1989). Z. C. Wu, D. Jelski, T. F. George, Chem. Phys. Lett. **137**, 291 (1987).
[16] N. B. Hannay, T. H. Geballe, B. T. Matthias, K. Andres, P. Schmidt, and D. MacNair, Phys. Rev. Lett. **14**, 225 (1965). I. T. Belash, O. V. Zharikov, and A. V. Palnichenko, Synthetic Metals **34**, 455 (1989).
[17] R. Al-Jishi, Phys. Rev. B **28**, 112 (1983).
[18] G. Grimvall, *The Electron-Phonon Interaction in Metals*, (North-Holland, Amsterdam, 1981). P. B. Allen and B. Mitrović, in *Solid State Physics* Vol. 37, p. 1, Ed. by H. Ehrenreich, F. Seitz and D. Turnbull (Academic Press, New York, 1982).
[19] P. Horsch and H. Rietschel, Z. Physik B **27**, 153 (1977).
[20] K. H. Benneman and J. W. Garland, in *Superconductivity in d- and f- band metals*, Ed. by D. H. Douglass, p. 103 (American Institute of Physics, New York, 1972).

Photoelectron Spectroscopy of Alkali Metal Cluster Anions

J.G. Eaton, L.H. Kidder, H.W. Sarkas, K.M. McHugh, and K.H. Bowen
Department of Chemistry
The Johns Hopkins University
Baltimore, MD 21218 USA

ABSTRACT. We present the photoelectron (photodetacment) spectra of K^-_{2-19} recorded using 2.540 eV photons. These spectra are highly structured, providing a wealth of information about the electronic structure of alkali metal clusters, including adiabatic electron affinities and electronic splittings for each size. Spectral assignments have been made with guidance from theory. For the smaller cluster sizes, detailed assignments of the spectra were aided by *ab initio* quantum chemical calculations. For the larger clusters, assignments were guided by a simple shell model. A substantial correlation is found between the predicted shell model energy levels and the observed patterns and spacings in the spectra.

1. Introduction

Among metal cluster anions, alkali metal cluster anions offer relative simplicity, and their photoelectron spectra exhibit unusual clarity. Alkali atoms are hydrogen-like, each with a s^1 valence electron and a rare gas electron core configuration. Bulk alkali metals are the simplest of metals, being essentially free electron metals. Because of these simplifying characteristics at the extrema of size, alkali metal clusters in the vast size regime between a single atom and the condensed phase offer unique opportunities for better understanding the electronic properties of metal aggregates generally.

Interest in alkali metal clusters has generated a rich literature pertaining not only to theoretical[1-11] and experimental[12-20] studies of neutral and cationic alkali clusters, but also to a lesser extent to theoretical[21-25] and experimental[26,27] work on alkali cluster anions. Negative ion photoelectron studies of metal cluster anions comprised of s^1 atoms have included work on metal cluster anions of the coinage metals by Lineberger[28,29], Meiwes-Broer[30], and Smalley[31] and on small alkali cluster anions by our group[27]. Here, we extend our previous work on alkali cluster anions and present the phototelectron (photodetachment) spectra of potassium cluster anions, K^-_n, with n = 2 -19.

2. Experimental

Negative ion photoelectron spectroscopy is conducted by crossing a mass-selected beam of negative ions with a fixed-frequency photon beam and energy analyzing the resultant photodetached electrons. The main elements of our negative ion photoelectron

P. Jena et al. (eds.), Physics and Chemistry of Finite Systems: From Clusters to Crystals, Vol. I, 493–507.

spectrometer are (a) an ion beam line along which negative ions are formed, transported, and mass-selected by an **E** × **B** Wien filter, (b) an argon ion laser which is operated intracavity in the ion/photon interaction region, and (c) a magnetically shielded, high resolution hemispherical electron energy analyzer, located below the plane of the crossed ion and photon beams. A particularly important attribute of this technique is its ability to size-select cluster anions before photodetachment so that photoelectron spectra of individual cluster anion sizes can be recorded.

To generate beams of potassium cluster anions, we use a heated supersonic expansion ion source. In this device, the stagnation chamber is divided into separately heated oven reservoir and nozzle channel sections. The alkali metal is heated in the oven to a temperature corresponding to several hundred torr equilibrium vapor pressure. The vapor is then coexpanded with several hundred torr of argon into high vacuum through the nozzle, which is maintained ~50 degrees hotter than the oven. Finally, a negatively biased hot filament injects low energy electrons directly into the expanding jet in the presence of axial magnetic fields, producing potassium cluster anions.

3. Results

The photoelectron spectra of K^- and the potassium cluster anions, K^-_{2-19} are presented in Figure 1. These spectra were recorded with 2.540 eV photons. The photoelectron spectra of alkali cluster anions are highly structured, an attribute that derives both from the inherent properties of the alkalis and from the available electron energy resolution (~30 meV) of our spectrometer. The individual peaks in these spectra arise due to photodetachment transitions between the ground electronic state of a given cluster anion and the ground and various energetically-accessible excited electronic states of its corresponding neutral cluster. Thus, one learns two main kinds of information from an analysis of these spectra, ie., adiabatic electron affinity values and electronic structure data, both as a function of cluster size. Adiabatic electron affinities for metal clusters are expected to increase with cluster size, reaching the work function of the bulk metal at infinite cluster size. There are two ways of looking at the electronic structure information in these spectra. In the picture alluded to above, the spacings between peaks in a given spectrum reflect the energy splittings between the various electronic states of the neutral cluster, albeit at the geometry of its corresponding cluster anion. In the other view, which is equivalent to the first picture at the Koopmans' theorem level of approximation, the peaks in a given spectrum are ascribed to electrons being removed or plucked from the occupied energy levels of the cluster anion. In this picture, the peak spacings and patterns directly reflect the upper portion of the occupied energy level structure of the cluster anion. In either picture, however, the observed peak structure in these spectra is a manifestation of how the electronic structure of the alkali metal cluster system is evolving as a function of cluster size. Eventually at some larger cluster size, this electronic structure is expected to organize itself into the essentially structureless valence band of bulk potassium metal.

4. Electron Affinities vs. Cluster Size

Adiabatic electron affinities (EA's) were extracted from the spectra and are presented as a function of cluster size in Figure 2. These results exhibit largely an odd-even alternation of

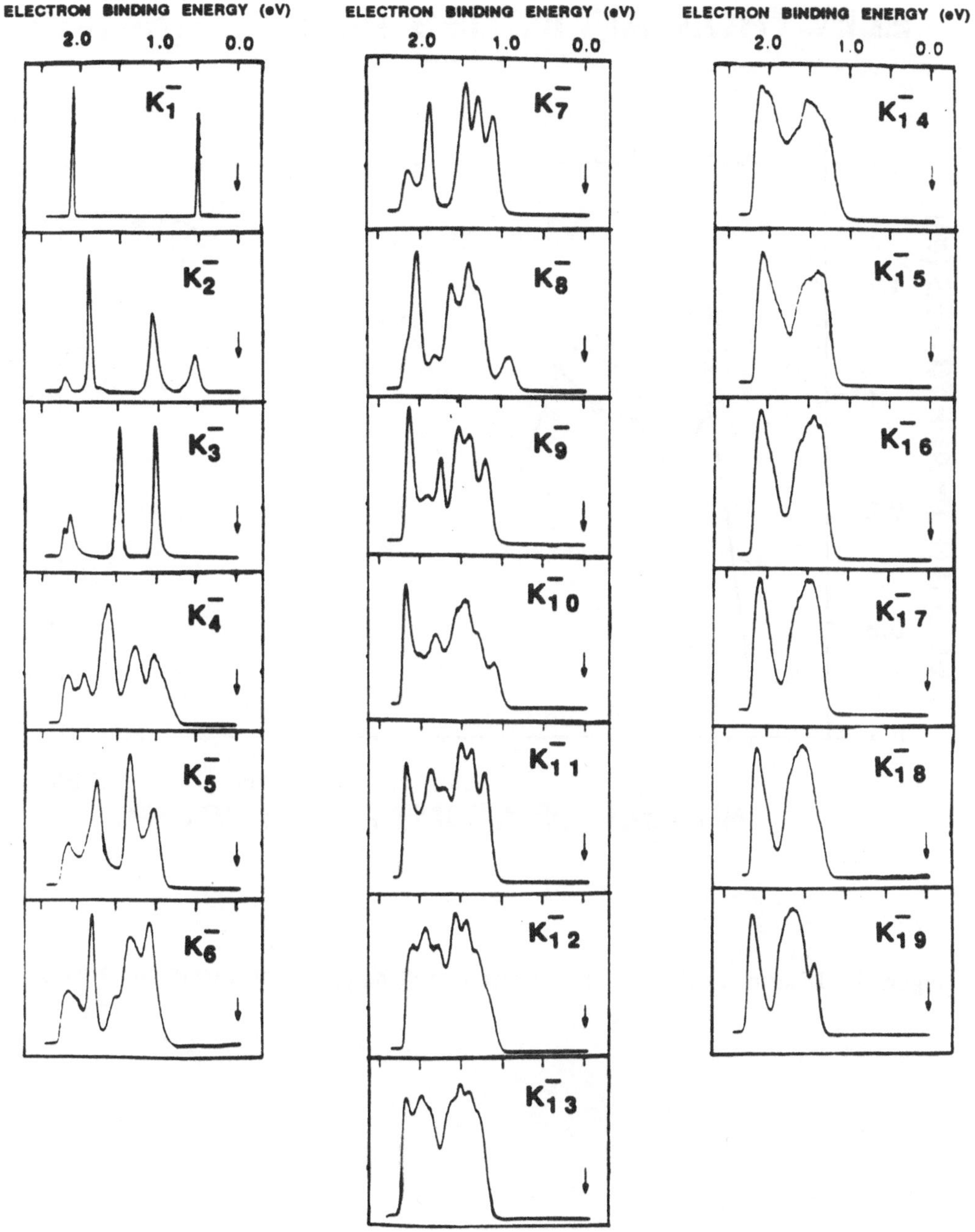

Figure 1. The photoelectron spectra of K^-_{1-19} recorded with 2.540 eV photons.

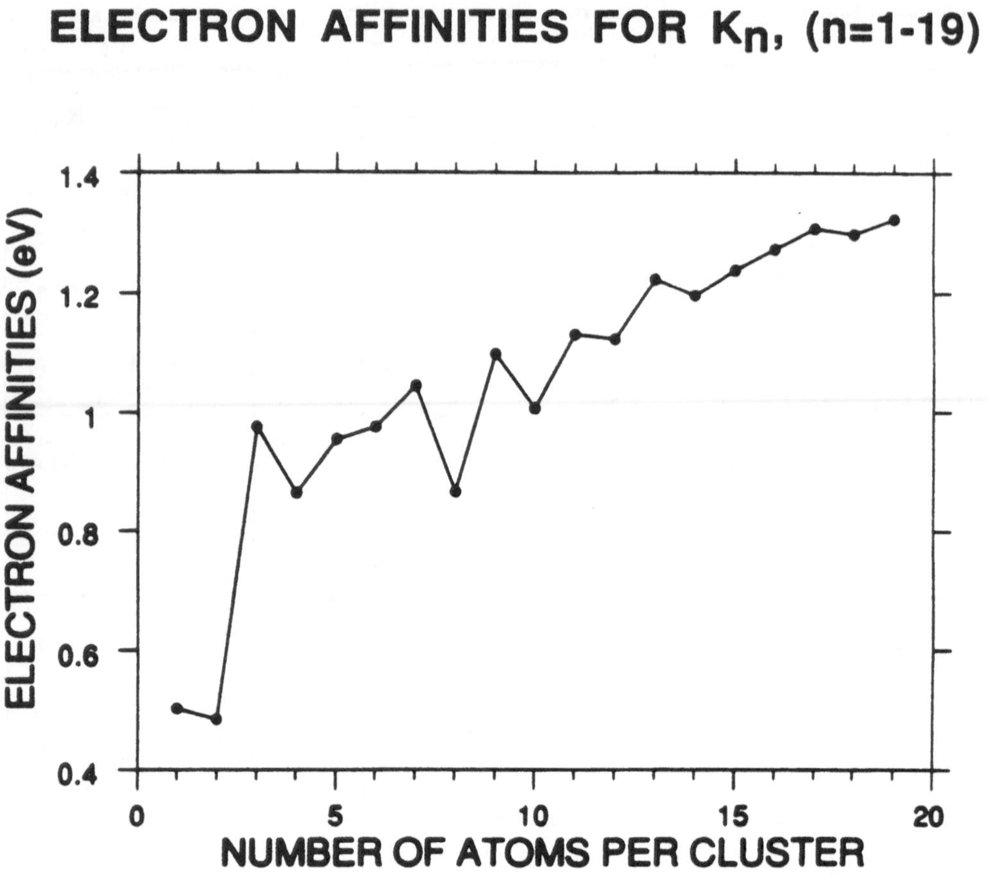

Figure 2. Measured adiabatic electron affinities of $K_{1\text{-}19}$ as a function of cluster size.

electron affinities with cluster size. Using *ab initio* quantum chemistry methods, Bonačić-Koutecký[4] has calculated the adiabatic electron affinities for potassium clusters from sizes n = 2-6. The quantitative agreement of her calculations with our experimental results is rather good. For theoretical guidance at larger potassium cluster sizes, we turned to a qualitative comparison with the predictions of simple shell models. Figure 3 compares our experimentally-determined EA vs. n trend with the qualitative EA vs. n trend predicted by an ellipsoidal shell model. The latter plots the scaled energy of the highest occupied energy level of the cluster containing n + 1 electrons as a function of cluster size, n. Since a large energy for the highest occupied energy level in the anion implies a relatively low electron affinity, we have inverted the energy scale on our experimental EA vs. n plot to facilitate comparison of these two graphs. It is evident that this shell model reproduces much of the EA vs. n trend found experimentally. Note that an ellipsoidal shell model predicts "peaks" in the EA vs. n trend in Fig. 3 at n = 8, 10, 14, and 18, and that we observe these experimentally. Additionally, however, we also see "peaks" at n = 4 and 12 which are not predicted by this shell model. (Note, of course, that "peaks" in the upper panel of Fig. 3 appear as "dips" in Fig. 2.)

5. Electronic Structure and Spectral Assignments

The photoelectron spectra of alkali metal cluster anions are laden with information about electronic structure, yet their interpretation is not obvious by inspection. Thus, our primary goal at this stage has been to assign these spectra in some rudimentary but useful way. In order to do this we need theoretical support. At relatively small alkali cluster anion sizes, the *ab initio* calculations of Bonačić-Koutecký et al. have been an invaluable guide. Two years ago Bonačić-Koutecký[21] modelled our photoelectron spectra of sodium cluster anions over the size range n = 2-5, and very recently she and her group performed similar calculations on potassium cluster anions ranging in size from n = 2-6. This was done by calculating the photodetachment transitions between several possible (energetically close) geometries of a given cluster anion and the ground and energetically-accessible excited electronic states of its corresponding neutral cluster, with the latter being taken in each case at the geometry of the particular isomer of the cluster anion in question. That is, high level quantum mechanical calculations were separately performed on both the cluster anion and its corresponding neutral cluster, followed by the computation of vertical photodetachmment transitions between them. By comparing the photoelectron spectra predicted in this way with a given experimentally observed alkali cluster anion photoelectron spectrum, a wealth of valuable information was obtained about the geometry (or isomeric geometries) of the alkali cluster anion and about the states involved in each of the observed photodetachment transitions. The success of this approach effectively led to an assignment of the spectra in that it identified the transitions which were involved in each observed spectral peak.

As larger cluster sizes are considered beyond the present range of *ab initio* quantum chemistry treatments, however, one must rely on some other form of theoretical guidance. In our earlier work with alkali cluster anions, we had noticed some intriguing correlations between the peak patterns and spacings of our spectra and the predictions of simple shell models. Consider K_7^- for example, an eight valence electron system. Imagine putting two of these electrons into a "s-like" shell and then the remaining six into a "p-like" shell. Further imagine that in such a small system structural effects (or perhaps crystal field

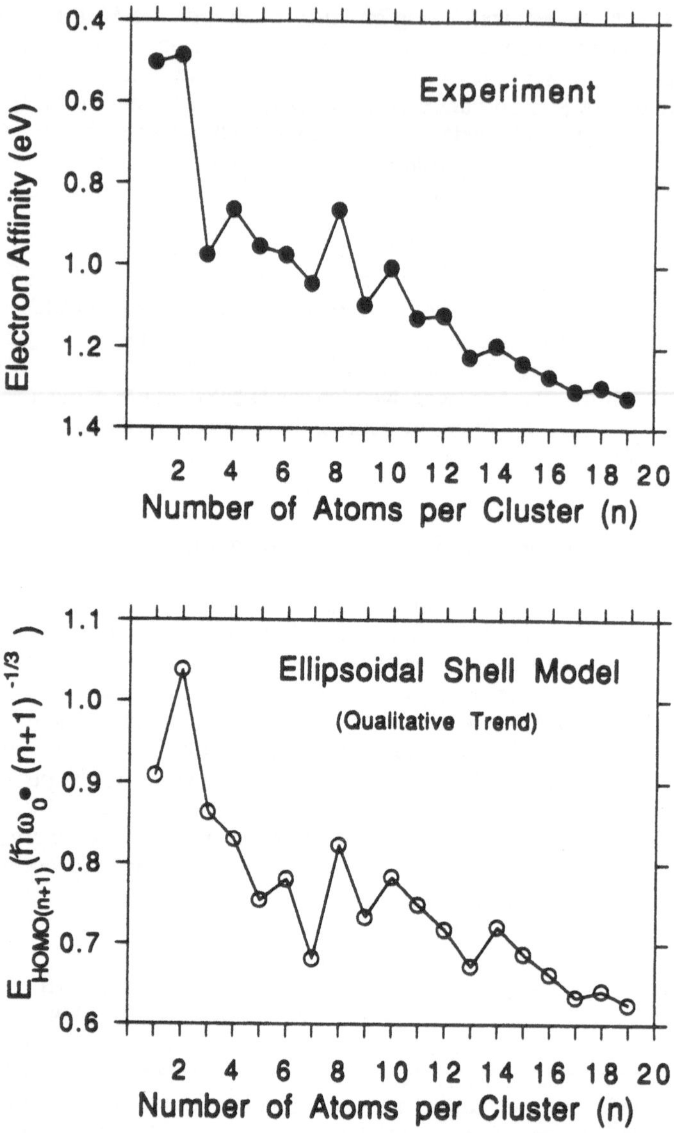

Figure 3. Comparison of the experimentally determined EA vs. n trend for $K_{n=1\text{-}19}$ (top) with the qualitative EA vs. n trend predicted by the ellipsoidal shell model (bottom). In the latter, the ellipsoidal shell model energy of the highest occupied level for the n+1 electron system is plotted as a function of cluster size, n. Note that the energy scale of the top plot has been inverted to facilitate comparison.

effects) lift the degeneracies of the three p sub-levels, leaving two electrons in each of three closely spaced p levels. Next, imagine removing each of these eight electrons from the cluster anion to their common ionization continuum limit, the energy required to do this in each case being its electron binding energy. That predicts a photoelectron spectrum with three closely spaced peaks at low electron binding energy, followed by a gap, followed by another peak at higher electron binding energy. This is what is observed, and furthermore the 1s-1p splitting is quantitatively close to the prediction of the Clemenger-Nilsson[6,8] model for K_8.

With these correlations to encourage us, we set out to see if a simple shell model could work well enough to aid in the assignment of our spectra. To calculate energy levels from which a predicted shell model "stick spectrum" could be generated, we used a program written by Saunders and Knight[20,32]. Essentially, this calculation treats a three dimensional harmonic oscillator having a single perturbation parameter. In addition to its relative simplicity, it has the advantage of easily being able to accommodate triaxial distortions. Only two parameters had to be fixed, one to set the energy scale and one to set the magnitude of the perturbation. These same two parameter values were then used for calculations on all cluster anion sizes. In this model the value of one quanta of energy is equal to the 1s-1p splitting in the closed shell system with eight valence electrons, and we set it in accordance with the value of the tentatively assigned 1s-1p splitting in our spectrum of K_7^-. Also in this model, the value of the perturbation parameter is equal to the 1d-2s splitting in the closed shell system with twenty valence electrons, and we chose it in accordance with the tentatively assigned 1d-2s splitting in the spectrum of K_{19}^-. Basically, this is a triaxial shell model parameterized for potassium cluster anions.

This approach allowed us to generate predictive "stick spectra", with distinctive spectral patterns and quantitative splittings, for first approximation comparisons with the observed photoelectron spectra. Figures 4-7 present these calculated "stick-spectra" (along with their level designations and electron occupations) as insets on the photoelectron spectra of K_n^- over the cluster size range, n = 4-19. The correlation between the shell model's "stick spectrum" and the actual spectrum for n = 4 is rather good. For n = 5, however, the correlation is not as good. The predicted 1p peak actually appears as two peaks in the real spectrum. Conceivably, the two degenerate 1p sub-levels could be split by some interaction not in the model to yield the two observed peaks. If so, the predicted value of the 1s-1p splitting would agree fairly well with that in the photoelectron spectrum. For n = 6, the correlation is somewhat more convincing. Two peaks deriving from 1p levels are predicted, but the lifting of the degeneracy of two of the 1p sub-levels would result in three 1p peaks. This is consistent with the observed spectrum. The case of n = 7 has been previewed above. The observed triplet of peaks is suggestive of the 1p level's three-fold degeneracy having been lifted. The case of n = 8 is rather convincing. The above kinds of arguments about 1p sub-levels and peaks are applicable here, but in addition the peak for a 1d level appears on cue at the predicted energy due to the opening of a new shell. Inspection of the n = 9 case is also fairly convincing if one allows the doubly degenerate 1p level to split into a doublet of peaks in the spectrum. The n = 10 case is consistent with the model but by itself is not convincing. The n = 11 case has the predicted number of peaks in the combined 1p and 1d manifolds. In most cases, the suspected 1s-1p splittings observed in the spectra are in reasonable accord with those predicted by the model. On the other hand, the model often does poorly in predicting the lifting of 1p sub-level degeneracies.

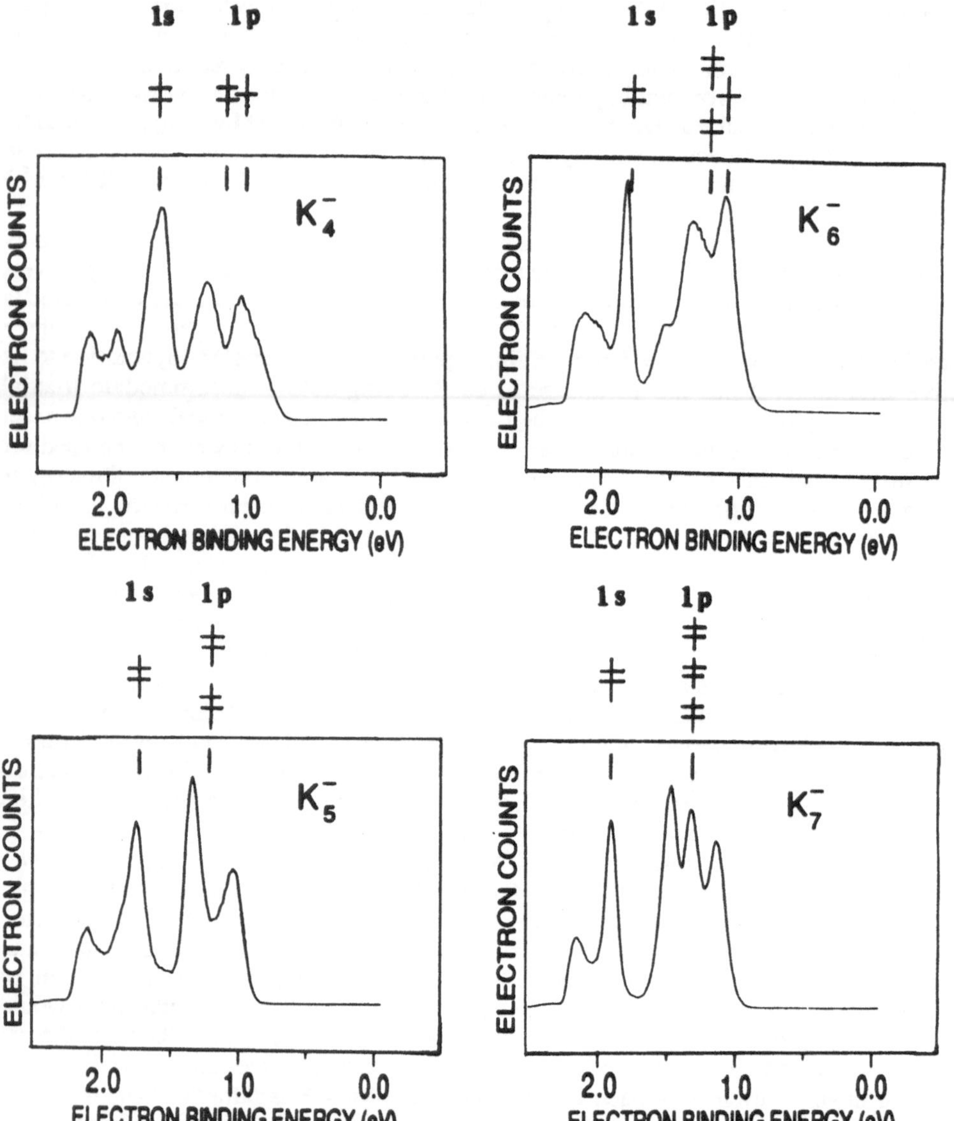

Figure 4. Comparisons of calculated shell model "stick spectra" with the corresponding observed photoelectron spectra of K_n^- (n=4-7). The "stick spectra" are shown as insets on each photoelectron spectrum. Designations and electron occupations for individual levels are shown above each "stick spectrum".

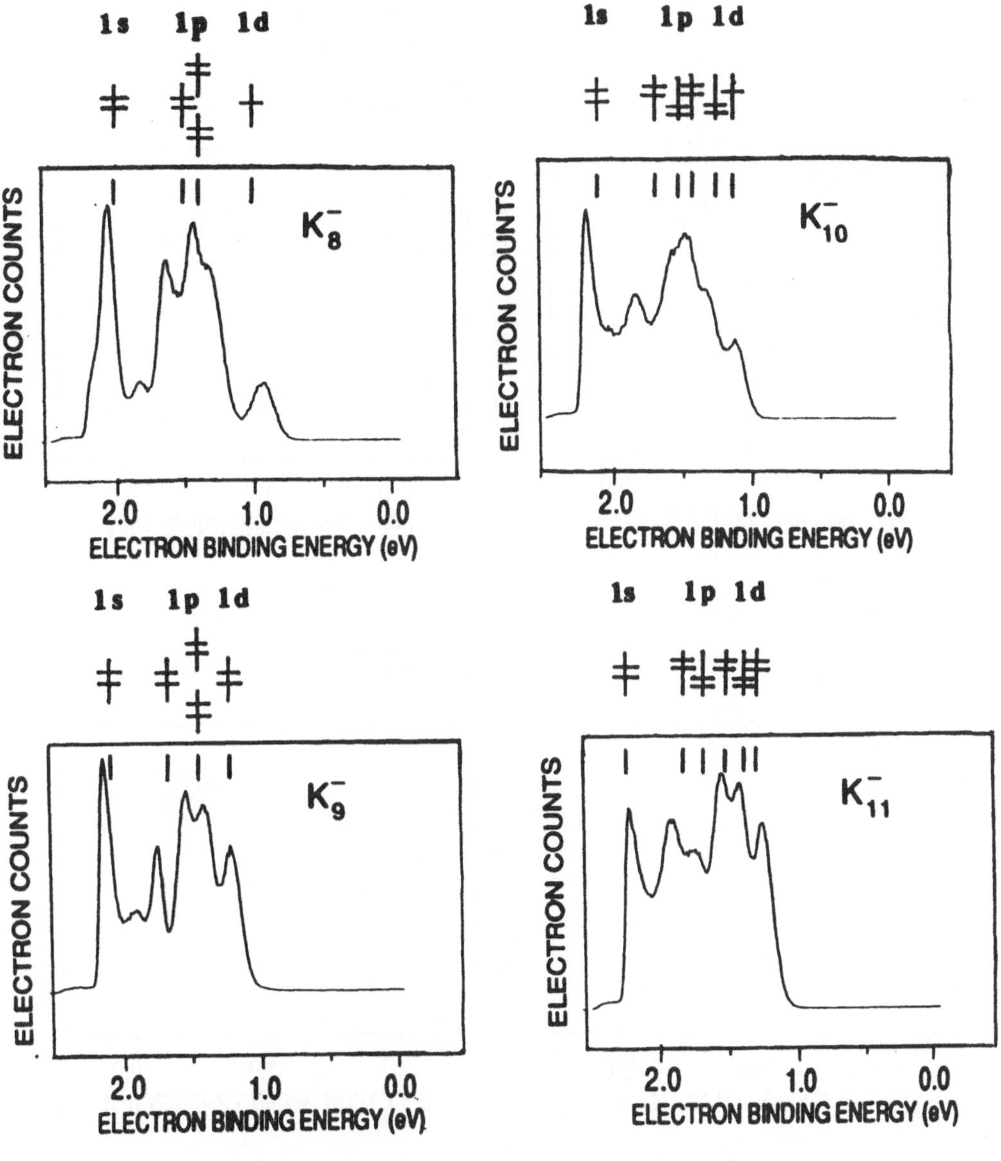

Figure 5. Comparisons of calculated shell model "stick spectra" with the corresponding observed photoelectron spectra of K_n^- (n=8-11). The "stick spectra" are shown as insets on each photoelectron spectrum. Designations and electron occupations for individual levels are shown above each "stick spectrum".

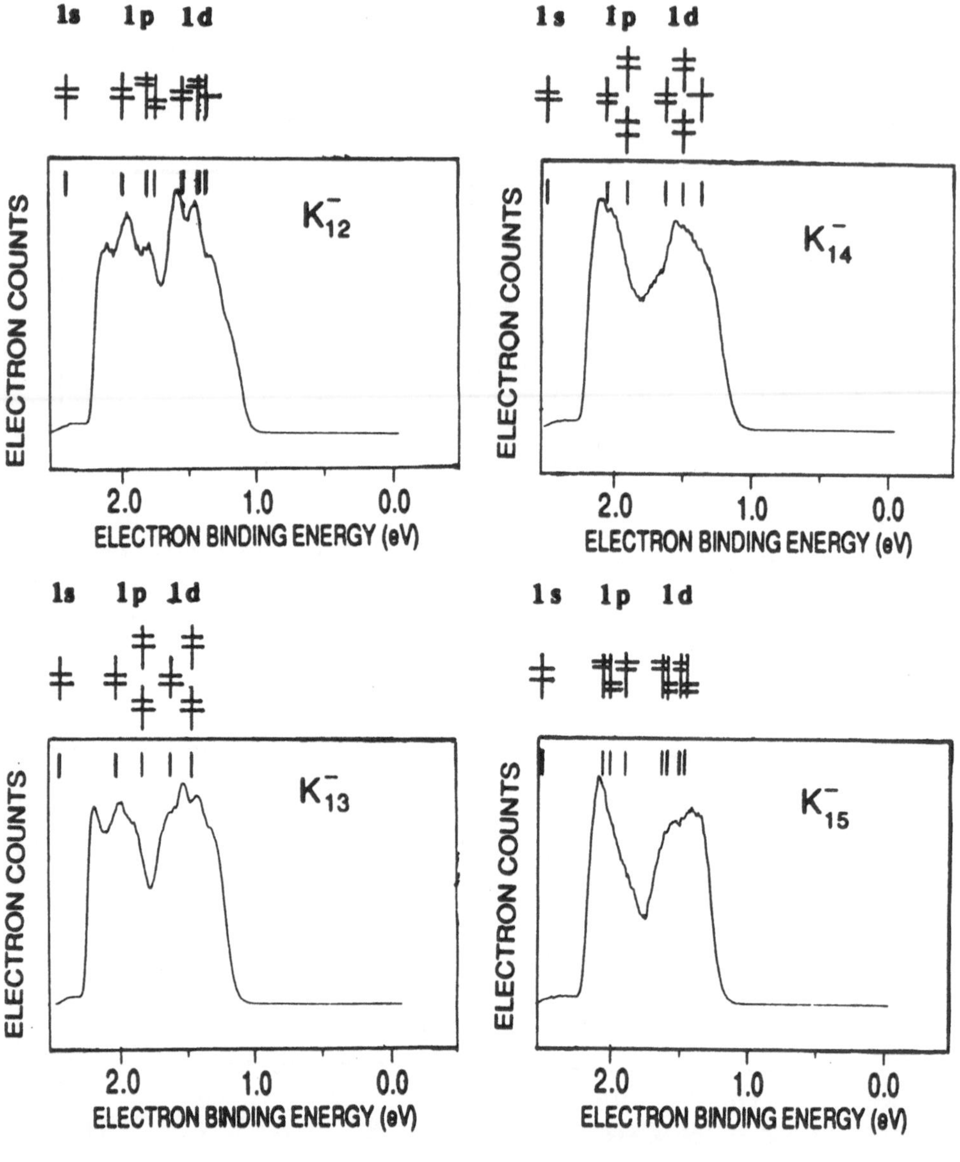

Figure 6. Comparisons of calculated shell model "stick spectra" with the corresponding observed photoelectron spectra of K_n^- (n=12-15). The "stick spectra" are shown as insets on each photoelectron spectrum. Designations and electron occupations for individual levels are shown above each "stick spectrum".

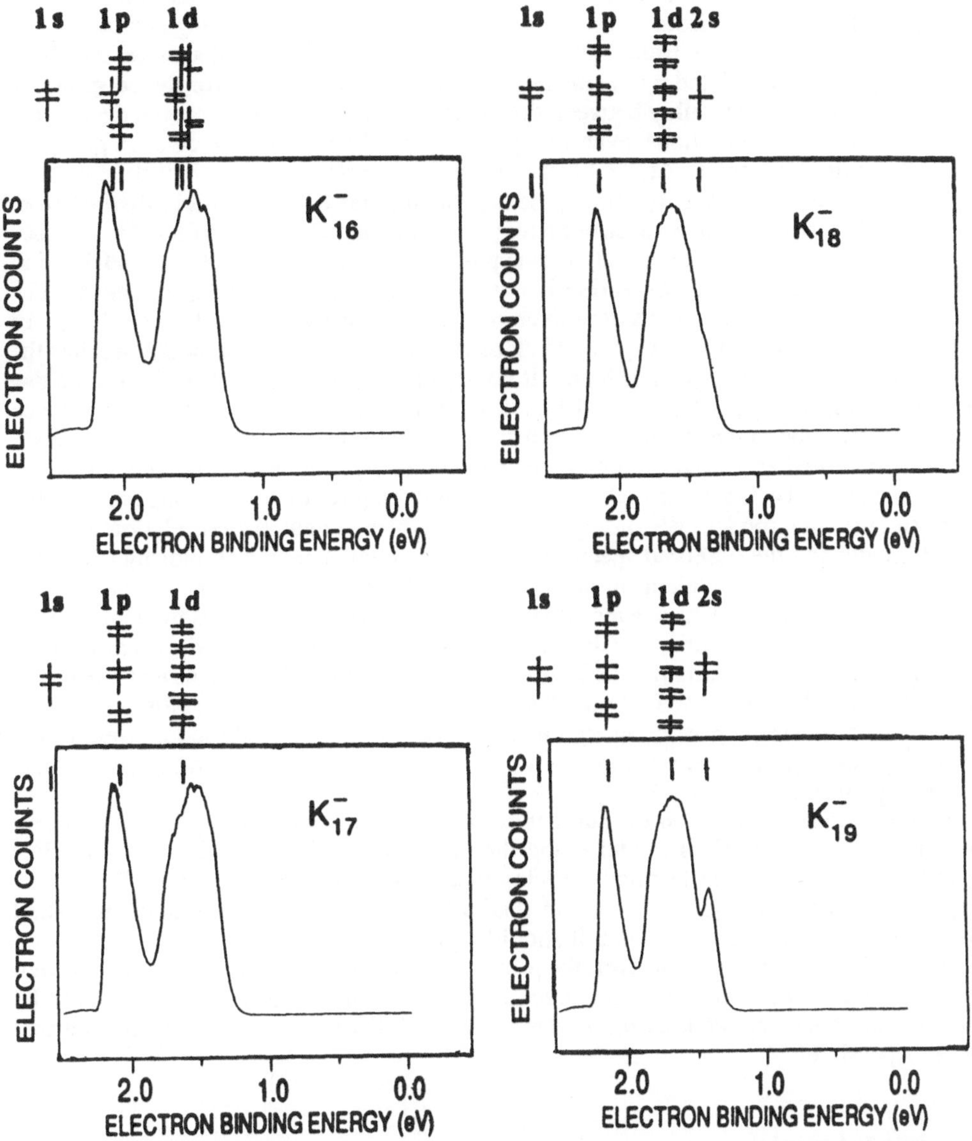

Figure 7. Comparisons of calculated shell model "stick spectra" with the corresponding observed photoelectron spectra of K_n^- (n=16-19). The "stick spectra" are shown as insets on each photoelectron spectrum. Designations and electron occupations for individual levels are shown above each "stick spectrum".

At n = 12, we lose sight of the 1s peak, ie., it shifts out beyond our photon range. Also here, 1p and 1d manifolds start to segregate into separate groupings of peaks, the beginning of an important trend. Within the 1p grouping of peaks there are three discernible peaks as predicted by the model. The same situation holds for the 1d grouping of peaks. By n = 13, the valley between the 1p and 1d groupings has become deeper and the two groupings more obvious, even though the spectral sub-structure on each grouping is more difficult to interpret. At n = 14, the spectral sub-structure on each of the two peak groupings is practically gone, but the 1p peak grouping has become clearly distinct from the 1d grouping, resulting in a spectrum having two broadened peaks. At n = 15, the pace of 1p and 1d segregation quickens. This is apparent in the predictions of the model and in the observed spectrum, and the valley between the two peak groupings deepens. The n = 16 case continues the trend as does the n = 17 case. The model predicts n = 17 (an 18 valence electron system) to be a spherical closed shell. By n = 18, the model predicts the appearance of a singly occupied 2s level. It faintly presents itself as a shoulder on the low electron binding energy side of the 1d shell peak grouping. At n = 19 (a 20 valence electron system), the model again predicts a spherical closed shell system. Now, however, the doubly occupied 2s level pops out dramatically as a new peak in the spectrum. The size range between n = 14 and 19 provides the most convincing set of correlations between the spectra and the shell model. We observe that, all else being equal, the correlation between the shell model and the observed spectra tends to be better (1) near shell closings, ie., near the 8, 18, and 20 valence electron systems, and (2) for larger cluster anions than for smaller ones. Indeed, both trends are what one would expect from a shell model for metal clusters. Viewing the evidence as a whole, we are led to the conclusion that there is something to a first approximation assignment of these photoelectron spectra in terms of a simple shell model. This is the first time that shell models have been successfully put to the test at the level of single particle energy levels. Always before, shell models were tested via consequential properties such as EA vs n trends, IP vs n trends, and mass spectral abundance patterns.

Shell models have been the subject of intense interest in cluster physics in recent years, the most famous shell model being the celebrated jellium model. Shell models generally arise as a result of fermions moving in attractive smooth potentials. Thus, even though the Saunders/Knight shell model that we have used here is not a jellium shell model, it is reasonable to expect that it should reproduce some of the gross features of a jellium calculation. To date, however, the predictions of a true jellium calculation have not been compared to the photoelectron spectra of potassium cluster anions. It would be interesting to do so, and we look forward to the availability of theoretical results that will make this possible.

Acknowledgements

It is a pleasure to acknowledge valuable conversations concerning this work with V. Bonačić-Koutecký, W. Knight, W. Saunders, D. Lindsay, J. Koutecky, W. Ekardt, M. Kappes, W. deHeer, E. Poliakoff, J. Pacheco, C. Yannouleas, B. Judd, L. Madansky, and K.-H. Meiwes-Broer. This work was supported by the (U.S.) National Science Foundation under grant number CHE-9007445.

References

1. Konowalow, D.D., and Rosenkrantz, M.E. (1982) 'Electronic structure and spectra of light alkali diatomic molecules and their molecular cations', in J.L. Gole and W.C. Stwalley (eds.), Metal Bonding and Interactions in High Temperature Systems, American Chemical Society, Washington D.C., pp. 3-17.

2. Cocchini, F., Upton, T.H., and Andreoni, W. (1988) 'Excited states and Jahn-Teller interactions in the sodium trimer', J. Chem. Phys., 88, 6068-6077.

3. Martins, J.L., Car, R., and Buttet, J. (1983) 'Electronic properties of alkali trimers', J. Chem. Phys., 78, 5646-5655.

4. Bonačić-Koutecký V., Fantucci, P., and Koutecký, J. (1991) 'Quantum chemistry of small clusters of elements of groups Ia, Ib, and IIa: Fundamental concepts, predictions, and interpretation of experiments', Chem. Rev., 91, 1035-1108.

5. Rao, B.K., Khanna, S.N., and Jena, P. (1987) 'Evolution of the electronic and structural properties of microclusters', Phys. Rev. B, 36, 953-960.

6. Clemenger, K. (1985) 'Ellipsoidal shell structure in free-electron metal clusters', Phys. Rev. B, 32, 1359-1362.

7. Cohen, M.L., Chou, M.Y., Knight, W.D., and de Heer, W.A. (1987) 'Physics of metal clusters', J. Phys. Chem., 91, 3141-3149.

8. de Heer, W.A., Knight, W.D., Chou, M.Y., and Cohen, M.L. (1987) 'Electronic shell structure and metal clusters', Solid State Physics, 40, 93-181.

9. Röthlisberger, U., and Andreoni, W. (1991) 'Structural and electronic properties of sodium microclusters (n=2-20) at low and high temperatures: New insights from *ab initio* molecular dynamics studies', J. Chem. Phys., 94, 8129-8151.

10. Ekardt, W. (1984) 'Work function of small metal particles: Self-consistent spherical jellium-background model', Phys. Rev. B, 29, 1558-1564.

11. Ekardt, W., and Penzar, Z. (1988) 'Self-consistent Sommerfield droplet as a simple model for an accurate prediction of the electronic properties of small metal particles', Phys. Rev. B, 38, 4273-4276.

12. Ross, A.J., Crozet, P., Effantin, C., d'Incan, J., and Barrow, R.F. (1987) 'Interactions between the A(1) $^1\Sigma_u^+$and b(1) $^3\Pi_u$ states of K_2', J. Phys. B: At. Mol. Phys., 20, 6225-6231.

13. Heinze, J., Schöhle, U., Engelke, F., and Caldwell, C.D. (1987) 'Doppler-free polarization spectroscopy of the $B^1\Pi_u$–$X^1\Sigma_g^+$ band system of K_2', J. Chem. Phys., 87, 45-53.

14. Gole, J.L. (1986) 'The gas-phase characterization of the molecular electronic structure of small metal clusters and cluster oxidation', in M. Moskovits (ed.), Metal Clusters, Wiley, New York, pp. 131-184.

15. Kappes, M.M., Kunz, R.W., and Schumacher, E. (1982) 'Production of large sodium clusters (Na_x $x \leq 65$) by seeded beam expansions', Chem. Phys. Lett., 91, 413-418.

16. Pollack, S., Wang, C.R.C., and Kappes, M.M. (1991) 'On the optical response of Na_{20} and its relation to computational prediction', J. Chem. Phys., 94, 2496-2501.

17. Brechignac, C., Cahuzac, Ph., and Roux, J. Ph. (1987) 'Photoionization of potassium clusters: Neutral and ionic cluster stabilities', J. Chem. Phys., 87, 229-238

18. Lindsay, D.M., Herschbach, D.R., and Kwiram, A.L. (1976) 'E.S.R. spectra of matrix isolated alkali atom clusters: Evidence for chemically bound Na_3', Mol. Phys., 32, 1199-1213.

19. Delacretaz, G., Grant, E.R., Whetten, R.L., Woste, L., and Zwanziger, J.W. (1986) 'Fractional quantization of molecular pseudorotation in Na_3', Phys. Rev. Lett., 56, 2598-2601.

20. Selby, K., Kresin, V., Masui, J., Vollmer, M., de Heer, W.A., Scheidemann, A., and Knight, W.D. (1991) 'Photoabsorbtion spectra of sodium clusters', Phys. Rev. B, 43, 4565-4572.

21. Bonačić-Koutecký V., Fantucci, P., and Koutecký, J. (1989) '*Ab initio* configuration-interaction study of the photoelectron spectra of small sodium cluster anions', J. Chem. Phys., 91, 3794-3795.

22. Rao, B.K., and Jena, P. (1988) 'Electronic localization in anionic and cationic clusters', Int. Jour. of Quant. Chem.: Quant. Chem. Symp., 22, 287-296.

23. Lindsay, D.M., Chu, L., Wang, Y., George, T.F. (1987) 'The Huckel model for small metal clusters. III. Anion structures and HMO electron affinities', J. Chem. Phys., 87, 1685-1689.

24. Ortiz, J.V. (1988) 'Electron binding energies of anionic alkali metal triatomics from partial fourth order electron propagator theory calculations', J. Chem. Phys. 89, 6353-6356.

25. Gole, J.L., Childs, R.H., Dixon, D.A., and Eades, R.A. (1980) 'The electronic structure of the lithium trimer anion and cation', J. Chem. Phys., 72, 6368-6375.

26. Lelyter, M., and Joyes, P. (1974) 'Mise en evidence do la stability c'ions polyatomiques par emission ionique secondaire: Cas de l'ion Al_2^-', J. Phys (Paris), 35, L85-L88.

27. McHugh, K.M., Eaton, J.G., Lee, G.H., Sarkas, H.W., Kidder, L.H., Snodgrass, J.T., Manaa, M.R., and Bowen, K.H. (1989) 'Photoelectron spectra of alkali metal cluster anions: Na_{2-5}^- , K_{2-7}^- , Rb_{2-3}^- , and Cs_{2-3}^-', J. Chem. Phys., 91, 3792-3793.

28. Ho, J., Ervin, K.M., and Lineberger, W.C. (1990) 'Photoelectron spectroscopy of metal cluster anions: Cu_n^- , Ag_n^-, and Au_n^- ', J. Chem. Phys., 93, 6987-7002.

29. Leopold, D.G., Ho, J., and Lineberger, W.C. (1987) 'Photoelectron spectroscopy of mass-selected metal cluster anions. I. Cu_n^-, n=1-10', J. Chem. Phys., 86, 1715-1726.

30. Ganteför, G., Gausa, M., Meiwes-Broer, K.H., and Lutz, H. (1988) 'Ultraviolet photodetachment spectroscopy on jet cooled metal-cluster anions', Farad. Disc. Chem. Soc., 86, 1-12.

31. Pettiette, C.L., Yang, S.H., Craycraft, M.J., Conceicao, J., Laaksonen, R.T., Chesnovsky, O., and Smalley, R.E. (1988) 'Ultraviolet photoelectron spectroscopy of copper clusters', J. Chem. Phys., 88, 5377-5382.

32. Saunders, W.A. (1986) Ph. D. Thesis, University of California, Berkeley.

ELECTRONIC MECHANISM FOR STABILITY OF QUASICRYSTALS

TAKEO FUJIWARA
Department of Applied Physics
University of Tokyo
Bunkyo-ku, Tokyo 113
Japan

ABSTRACT. The electronic structures are calculated for several crystalline approximants of quasicrystals, i-AlMn, i-AlCuLi and d-AlFe. Their densities of states show a large depression called a pseudogap at the Fermi energy and the origin of the pseudogap is attributed to the electron scattering by the lattice structure. The pseudogap enhances a stability of the cohesion, which is known as the Hume-Rothery mechanism. Cohesive energies for several models of AlCuLi are discussed.

1. INTRODUCTION

Since the discovery of quasicrystals in 1984 [1,2], experimental and theoretical studies have been accumulated on their structures. There are two different models for icosahedral quasicrystals, one is the quasiperiodic lattice models [2], e.g. Penrose tilings, and the other is the random tiling model [3] which stands upon an idea that the ground state should be a periodic structure. Even in the quasiperiodic lattice model, actual atomic structures of respective quasicrystals have not been definitely determined.

Recently, there have been discovered many kinds of quasicrystals and several related crystalline materials. It is very common that some crystalline structures are stabilized at the stoichiometry near which the quasicrystalline phase can exist, e.g. crystalline R-phase AlCuLi of quasicrystalline AlCuLi. In these cases, the local atomic environment of a quasicrystalline structure may be very close to that of crystalline one. From this point of view, there are two different types of icosahedral quasicrystals, one is the Mackay icosahedron type and the other is the triacontahedron type.[4] Typical examples of the former Mackay icosahedron type are i-AlMnSi and i-AlCuFe, where a unit of packing is an icosahedral cluster and *glue atoms* tie up these clusters. An example of the latter triacontahedron type is i-AlCuLi where the fundamental unit is a triacontahedral atomic cluster sharing several atoms with each other and the packing does not require any glue atom. From the viewpoint of the crystallography (i.e. according to extinction of some diffraction spots), there are three different types of icosahedral symmetry, i.e. simple, face center and body center. There are, other than the icosahedral symmetry, several different symme-

P. Jena et al. (eds.), Physics and Chemistry of Finite Systems: From Clusters to Crystals, Vol. I, 509–518.

try observed in experiments, decagonal (AlFe, AlMn, AlCoNi etc) and octagonal (CrNiSi, VNiSi).

Some physical properties in crystalline compounds of close stoichiometry are very similar to those in quasicrystalline compounds, e.g. atomic density or electronic specific heats. In some cases such as MgGaZn (triacontahedron type), several different crystalline phases can coexist with the quasicrystalline phase, sharing some growth direction [5], and this reveals a possibility of thermodynamic coexistence. These crystalline compounds are called ***crystalline approximants***, whose lattice can be specified generally by an approximant rational number of the irrational number defining the quasiperiodicity of the quasiperiodic lattice. For example, two-dimensional decagonal and three-dimensional icosahedral lattices are specified by the golden ratio $\tau = (\sqrt{5}+1)/2 = 1.6180\cdots$. The golden ratio can be expressed by a continued fraction as

$$\tau = 1 + \cfrac{1}{1 + \cfrac{1}{1 + \cfrac{1}{1 + \cdots}}}. \tag{1}$$

The optimal approximant of the golden ratio is given as a series of

$$1/1, \quad 2/1, \quad 3/2, \quad 5/3, \quad \cdots, \tag{2}$$

which is a truncation of the infinite continued fraction (1) at certain finite step. The three-dimensional icosahedral quasilattice is defined by a projection of hypercubic lattice points of a six-dimensional space within a strip region onto a three-dimensional incommensurate subspace $\mathbf{E}_{\|}$ [6];

$$\mathbf{E}_{\|} = \sum_{i=1}^{3} t_i \mathbf{r}_i,$$

$$\begin{cases} \mathbf{r}_1 = (1, \tau, 0, -1, 0, \tau) \\ \mathbf{r}_2 = (0, 1, \tau, 0, -\tau, -1) \\ \mathbf{r}_3 = (\tau, 0, 1, \tau, 1, 0), \end{cases} \tag{3}$$

where $-\infty < t_i < \infty$. On the other hand, the crystalline approximants can be defined by a similar projection of six-dimensional hypercubic lattice points to the three-dimensional subspace commensurate to the original six-dimensional hypercubic lattice, which is simply obtained by a replacement

$$\begin{cases} \mathbf{r}_1 \to \mathbf{r}'_1 = (p_1, q_1, 0, -p_1, 0, q_1) \\ \mathbf{r}_2 \to \mathbf{r}'_2 = (0, p_2, q_2, 0, -q_2, -p_2) \\ \mathbf{r}_3 \to \mathbf{r}'_3 = (q_3, 0, p_3, q_3, p_3, 0), \end{cases} \tag{4}$$

where p_i and q_i are rational numbers of the series (2).[7,8] The observed crystalline approximant is $p_i = q_i = 1/1$ ($i = 1 \sim 3$) for AlCuLi and AlMnSi and $p_i = q_i = 3/2$

for AlCuFe. Several different crystalline approximants are observed to coexist in MgGaZn, corresponding to $p_i = q_i = 1$, $p_i = q_i = 2/1$, $p_i = q_i = 3/2$, $p_1 = q_1 = 3/2$ and $p_2 = q_2 = p_3 = q_3 = 2/1$ etc. [5]. In Mg(Ga,Al)Zn, observed crystalline approximant is $p_1 = q_1 = 3/2$, $p_2 = q_2 = p_3 = q_3 = 1/1$ [9].

For several years, it has been believed that quasicrystals cannot avoid intrinsic structural defects called phasons. Phason is defined as a phase mismatch of actual three-dimensional lattice with the ideal three-dimensional quasiperiodic lattice, which is introduced into actual systems for a local stability. A recent discovery of phason free icosahedral quasicrystals is, therefore, very important for analyzing the stability of quasicrystals.[10] The discovery of the irreversible phase transformation between quasicrystalline and crystalline phases in some specific narrow range of stoichiometry of AlCuFe is very important and makes the world more complex.[11] The random tiling model was presumably encouraged by the discovery of the transformation of crystalline and quasicrystalline phases. The ground state at low temperatures, in the random tiling model, is assumed to be a crystalline lattice and the phason-phonon coupling stabilizes the quasicrystalline phase at high temperatures by an entropy term. However, the phase transformation is not common in all stable phason free quasicrystals but very specific for quasicrystals of some specific stoichiometry, e.g. $Al_{63.5}Cu_{24}Fe_{12.5}$.[12]

Recently, physical properties have received much attention, especially on the electronic properties after a discovery of perfect (i.e. phason free) icosahedral quasicrystals. Some difference of X-ray photoelectron spectra between approximant crystals and quasicrystals has been observed, e.g. direct observation of the pseudogap and slightly enhanced depression of the electronic density of states in quasicrystals.[13] The observed temperature dependent coefficient of specific heat is almost the same in approximant crystals and quasicrystals. The electron/atom ratio is almost constant in a range between 2.0 and 2.2 in stable quasicrystals.[14] Most striking difference between approximant crystals and quasicrystals have been reported on the electric transport, absolute values of resistivity and its temperature dependence.[15] High qualified samples of quasicrystals shows anomalous electric resistivity, much higher than those in crystalline and amorphous phases. For example, the electric resistivity in icosahedral AlCuLi ($\sim 870\mu\Omega$cm) is six times larger than in crystals, though the electron density at the Fermi energy is almost the same.[15] The electric resistivity of AlCuRu has been reported to be $30000\mu\Omega$cm.[16] In MgGaZn or AlCuLi, the resistivity changes according to the phase transformation under pressure or thermal treatment and that in quasicrystalline phase shows the highest values.[15] The temperature dependence of the resistivity in normal metals is usually due to scattering by phonons and the resistivity increases with increasing temperatures. The exceptions may be the cases of the s-d scattering and the Anderson localization, but the possibility of s-d scattering can be excluded here. The resistivity in several quasicrystals decreases with increasing temperatures. Magnetoresistance, the Hall coefficient and the thermoelectric power show also some anomaly.[15,16,17] Several recent reports on the electric resistivity tell similar anomalous behavior of resistivity in higher approximant crystals which may be determined on a balance between the lattice periodicity and the mean free path and electrons cannot recognize the difference of quasicrystals and their higher approximants. Another interesting experimental result is the anisotropy of temperature dependence in decagonal AlCuCoSi.[18] The

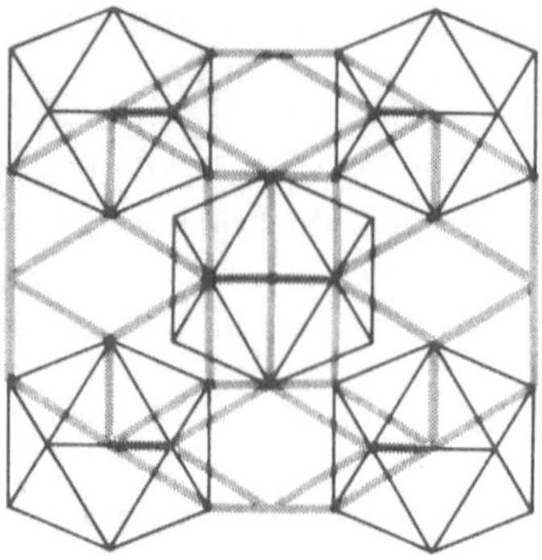

Figure 1. Idealized α-AlMnSi structure consisting of Mackay icosahedra projected onto the (100)-plane. The shaded lines show five rhombic dodecahedra.

absolute value of resistivity is $280\mu\Omega$cm in the quasiperiodic plane (ab-plane) and decreases with increasing temperatures. On the other hand, the absolute value of resistivity along the periodic direction (c-axis) is $65\mu\Omega$cm and increases with increasing temperatures. This strong anisotropy may be hardly explained by the Anderson localization. Fujiwara [19], and Fujiwara and Yokokawa [20] first calculated the electronic structure in several crystalline approximants, e.g. α-AlMn, R-AlCuLi and $Al_{13}Fe_4$. Recently, Hafner and Krajci calculated the density of states in approximants of AlMgZn.[21]

In the present paper, I present our theoretical calculations of the electronic structures in crystalline approximants of icosahedral (AlMn[19], AlCuLi[19,20]) and decagonal (AlFe[20]) quasicrystals and show the calculated cohesive energy. The Linear Muffin-Tin Orbital method, with the Atomic Sphere Approximation (LMTO-ASA), has been used throughout the calculations, because this method is appropriate for densely packed systems of a few hundred atoms in a unit cell. These results are themselves interesting and can be compared with recent experimental results.

2. CRYSTALLINE APPROXIMANT

Quasiperiodic lattice structure defined by the irrational number does not specify atomic positions and several variations of atomic positions are possible. The projection method for quasiperiodic lattices can be generalized to atomic positions in quasicrystals. Some decoration of the quasiperiodic lattice (e.g. Penrose lattice) can be defined by using a projection of lattice points, edge centers and body centers of the six-dimensional hypercubic lattice to the physical space through several occupation domains.[22] Yamamoto determined the atomic positions in lower order crystalline approximants of AlMnSi and AlCuLi, whose lowest order approximant is tested with the observed (1/1) approximants.[23] Another systematic models are proposed by Henley and Elser.[7] In the lowest order (1/1) crystalline approximants, e.g. α-AlMnSi, R-AlCuLi, $(Al,Zn)_{49}Mg_{32}$, the whole system can be uniquely divided into prolate rhombohedra (PR) and rhombic dodecahedra (RD) and all oblate

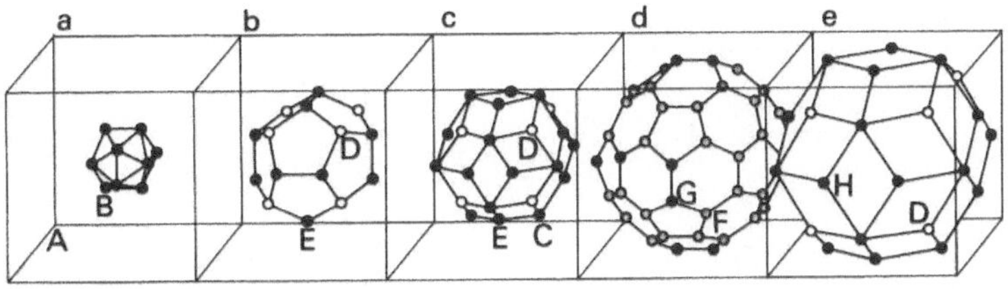

Figure 2. Shell structure of R-AlCuLi.

rhombohedra (OR) are absorbed by the regrouping into RD of two PR and two OR. In the Henley-Elser model of icosahedral quasicrystals, neighboring pairs of OR are grouped with accompanied two PR to form RD. PR and RD are decorated in the same way as the (1/1) approximant. The decoration of remaining OR is in accord with neighboring clusters.

For selfconsistent electronic calculations, a possible size of the unit cell in the models is limited by a computer facility and we use here real crystalline structures, i.e. the lowest (1/1) approximant for AlMn, AlCuLi and AlFe.

2.1. α-AlMn

The unit cell of α(AlMnSi) is shown in Figure 1. Large holes are at (0,0,0) and $(\frac{1}{2},\frac{1}{2},\frac{1}{2})$, which are the vertex positions of a RD connected by the longest body diagonal. An icosahedral cluster consists of 12 Al(α), 12 Mn, 24 Al(β1) and 6 Al(β2) around this hole (MI center), called the 54-atom Mackay icosahedron (MI). The Mn are on the vertices of a MI. The Al(α) is at the midpoint between a hole (MI center) and a Mn and form an inner icosahedron. Two adjacent MIs (in the (111) direction) share a distorted Mn octahedron , connected by glew Al atoms. The Al(β1) are on the edges of a MI at the midpoint between two Mn atoms and are shared with a Mn octahedron. The Al(β2) are each on one of other six edges of a MI at the midpoint between two Mn atoms. All γ-sites (24 sites) and half of the δ-sites (6 sites in one bcc-sublattice) are occupied by Al or Si atoms, which are glue atoms for a MI pair in the (100) direction. The other half of the δ-sites (6 sites in the other bcc-sublattice) are unoccupied. Actual α-AlMnSi is a simple cubic crystal of $Pm3$, containing 138 atoms in a unit cell. A Mn-Mn nearest neighbor contact is not preferable.

2.2. R-AlCuLi

There are eight classes of atomic sites, A~H in R-AlCuLi. Figure 2 shows the R-AlCuLi bcc structure: (a) first an icosahedron of B centered at a vacant A site (0,0,0) or $(\frac{1}{2},\frac{1}{2},\frac{1}{2})$, (b) a pentagonal dodecahedron of 8 D and 12 E above the faces of the icosahedron of B, (c) a smaller triacontahedron (s-TC) of 12 C (on face-centers of (b)) with 8 D and 12 E appearing already in (b), (d) a distorted truncated triacontahedron (t-TC) of 48 F (a hexagonal face of Fs is shared with a nearest neighbor t-TC,) and 12 G (shared with a next nearest t-TC), (e) a larger triacontahedron (l-TC) of 24 H and 8 D (appearing already in neighboring clusters in a (b) shell). The structure

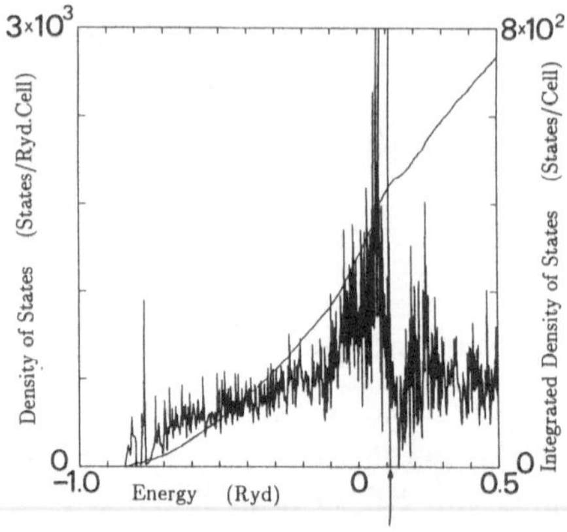

Figure 3: Density of states and integrated density of states in α-AlMnSi. Fermi energy is indicated by the vertical line and arrows.

does not contain any glue atom once it is regarded as a packing of touching and overlapping l-TCs of 136 atoms. Sites A, centers of icosahedra of B, are vacant. Sites D, E and H are occupied by Li atoms, and sites G by Al. Sites B, C and F are occupied randomly by Al and Cu and their populations (Al/Cu) are 88.6/11.4, 53.8/46.2 and 88.6/11.4, respectively. The R-AlCuLi is $Im3$ of 80 atoms in a bcc unit cell.

2.3. $Al_{13}Fe_4$

An approximant crystal $Al_{13}Fe_4$ for AlCoNi is the base centered monoclinic structure $C2/m$ of 51 atoms in a unit cell. The atomic structure of decagonal AlFe may be constructed by locating Fe atoms on the vertex positions of a two-dimensional Penrose lattice when the structure is projected along the c-axis. Two penetrating icosahedral clusters bridge a gap between these layers.

3. ELECTRONIC STRUCTURES

3.1. α−AlMn

Because favorable positions of Si atoms in α-AlMnSi have not been experimentally determined, we use a model of a unit cell of 114 Al and 24 Mn where Al atoms are substituted for all Si atoms. The total density of states is shown in Figure 3. The DOS consists of a set of spiky peaks; every band is highly dispersionless. The deep tail of the DOS is mainly of Al 3s character. The Mn 3d state is split into two at E_F and two peaks are at $E \simeq 0.0 \sim 0.1$Ry and at 0.22Ry. This splitting and opening of a pseudogap at $E \simeq 0.14$Ry causes a deviation of the high DOS peak from E_F and stabilizes the system. With a few Si atoms substituted for Al atoms, the Fermi energy would shifts towards the minimum of the pseudogap and the stability could

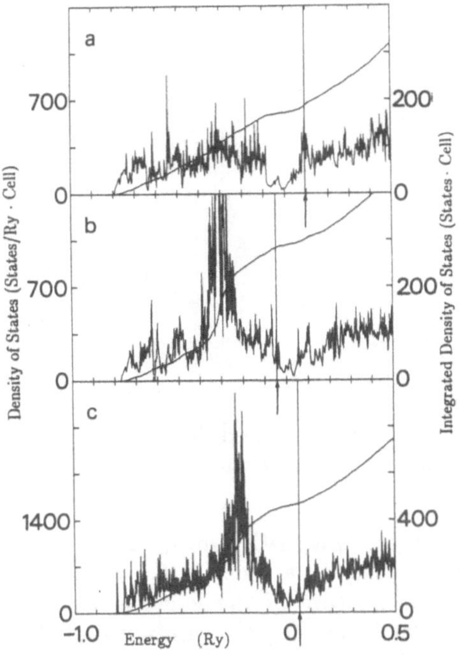

Figure 4: Density of states and integrated density of states in R-AlCuLi. (a) $Al_{54}Li_{26}$ (bcc $Im3$), (b) $Al_{42}Cu_{12}Li_{26}$ (bcc $Im3$) and (c) $Al_{96}Cu_{12}Li_{52}$ (sc $Pm3$). Fermi energy is indicated by the vertical line and arrows.

be enhanced.

The minimum of the pseudogap measured from the bottom of the valence band corresponds to a momentum $k = 1.75 \text{Å}^{-1}$, if we assume a free electron parabolic band. The value $2k$ fairly coincides with a momentum transfer Q of strong diffraction peak, the (211111) spot with $Q = 2.896 \text{Å}^{-1}$ or (221001) with $Q = 3.043 \text{Å}^{-1}$. There can be found several deep s,p-bonding bands of the glue Al atoms.

3.2. R-AlCuLi

We use four different models: $Al_{54}Li_{26}$ of $Im3$ with all C-sites occupied by Al atoms, $Al_{42}Cu_{12}Li_{26}$ of $Im3$ with all C-sites occupied by Cu atoms, $Al_{84}Cu_{24}Li_{52}$ of $Pm3$ symmetry with half of F-sites on one bcc sublattice occupied by Al atoms and the other half on the other bcc sublattice by Cu atoms, $Al_{96}Cu_{12}Li_{52}$ of $Pm3$ symmetry with half of C-sites on one bcc sublattice occupied by Al atoms and the other half on the other bcc sublattice by Cu atoms.

The total DOS and the integrated DOS of three among four are shown in Figure 4. Just as in α-AlMn, the DOS consists of a set of spiky peaks and the deep tail of the DOS originates from Al and Li s-states. The main narrow peak at $E \simeq$

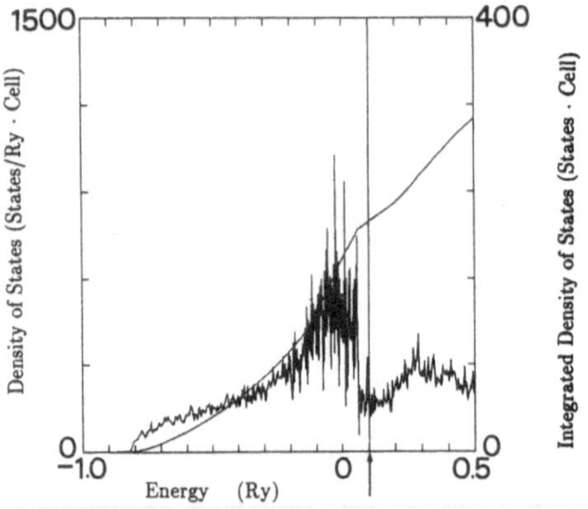

Figure 5: Density of states and integrated density of states in $Al_{13}Fe_4$. Fermi energy is indicated by the vertical line and arrows.

$-0.3 \sim -0.1$Ry is Cu 3d-states. The DOS shows a clear opening of a pseudogap at the Fermi energy. Substituting Al atoms for Cu atoms, (1) the position of the pseudogap measured from the bottom of the valence band does not change at all, but (2) the Fermi energy shifts from the true minimum of the pseudogap according to the Cu concentration, and (3) the Cu d-states do not contribute to the formation of the pseudogap and the main role of Cu atoms is a source of valence electrons. The Fermi energy measured from the bottom of the band corresponds to a momentum $k = 1.64\text{Å}^{-1}$ in $Al_{96}Cu_{12}Li_{52}$. Coincidence is again observed between $2k$ and Q of strong diffraction peaks in an icosahedral AlCuLi, i.e. the (221001) spot with a momentum transfer $Q = 2.793\text{Å}^{-1}$ and the (222100) with $Q = 3.189\text{Å}^{-1}$. The cohesive energies of four systems, i.e. $Al_{54}Li_{26}$ of bcc $Im3$, $Al_{42}Cu_{12}Li_{26}$ of bcc $Im3$, $Al_{84}Cu_{24}Li_{52}$ of sc $Pm3$ and $Al_{96}Cu_{12}Li_{52}$ of sc $Pm3$ are -14.828Ry, -21.327Ry, -19.218Ry and -18.951Ry (per simple cubic unit cell of 160atoms), respectively. The $Al_{42}Cu_{12}Li_{26}$ of $Im3$ is the most stable, and this means a observed trend for the actual R-AlCuLi ($Al_{88.7}Cu_{19.3}Li_{52}$) stabilized by electron donors of Cu atoms.

3.3. $Al_{13}Fe_4$

The total DOS and the integrated DOS of $Al_{13}Fe_4$ are shown in Figure 5. The DOS consists of a set of spiky peaks and the deep tail of the DOS originates from Al s-states. The main peak at $E \simeq -0.1 \sim 0.0$Ry is Fe 3d-states which extend to higher energies. The pseudogap opens at the E_F though the density is rather high. The remaining states in the pseudogap are mainly the Fe 3d-states. The $E - \mathbf{k}$ relation near the E_F region shows a very strong anisotropy. This fact is surely one of reasons of the strong anisotropy of the resistivity in AlCoNi and AlCuCoSi but presumably not all. The position of the pseudogap is at about $k = 1.73\text{Å}^{-1}$ and a reasonable coincidence is found between $2k_F$ and a strong scattering wavevector $Q = 3.017\text{Å}^{-1}$ in $Al_4Mn_{0.72}Fe_{0.28}$.

4. DISCUSSIONS

The clusters, either icosahedra or triacontahedra, are generally stable constituent units of icosahedral quasicrystals and crystalline approximants. No atoms are experimentally observed at the centers of triacontahedra in AlCuLi and the Mackay icosahedra in AlMn, which can be founded by the theoretical calculations. When atoms are located at these centers, we could not obtain a stable solution. Deep bonding states can be observed in all cases, which shows that the stability contribution is partly due to formation of bonding states through touching atoms of neighboring clusters.

The existence of the pseudogap at the Fermi energy is generic in these systems; crystalline approximants of realistic quasicrystals. Considering the existence of the pseudogap even in AlLi and the similarity of the crystalline structures between AlCuLi and AlMn, we believe that the pseudogap here could not be due to the antiresonance of the transition metal d-band in the Al sp-band. In contrast to the case of R-AlCuLi, the pseudogaps are rather narrow in α-AlMn and $Al_{13}Fe_4$ and the transition metal d-states smear out the pseudogap. The values of the density of states at the Fermi energy is twice or three times larger, which may relate to the metastability of these systems.

The relation $Q \simeq 2k_F$ and the pseudogap at the Fermi energy suggest a mechanism of opening a gap attributed to a touching of the Fermi surface at the effective Brillouin zone and enhancement of the cohesive energy. This band mechanism of enhancement of cohesive energy is known as the Hume-Rothery mechanism. This mechanism might work in a quasicrystalline case more efficiently because the effective Brillouin zone should be almost spherical in its polyhedron shape.

One can estimate the density of states at the Fermi energy from the coefficient of the linearly temperature dependent term of the specific heat. The typical example of the similarity is in AlCuLi, 0.13states/eV · atom (i-phase), and 0.12states/eV · atom (R-phase) [24]. They agree well with the present calculated result, 0.12states/eV · atom. The free electron value is estimated to be 0.33states/eV · atom and is three times larger than the above values, which suggests the existence of the strongly enhanced pseudogap of the density of states. The DOS depression or the pseudogap in both approximant and quasicrystalline AlCuLi has been observed by photoemmision spectroscopy experiments.[13] A large difference of resistivity in crystalline and quasicrystalline AlCuLi cannot be explained simply by the difference of the density of states.

In conclusion, we presented the electronic densities of states in several crystalline approximants of quasicrystals, showing the universality of the pseudogap in a quasicrystal at the Fermi energy. The origin of the pseudogap was argued as the Hume-Rothery mechanism and the existence of the pseudogap causes an enhancement of cohesive energies. The stability trend towards the observed Cu contents was demonstrated in several artificial atomic contents of R-AlCuLi. The calculation is now being extended to crystalline approximants with successive larger unit cells, which could give the key character for stabilizing the larger periodicity towards quasiperiodic systems.

References

1 D.Shechtman, I.Blech, D.Gratias and J.W.Cahn, Phys. Rev. Lett. **53**, 1951 (1984).
2 D.Levine and P.J.Steinhardt, Phys. Rev. Lett. **53**, 2477 (1984).
3 C.L.Henley, Comments Cond. Mat. Phys. **13**, 59 (1987).
4 C.L.Henley, in *Quasicrystals*, edited by T.Fujiwara and T.Ogawa (Springer-Verlag, Heidelberg, 1990).
5 F.Spaepen, L.C.Chen, S.Ebalard and W.Ohashi, *Proc. of the Adriatico Research Conference on Quasicrystals*, edited by M.V.Jarić and S.Lundqvist (World Scientific, Singapore 1990).
6 V.Elser, Acta Cryst. **A42**, 36 (1986).
7 V.Elser and C.L.Henley, Phys. Rev. Lett. **55**, 2883 (1985); C.L.Henley and V.Elser, Phil. Mag. **B53**, L59 (1986).
8 H.Tsunetsugu, T.Fujiwara, K.Ueda, and T.Tokihiro, J. Phys. Soc. Jpn. **55**, 1420 (1986).
9 K.Edagawa, K.Suzuki, M.Ichihara, S.Takeuchi, A.Kamiya and U.Mizutani, Phil. Mag. Lett. to be published.
10 A.P.Tsai, A.Inoue and T.Masumoto, J. Mater. Sci. Lett. **7**, 322 (1988).
11 A.I.Goldman, J.E.Shield, C.A.Guryan and P.W.Stephen, *Proc. of the Adriatico Research Conference on Quasicrystals*, edited by M.V.Jarić and S.Lundqvist (World Scientific, Singapore 1990); P.Bancel, Phys. Rev. Lett. **63**, 2741 (1989), *ibid* **64**, 496 (1990).
12 A.Waseda, K.Kimura, H.Ino and K.Edagawa, in *Quasicrystals* (*Proc. China-Japan Seminars*), edited by K.H.Kuo and T.Ninomiya, (World Scientific, Singapore 1990).
13 H.Matsubara, S.Ogawa, T.Kinoshita, K.Kishi, S.Takeuchi, K.Kimura and S.Suga, Jpn. J. Appl. Phys. **30**, L389 (1991).
14 A.Inoue, in *Quasicrystals*, edited by T.Fujiwara and T.Ogawa (Springer-Verlag, Heidelberg, 1990).
15 For a comprehensive review article, see K.Kimura and S.Takeuchi (experiments), and T.Fujiwara and H.Tsunetsugu (theory), to be published in *Quasicrystals: The State of the Art* (World Scientific, Singapore 1991).
16 B.D.Biggs, S.J.Poon and N.D.Munirathnam, Phys. Rev. Lett. **65**, 2700 (1990).
17 S.J.Poon, to be published in Advances in Physics.
18 D.L.Zhang, Y.P.Wang, L.Lu, X.M.Wang, S.C.Cao and S.Y.Lin, in *Quasicrystals* (*Proc. China-Japan Seminars*), edited by K.H.Kuo and T.Ninomiya, (World Scientific, Singapore 1990).
19 T.Fujiwara, Phys. Rev. **B40**, 942 (1989); T.Fujiwara, J. Non-Cryst. Solids, **117/118**, 844 (1990).
20 T.Fujiwara and T.Yokokawa, in *Quasicrystals*, edited by T.Fujiwara and T.Ogawa (Springer-Verlag, Heidelberg, 1990); T.Fujiwara and T.Yokokawa, Phys. Rev. Lett. **66**, 333 (1991).
21 J.Hafner and M.Krajci, private communications.
22 K.N.Ishihara and P.H.Singu, J.Phys.Soc.Jpn. **55**, 1795 (1986).
23 A.Yamamoto, in *Quasicrystals*, edited by T.Fujiwara and T.Ogawa (Springer-Verlag, Heidelberg, 1990).
24 K.Kishi, K.Kimura, T.Hashimoto, and S.Takeuchi, J. Phys. Soc. Jpn. **59**, 1158 (1990).

SUPERCONDUCTIVITY IN GRANULAR FILMS

B. WEITZEL and H. MICKLITZ(*)
Experimentalphysik IV
Ruhr-Universität Bochum
Universitätsstr. 150
D-4630 Bochum, F.R.G.

ABSTRACT. Granular systems built from well-defined metal clusters, which have been produced with the help of a cluster source, give the opportunity to study systematicly the influence of the cluster size on the superconducting transition temperature T_C. Bi- and In-clusters show an enhancement of T_C compared to bulk material. In the case of Bi, which is *not* superconducting in the bulk material, the occurence of superconductivity is explained by an enhancement of the electron density of states $N(\epsilon_F)$ at the cluster surface. Pb-cluster show a decrease in T_C compared to bulk Pb.

1. Introduction

Granular superconductors have been studied since many years [1]. The discovery of high-T_C superconductivity in oxides with its original granular structure renewed the interest in this field. Some special features of granular superconductors are directly connected with their structure: the intergrain junctions result in a transformation of a type-I bulk superconductor to a strong type-II one, with large values of critical field H_{C2}, and in a strong depression of the critical current j_c. We will *not* discuss these problems in this paper. Furthermore, we will *not* talk about granular films *below* the so-called percolation threshold, where we essentially have *isolated* metal clusters. In this regime so-called *quantum-size* effects (QSE) due to a finite splitting of electronic levels may occur. These QSE together with a weak Josephson coupling between the clusters generally will lead to a decrease in the superconducting transition temperatur T_C and finally in a break down of superconductivity for very small, isolated metal clusters [2]. We will concentrate here on a rather unexpected feature, namely, the suprising observation that in most of the granular superconductors, built from metal clusters embedded in a dielectric medium, far *above* the percolation threshold, T_C is enhanced [3]. While non-phononic mechanism have been proposed to explain this effect [4], it was soon realized that surface (or interface) effects very likely are responsible for this T_C-enhancement [5]. If this is discussed within the framework of ordinary BCS-theory, it indicates an increase in the electron-phonon coupling constant λ_S occuring at the metal cluster surface. λ_S can be written as $\lambda_S \propto N(\epsilon_F) < g^2 > [< \omega^2 >]^{-1}$, with $N(\epsilon_F)$ being the electron density of states at the Fermi energie ϵ_F, $< g^2 >$ a Fermi surface momentum average over the electronic matrix elements and $< \omega^2 >$ an average over the phonon frequencies. Thus, T_C-enhancement is obtained either by an increase of $N(\epsilon_F)$ or $< g^2 >$ or by a decrease in $< \omega^2 >$. Strongin et.al. [5] assumed the occurence of soft

P. Jena et al. (eds.), Physics and Chemistry of Finite Systems: From Clusters to Crystals, Vol. I, 519–525.

surface phonons, leading to a decrease in $< \omega^2 >$, to be the dominating reason for the T_C–enhancement observed in granular Al, In, Sn and Zn. The large T_C–enhancement in granular Al ($T_C \simeq 6.6$ K compared to bulk Al with $T_C \simeq 1.16$ K [6]), however, is above the McMillan limit [7], i.e. seems to be too high to be explained by surface phonons alone. Another interesting idea for the T_C–enhancement in inhomogeneous (percolating) superconductors quite general, very recently was brought up by Fontana and Cassi [8]: they argue that the fractal character of a percolation cluster and the connected fractal dynamics may lead to a rescaling of the Debye frequency and thus to an enhancement in T_C. In order to add more reliable experimental data to this still puzzling problem of T_C–enhancement in granular metals, we have decided to study granular systems with a better defined granular structure. This was accomplished by using a cluster source for the *in–beam* preperation of well–defined metal clusters and deposition of these clusters at low temperatures on a substrate [9]. A similar technique recently has been used by Riefer et.al. [10].

2. Experimental set-up

Figure 1 shows a schematic experimental set–up. The metal cluster source, working with the so–called inert gas aggregation technique is described elsewhere [11]. A liquid helium cooled cryopump ($\simeq 10$ K) removes most of the aggregation gas (Ar) from the cluster beam. The cluster source is attached to a rotatable, variable temperature helium cryostate ($2 \lesssim T \lesssim 300$ K) allowing the deposition of the metal cluster beam together with the matrix gas (Kr, Xe, Ge) beam at low temperature on a sapphire substrate which was contacted before by an evaporated gold electrode system and mounted at the cold finger of the helium cryostat.

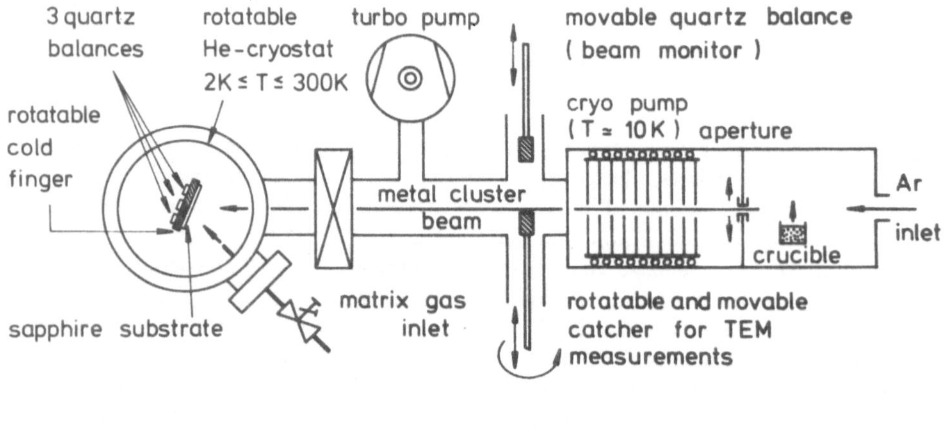

Figure 1. Experimental set–up (schematic).

Since metal and matrix gas beam were *not* parallel, a composition gradient was created along the substrate in order to measure simultaneously samples with different compositions x in one sample preparation process. Deposition rates and beam profiles were measured with the help of three quartz balances mounted on the rotatable cold finger. After adjustment of the individual deposition rates the cold finger was rotated for 180° for film deposition onto the sapphire substrate. Metal cluster size was determined with the help of transmission electron microscopy (TEM). For this purpose a thin ($\simeq$ 100 Å) carbon foil *catcher*, hold at $\simeq$ 300 K, could be brought for a short time ($<$ 0.5 s) into the metal cluster beam (see figure 1). A short cluster deposition time is necessary in order to prevent cluster coagulation on the carbon film which would lead to a changed size distribution. The size distribution of the clusters, collected on this film, was measured after removal of the carbon film at the end of each experiment.

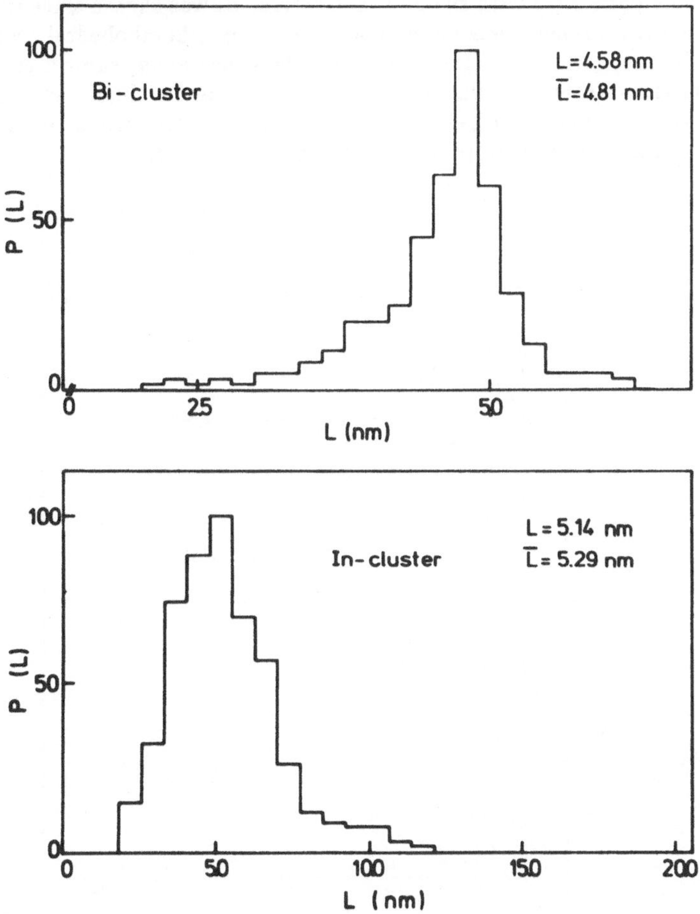

Figure 2. Typical size distribution of Bi–clusters (mean cluster size $\overline{L}$ = 4.81 nm, peak maximum L = 4.58 nm) and In–clusters (mean cluster size $\overline{L}$ = 5.29 nm, peak maximum L = 5.14 nm), determined with the help of TEM.

Figure 2 shows two typical cluster size distributions for Bi– and In–clusters, respectively. The deposition temperature for the granular films was $\simeq$ 40 K, in order to prevent condensation of the cluster aggregation gas (Ar) which could not be totally removed from the metal cluster beam by the liquid helium cooled cryopump (integrated in the cluster source) and the succeeding turbomolecular pump (see figure 1). The typical film thickness was $\simeq$ 150 nm. The two major advantages of this preparation method for granular films are: (i) the cluster size distribution is rather narrow and (ii) mean cluster size $\overline{L}$ and composition x can be varied independently.

3. Experimental results

Experiments were performed with Bi– and In–clusters. Bi was chosen since it is well known that small structural changes make non superconducting, rhombohedral semi–metallic Bi to a superconducting metal [12]. In–clusters, on the other hand, show a structural phase transition from the tetragonal bulk to the fcc structure for In–cluster size $\lesssim$ 5.5 nm [13]. The metal clusters were deposited either together with a matrix gas beam (Kr, Xe, Ge) or in the presence of a partial pressure of $\simeq 10^{-6}$ mbar of O_2 or H_2.

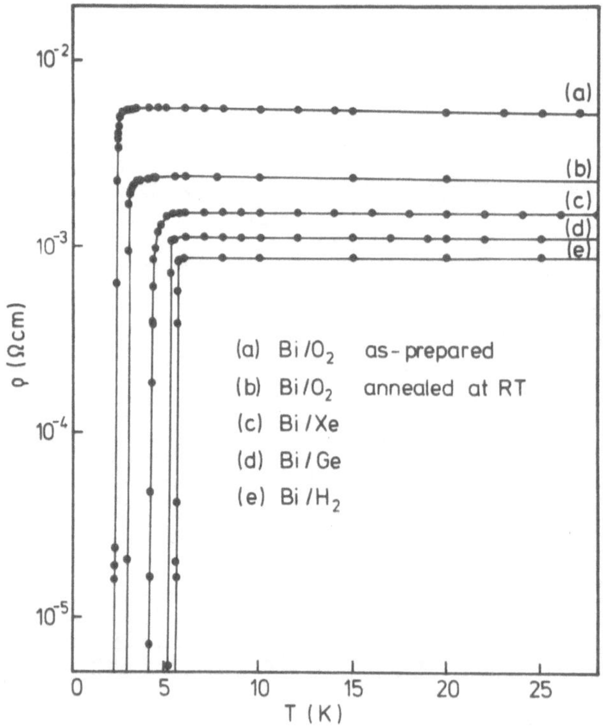

Figure 3. Temperature dependence of the resistivity of granular films built from Bi–clusters (mean size $\overline{L} \simeq$ 3.8 nm) embedded in different matrices (Xe, Ge) or with H_2 (O_2) gas adsorbed on the cluster surface.

Figure 3 shows the resistivity versus temperature curves for various granular Bi–films (mean Bi–cluster size $\overline{L} \simeq 3.8$ nm, metal volume fraction $\gtrsim 0.9$). The occurence of superconductivity in *all* films is clearly to be seen. There seems to be a relationship between the normal–state resistivity ϱ and the superconducting transition temperature T_C: films with the highest ϱ have the lowest T_C and vice versa. x–ray diffraction analysis of the Bi/O_2 films brought to room temperature revealed the rhombohedral bulk structure of the Bi–clusters ($\overline{L} \simeq 3.8$ nm). This finding is in agreement with that of Yokozeki and Stein [13]. The room temperature annealed Bi/O_2–films were still superconducting, with an even slightly increased T_C, when cooled down again to low temperatures (see figure 3). The influence of Bi–cluster size on T_C can be seen in figure 4: here we have plotted the change ΔT_C in T_C between the granular films (cluster size $\overline{L}$) and the bulk material as a function of $\overline{L}$. Bi–clusters show a linear increase in T_C [$\Delta T_C = (T_C)_{cluster} - (T_C)_{bulk} = (T_C)_{cluster}$, since bulk Bi with rhombohedral structure is *not* superconducting]. All experimental facts can be satisfactorly explained [12] by assuming that the Bi–cluster surface layer is superconducting due to a strongly enhanced density of states $N(\epsilon_F)$, while the cluster core is similar to bulk Bi semi–metallic and not superconducting. First experiments with In–clusters in the cluster size range $5 \lesssim \overline{L} \lesssim 8$ nm show an increase in T_C compared to bulk In (see figure 4). However, no difference in T_C between $\overline{L} \simeq 5$ nm and 8 nm was observed as one might expect from the changed crystallographic structure of the clusters (see above). Further experiments definetely are needed to clear this point.

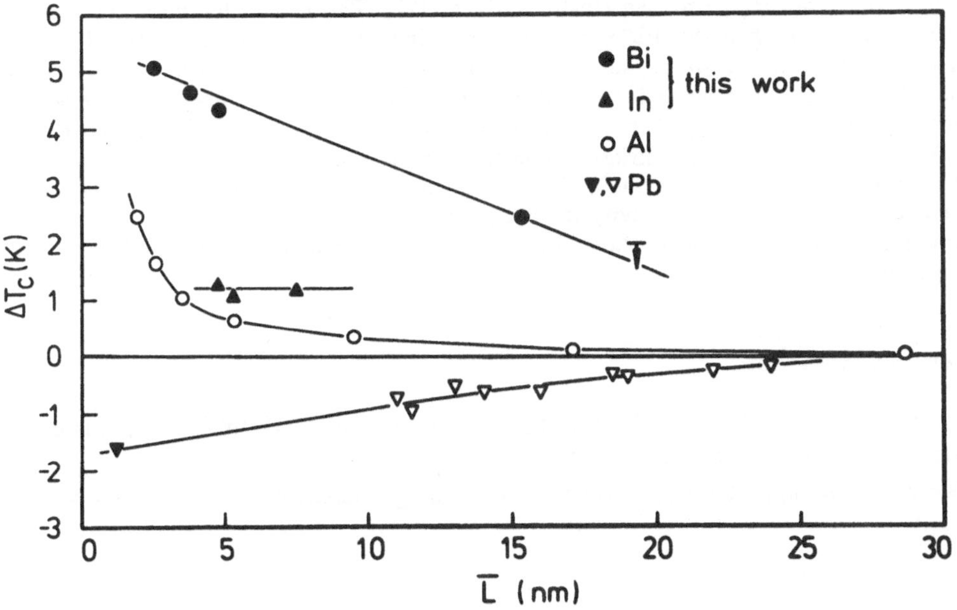

Figure 4. Change ΔT_C in T_C between the granular films (cluster size $\overline{L}$) and the bulk material as a function of $\overline{L}$. The data of granular Al (○) are from ref. [15] and the data of granular Pb [(▼) and (▽)] are from refs. [10] and [14], respectively.

Other experiments on granular films using a metal cluster source for the film preparation have been reported by Riefer et.al. [10]. They have studied granular Pb–films with a Pb–cluster size of $\overline{L} \simeq 1.2$ nm and observed a *decrease* in T_C compared to bulk Pb (see figure 4). This result essentially is in agreement with that of Krah and Baumann [14] who studied granular Pb films with $\overline{L}$ between 10 and 30 nm (see figure 4). However, since in both of these experiments the film resistivity ϱ is *not* given, it is difficult to decide wether part of this decrease in T_C is due to weak Josephson coupling between the clusters (see introduction).

We have also plotted in figure 4 the result on granular Al–films reported by Deutscher et.al. [15]. In these experiments the granular films were obtained by evaporation of Al in the pressure of O_2 on a warm substrate which allows Al–cluster formation due to surface migration. A strong but *not* linear increase of T_C with decreasing cluster size was observed.

4. Conclusion

The superconducting transition temperatures T_C of granular films differ from those of the bulk material. Both an increase as well as a decrease in T_C is observed depending on the system studied. While the increase in T_C with decreasing cluster size in granular Al has been assumed to be due to soft surface phonons [5], we have found convincing evidence that the linear increase in T_C with decreasing $\overline{L}$ in granular Bi is due to a strongly enhanced density of states $N(\epsilon_F)$ at the cluster surface [12]. The reason for both the observed decrease in T_C of granular Pb as well as the $\overline{L}$–independent increase in T_C of granular In is not yet clear. However, the already now existing data (see figure 4) indicate that the different possible reasons for a change in T_C (surface effects of phononic or electronic nature, structural effects, quantum–size effects QSE) are more or less important in the different systems. Surface or interface effects certainly play a major role. The phenomena of surface– or interface–superconductivity is an unsolved problem by itself. An example is the occurence of superconductivity in thin (~ 1 nm) Au–layers on Ge [16]. The study of granular systems with a well–defined metal cluster size seems to be a possibility to attack this problem.

Acknowledgement

Our thanks go to Prof.Dr. E. Hornbogen, Institut für Werkstoffe, Ruhr–Universität Bochum, for the TEM measurements.

This work was supported by the Deutsche Forschungsgemeinschaft.

References

(*) Address now: II. Physikalisches Institut, Universität zu Köln, Zülpicherstr. 77, D–5000 Köln, F.R.G.

[1] For a review see: Perenboom, J.A.A.J., Wyder, P. and Meier, F. (1981) 'Electronic Properties of Small Metallic Particles', Physics Reports **78**, 173 – 292.

[2] Anderson, P.W. (1959) 'Theory of Dirty Superconductors', J. Phys. Chem. Solids **11**, 26 – 30.

[3] For a review see: Abeles, B. (1976) 'Granular Metal Films', Applied Solid State Science **6**, 1 – 117.

[4] Ginzburg, V.L. (1964) 'On Surface Superconductivity', Phys. Lett. **13**, 101 – 102.

[5] Strongin, M., Kammerer, O.F., Crow, J.E., Parks, R.D., Douglass, D.H. and Jensen, M.A. (1968) 'Enhanced Superconductivity in Layerd Metallic Films', Phys. Rev. Lett. **21**, 1320 – 1323.

[6] Fontaine, A. and Meunier, F. (1972) 'Normal and Superconducting Properties of Metallic Grains (Al) Embedded in a Semiconductor (Ge): Experimental Evidence for an Additional Attractive Mechanism between Electrons', Phys. kondens. Materie **14**, 119 – 137.

[7] McMillan, W.L. (1968) 'Transition Temperature of Strong–Coupled Superconductors', Phys. Rev. **167**, 331 – 344.

[8] Fontana, M.P. and Cassi, D. (1989) 'T_C Enhancement in Inhomogeneous Superconductors and Fractal Dynamics', Il Nuovo Cimento **11D**, 783 – 789.

[9] Weitzel, B., Schreyer, A. and Micklitz, H. (1990) 'Metal–Insulator Transition and Superconductivity in Bi/Kr Composites Built from Well–Defined Bi–Clusters', Europhys. Lett. **12**, 123 – 128.

[10] Riefer, G., Höcketstaller, M., Jerger, Th., Nassal, B., Pflaum, R. and Recknagel, E. (1991) 'Superconducting Lead Cluster Films With Particle Size Dependent Reduction of T_C', Z. Phys. D**20**, 337 – 339.

[11] Frank, F., Schulze, W., Tesche, B., Urban, J. and Winter, B. (1985) 'Formation of metal clusters and molecules by means of the gas aggregation technique and characterisation of size distribution', Surf. Sci. **156**, 90 – 99.

[12] Weitzel, B. and Micklitz, H. (1991) 'Superconductivity in Granular Systems Built from Well–Defined Rhombohedral Bi–Clusters: Evidence for Bi–Surface Superconductivity', Phys. Rev. Lett. **66**, 385 – 388.

[13] Yokozeki, A. and Stein, G.D. (1978) 'A metal cluster generator for gas–phase electron diffraction and its application to bismth, lead, and indium: Variation in microcrystal structure with size', J. Appl. Phys. **49**, 2224 – 2232.

[14] Krah, W. and Baumann, F. (1981) 'Resistance and Superconductivity of Small Lead Particles', Surf. Sci. **106**, 373 – 381.

[15] Deutscher, G., Gershenson, M., Grünbaum, E. and Imry, Y. (1973) 'Granular Superconducting Films', J. Vac. Sci. Technol. **10**, 697 – 701.

[16] Dwir, B. and Deutscher, G. (1987) 'Superconductivity at Contact of Ultrathin Gold Films with Amorphous Germanium', in S.A. Wolf and V.Z. Kresin (eds.), Novel Superconductivity, Plenum Press, New York and London, pp. 23 – 28.

MÖSSBAUER STUDIES OF QUASICRYSTALS

R. A. DUNLAP and D. W. LAWTHER
Department of Physics, Dalhousie University
Halifax, Nova Scotia, Canada B3H 3J5

V. SRINIVAS
Low Temperature Physics Division
Tata Institute of Fundamental Research
Bombay, India 400005

ABSTRACT. The application of ^{57}Fe Mössbauer effect spectroscopy to the study of quasicrystals is overviewed. Paramagnetic quasicrystals of the composition Al-TM-M (TM = transition metal, M = metalloid) and Ti-Ni-Fe-Si, which are prepared by rapid solidification, and Al-Cu-Fe, which is prepared by equilibrium processing methods, are considered in terms of a distribution of quadrupole interactions.

1. INTRODUCTION

The first experimental evidence for the existence of aperiodic structures was provided in 1984 when Shechtman et al. [1] observed five fold symmetry in some rapidly solidified Al-Mn alloys. These materials, known as quasicrystals, possess long range rotational order but no translational order. For the past seven years, the manifestation of this novel atomic structure in the physical properties of quasicrystals has been the subject of extensive investigations. Mössbauer spectroscopy is an important experimental technique for the study of structural as well as magnetic and electronic properties of solids. Because of the inherent disorder in quasicrystalline materials and the unique atomic order, the analysis of Mössbauer spectra, either on the basis of multiple discrete probe environments or on the basis of distributed Mössbauer parameters, must be performed with suitable care.

The Mössbauer effect was first applied to quasicrystalline materials by Swartzendruber et al. [2] in 1985. These measurements were made on multiphase samples and resulting spectra were fit to a simple quasicrystalline structural model. Other, early Mössbauer studies [3-6] of icosahedral alloys had to contend with multiphase samples and a relatively poor understanding of the details of the atomic structure of quasiperiodic materials. Subsequent Mössbauer effect studies have benefitted from a growing knowledge of methods of preparing single phase quasicrystals [e.g. 7] and progress in understanding the microsctructure of these materials [e.g. 8] and their relationship with analogous crystalline phases [9, 10]. As well, the application of external magnetic fields have helped to elucidate the details of Mössbauer results [e.g. 11-12].

The literature on Mössbauer studies of quasicrystalline materials is extensive and the present treatment is not intended to be comprehensive. Rather, the present overview will concentrate on studies of paramagnetic quasicrystals which exhibit icosahedral symmetry.

P. Jena et al. (eds.), Physics and Chemistry of Finite Systems: From Clusters to Crystals, Vol. I, 527–536.

2. METASTABLE Al-BASED ALLOYS

2.1. Zero field studies

Virtually all Al-based metastable quasicrystals show a well resolved quadrupole doublet. Even in well-ordered, single phase quasicrystals with minimal strains, the component peaks of the doublet are measurably broader than the intrinsic source linewidth. This indicates the presence of Mössbauer probe atoms in more than a single well defined environment. Although the details of these environments and whether there is a continuous distribution of sites, or whether there are discrete classes of sites, have been considered in substantial detail, definitive conclusions may not be possible. In general, the analysis of broadened Mössbauer spectra presents problems. In the case of quasicrystals, this difficulty is compounded by some authors' attempts to draw general conclusions concerning Mössbauer results in quasicrystals of different compositions, or even different samples of the same composition. As the Mössbauer effect is a very sensitive method for studying defect structure, and the defect structure of many quasicrystals is a sensitive function of composition and processing conditions (see, e.g. Section 3), experimental results should not necessarily be construed as representative of more than a specific sample. Nonetheless, it has been found that many metastable quasicrystalline Al-based alloys show common or related Mössbauer features, and these are informative in formulating an overall picture of the structural and electronic properties of these materials. Space permits only a small number of these studies to be considered here.

Perhaps the simplest analyses of quasicrystalline Mössbauer spectra have utilized a combination of two symmetric doublets; these either have components with similar isomer shifts [2] or similar quadrupole splittings [13]. Early studies tended to require relative intensity ratios of the two components to be the golden ratio [2], although more recent Mössbauer studies have not put constraints on the relative line intensities.

The first analysis of a quasicrystalline Mössbauer spectrum on the basis of a distribution of quadrupole interactions was presented by Eibschutz et al. [5]. In this work, a shell model [14] for the quadrupole distribution, $p(\Delta)$, was modified by the introduction of the parameter n;

$$p(\Delta) \propto (\Delta/\sigma)^n \exp(-\Delta^2/(2\sigma^2)) \quad (1)$$

where σ is a fitted parameter. Other authors have used an expansion of $p(\Delta)$ in discrete components [15-17] which places fewer constraints on the form of $p(\Delta)$ and should allow for a more obective analysis of the shape of this function. In cases where a distribution of quadrupole splittings have been used to fit the spectrum, a correlation between the isomer shift, δ, and quadrupole interaction Δ of the form

$$\delta(\Delta) = \delta_0 + \alpha\Delta + \beta\Delta^2 + ... \quad (2)$$

has been included to account for any spectral asymmetry. Some other methods of expanding $p(\Delta)$ have been implemented; these include the Fourier expansion method of Window [18] and a folded Gaussian [19].

A typical example of an ^{57}Fe Mössbauer spectrum in an Al-based quasicrystal is illustrated in Figure 1. The results of various analyses of $p(\Delta)$ are illustrated in Figure 2. While all analyses give consistent results for the mean value of Δ, the most probable value of Δ and the width of $p(\Delta)$, there are some noticeable differences in $p(\Delta)$ for small Δ. In all cases except the shell model, the value of $p(0)$ is non-zero. In this case, at least, the shell model has placed an unrealistic constraint on the distribution function. This same feature has been seen for a number of other Al-based quasicrystals [20,21]. This is not to say that fits to

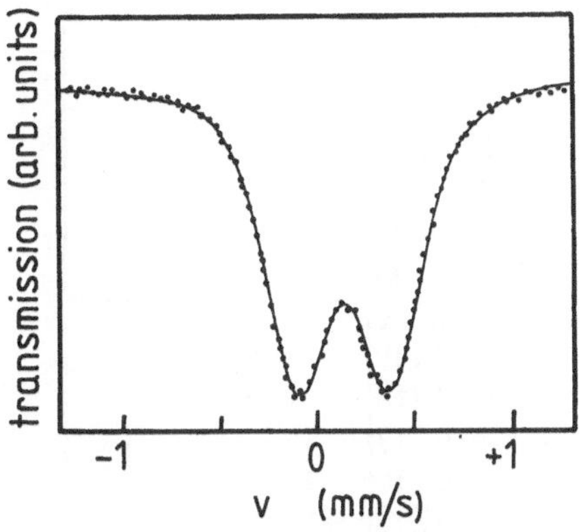

Figure 1: Room temperature ^{57}Fe Mössbauer effect spectrum of single phase icosahedral $Al_{80}Mo_9Fe_{11}$ (from ref. 20).

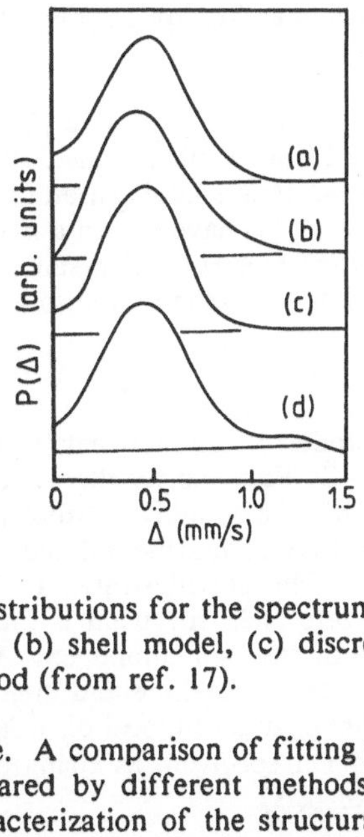

Figure 2: Quadrupole splitting distributions for the spectrum of Figure 1 as obtained by: (a) folded Gaussian model, (b) shell model, (c) discrete expansion method, and (d) Fourier expansion method (from ref. 17).

the shell model are uninformative. A comparison of fitting parameters for alloys of different compositions [19], or those prepared by different methods [21], can provide important information concerning the characterization of the structural and electronic details of the samples as long as the physical significance of the parameter values themselves is not over-interpreted. These results, as well, shed some suspicion on the interpretation of specific details of the shape of p(Δ) obtained by certain analysis methods.

A somewhat different approach to the extraction of useful information from Mössbauer spectra of paramagnetic quasicrystals has been taken by Schurer, van der Woude and co-workers [7,22,23], as they have sought to understand quasicrystal stability in terms of average values of the Mössbauer effect parameters. An example of these results is illustrated for the $Al_{80}Mn_{20-x}Fe_x$ series of quasicrystals in Figure 3. Both Δ and δ show maxima near a composition with x=9, although the trends are much clearer in δ than in Δ. This corresponds

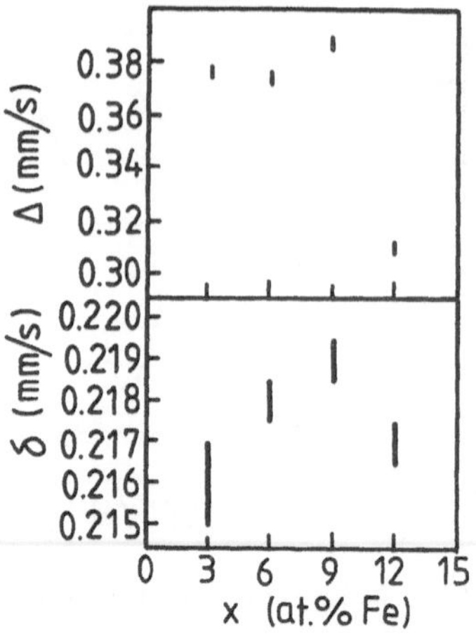

Figure 3: ^{57}Fe quadrupole splitting and isomer shifts relative to α-Fe obtained at room temperature for quasicrystalline $Al_{80}Mn_{20-x}Fe_x$ alloys as a function of composition. Data are from ref. [22].

to a density of valence electrons per atom, e/a, of 1.76 [22] and has been explained on the basis of alloy stabilization with the Hume-Rothery rules. Bancel and Heiney [24], as well as Friedel and Denoyer [25], have shown that a stabilization of the icosahedral phase is expected for e/a = 1.74. Similarly, Dunlap et al. [4] have shown a maximum in the isomer shift of Fe-doped binary Al-TM (TM = transition metal) quasicrystals near TM = Mn, and Lawther et al. [26] have shown that this composition corresponds well to a maximum in the alloy stability.

2.2. *In-field studies*

The first in-field Mössbauer studies of quasicrystals were reported by LeCaer et al. [12] for icosahedral Al-Mn-Fe and Al-Cr-Fe. Shortly thereafter, Eibschutz et al. [11] reported in-field studies of icosahedral Al-Mn-Fe-Si alloys. More recently, Brand et al. [27] have reported a more thorough in-field study of icosahedral and crystalline, Al-Mn-Si doped with Fe. Typical in-field Mössbauer results are illustrated in Figure 4. The quadrupole distribution as obtained from the zero field spectrum is illustrated in Figure 4b, and the distribution, including a separation of the sign component of Δ, is shown in Figure 4d. On the basis of such an analysis for several crystalline and quasicrystalline Al-Mn based alloys, Brand et al. [27] have concluded that changes in local disorder of paramagnetic quasicrystals is primarily manifested in the Mössbauer spectrum as changes in the relative proportion of negative and positive components of p(Δ).

3. STABLE Al-Cu-Fe QUASICRYSTALS

The existence of a thermodynamically stable quasicrystalline phase in the Al-Cu-Fe alloy system was first reported by Tsai et al. [28]. In the past four years, this alloy system has been studied extensively and, as Fe is a substantial component of these materials, Mössbauer effect measurements have been an important part of these investigations [e.g. 9,21,29]. Many early Mössbauer studies of these materials have yielded seemingly inconsistent results. It has been recognized, however, that such differences are the result of differences in local atomic

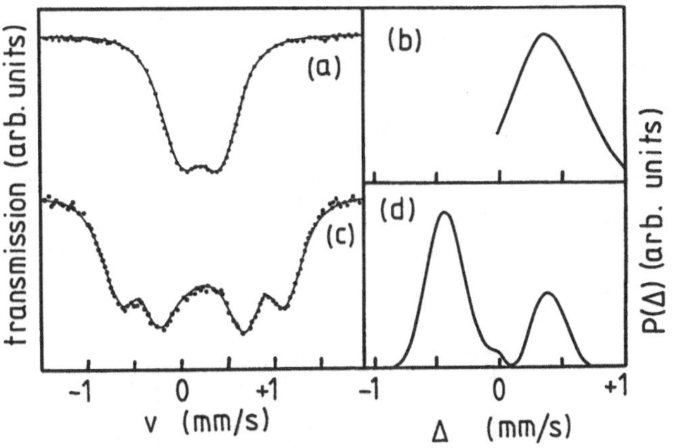

Figure 4: ^{57}Fe Mössbauer spectra obtained at 4.2 K for icosahedral $Al_{84}Mn_{10.8}Fe_{4.2}Si_1$: (a) without an external magnetic field, and (c) in an external field of 5 T. Corresponding distributions of the quadrupole splitting are shown in (b) and (d), respectively. Data are from ref. [27].

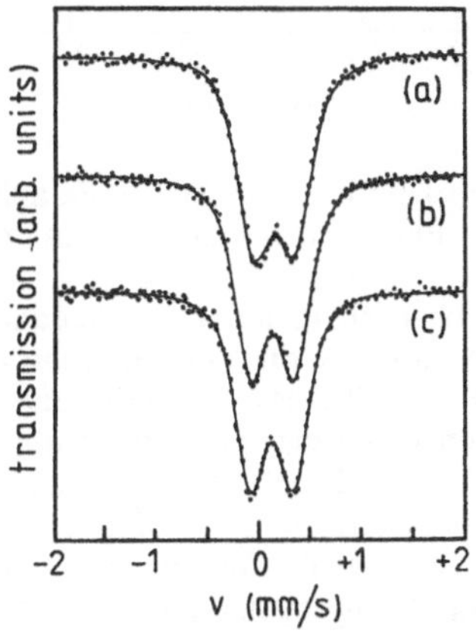

Figure 5: ^{57}Fe Mössbauer effect spectra obtained at room temperature for $Al_{63}Cu_{25}Fe_{12}$: (a) perfect quasicrystalline phase, (b) modulated quasicrystalline phase, and (c) rhombohedral crystalline phase. Solid lines are fits to distributions of quadrupole interaction, as described in the text.

structure [21,30]. It has, similarly, been recognized that magnetic properties are a sensitive function of local order in these materials [21,31] and that substantial differences in Mössbauer parameters and magnetic susceptibility can occur in alloys which, although prepared by different methods, show virtually identical X-ray diffraction patterns. Recently, Janot et al. [10] have proposed an interpretation of the phase diagram of the Al-Cu-Fe system that predicts the formation of distinct phases as a function of annealing temperature which show only minor differences in their diffraction pattern. They have recognized three phases: the perfect quasicrystalline phase, the modulated quasicrystalline phase, and the rhombohedral

Table I: Mean ^{57}Fe Mössbauer parameters of Al-Cu-Fe obtained at room temperature. Δ = quadrupole splitting, δ = isomer shift, and Γ = full width half maximum of spectral peaks. Values in mm/s ±0.005.

phase	Δ	δ	Γ
perfect quasicrystal	0.368	+0.233	0.358
modulated quasicrystal	0.407	+0.223	0.346
rhombohedral	0.405	+0.220	0.332

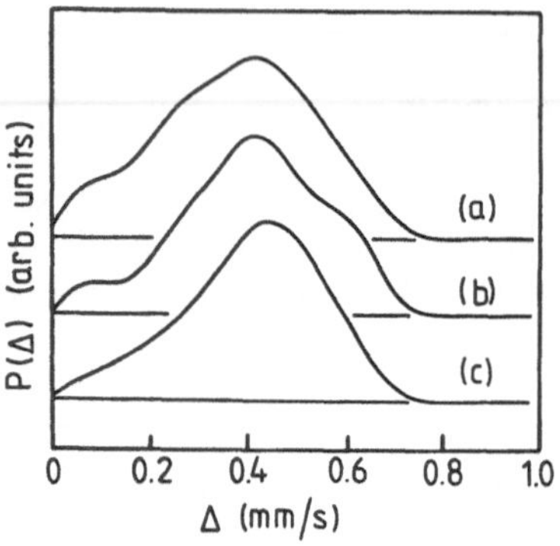

Figure 6: Quadrupole splitting distributions for Al-Cu-Fe alloys expanded as discrete points, as decribed in the text. Phases correspond to those of Figure 5.

phase, which have probably all been characterized as icosahedral alloys in many previous studies. This confusion forms a basis for the interpretation of apparent inconsistencies in Mössbauer parameters of these alloys. We have recently underaken Mössbauer studies of Al-Cu-Fe alloys [9, 32] in conjunction with other methods of characterizing their defect structure. Typical results are illustrated in Figure 5. Mean Mössbauer parameters obtained from these spectra are given in Table I. The general features of these results are as follow: The mean quadrupole splittings and isomer shifts of the modulated quasicrystal and rhombohedral phases are virtually identical. This suggests similar atomic environments in these two phases, as would be consistent with the model of Janot et al. [10]. The linewidths indicate, perhaps not unexpectedly, that the modulated quasicrystalline phase has more local disorder than the rhombohedral phase. The perfect quasicrystalline phase has an even larger linewidth, suggesting even more local disorder, and a smaller quadrupole splitting, indicating a more symmetric Fe environment.

Figure 6 shows that quadrupole splitting distributions of these three phases show similar features. All seem to show an extended tail with a small secondary peak on the low velocity side of the principal peak in p(Δ). The physical significance, if any, of this characteristic has not yet been determined.

The minor distinctions between the Mössbauer parameters of the three recognized phases of the Al-Cu-Fe system are of importance in the analysis of the results for Mössbauer studies of these alloys. In the case of samples which have not been well characterized, multiple phases may occur without yielding obvious anomalies in the X-ray diffraction patterns. The presence of more than one phase may produce spurious Mössbauer results and these may be manifested as anomalies in the linewidths, spectral symmetry and the shape of p(Δ).

4. Ti-BASED QUASICRYSTALS

A quasicrystalline phase in rapidly solidified Ti-based alloys was first reported by Kuo and co-workers [33,34]. Subsequently, the structural and physical properties of related Ti-based quasicrystals have been studied extensively by Kelton and co-workers [35,36] and Dunlap and co-workers [8,37-41]. Mössbauer effect studies have been reported for various $Ti_{56}Ni_{28-x}Fe_xSi_{16}$ alloys and typical results are illustrated in Figure 7. This spectrum is distinctly different from that seen in other paramagnetic quasicrystalline alloys, as only a single peak is observed. This indicates that the Fe sites in these alloys are much more symmetric than in Al-based quasicrystals. Although spectra for $2.5 \leq x \leq 20$, show no resolvable splitting, those for alloys with Fe content greater than 5 at% show substantial broadening of the Mössbauer resonance peak, indicating the presence of at least some sites involving a quadrupole splitting. The spectra were, therefore, fit to a distribution of quadrupole interactions and the results of this analysis for p(Δ) are shown in Figure 8. Two important features are seen in these results: (1) larger concentrations of Fe yield a broader distribution of quadrupole splittings, and (2) a shift of the most probable quadrupole splitting away from zero is observed for higher Fe content. The first feature presumably results from an increasing variety of probe sites within the alloy with increasing Fe content. The second feature, most likely, indicates that, for larger concentrations of Fe, some probe atoms enter less symmetric environments, but for small Fe concentrations there is a preference of Fe substitution for atoms at highly symmetric sites. This situation can be viewed in the context of the NMR measurements in icosahedral Ti-Mn by Jeong et al. [36], where the transition metal (i.e. Mn) sites are highly symmetric. Dunlap et al. [37] have suggested that in icosahedral Ti-Ni-Si small amounts of Fe substitute for the transition metal atoms (i.e. Ni) which decorate highly symmetric sites located at vertex positions of the prolate rhombohedra of the three dimensional Penrose tiling.

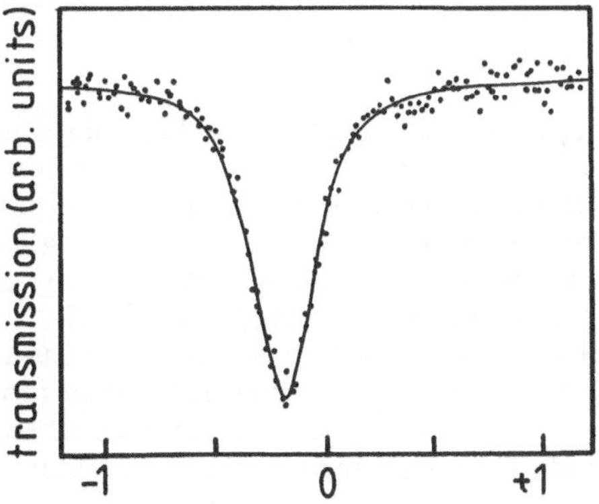

Figure 7: Room temperature ^{57}Fe Mössbauer effect spectrum for icosahedral $Ti_{56}Ni_{23}Fe_5Si_{16}$ (from ref. [8]).

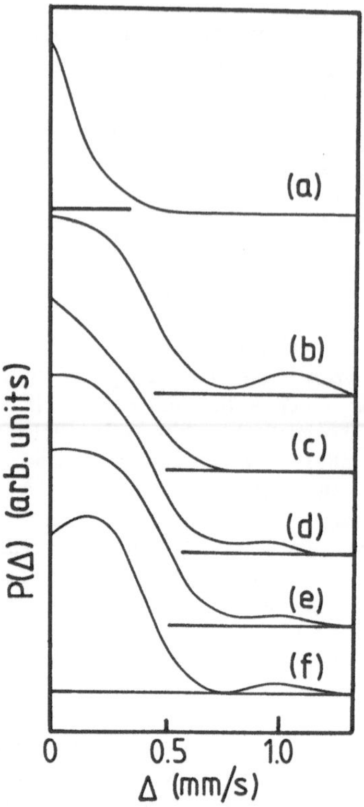

Figure 8: Room temperature ^{57}Fe quadrupole splitting distributions in $Ti_{56}Ni_{28-x}Fe_xSi_{16}$ quasicrystals as a function of x (data from ref. [40]): (a) x = 2.5, (b) x = 5.0, (c) x = 7.5, (d) x =10, (e) x = 15, and (f) x = 20.

5. CONCLUSIONS

Mössbauer effect studies using an ^{57}Fe probe have been important in providing an understanding of the physical properties of quasicrystals. Systematic changes in the Mössbauer parameters of alloys of differing composition have provided insight into the stabalization mecahnism responsible for icosahedral phase formation. Clear differences in local transition metal site symmetry has been seen between Al-based and Ti- based quasicrystals. The small or negligible quadrupole splitting in Ti-Ni-Fe-Si alloys shows the Fe sites to be highly symmetric. While quadrupole splitting distributions in Al-based quasicrystals have been the subject of some controversy, recent results on stable icosahedral Al-Cu-Fe and analogous crystalline phases have provided an understanding of the possible reasons for variations in results previously reported in the literature.

ACKNOWLEDGEMENTS

This work was supported by grants from the Natural Sciences and Engineering Research Council of Canada. The authors are grateful for comments and suggestions on various aspects of this work from Drs. R. C. O'Handley, M. E. McHenry, D. Bahadur, K. Dini, and D. J. Lloyd.

REFERENCES

1. Shechtman, D., Blech, I., Gratias, D., and Cahn, J. W. (1984), "Metallic phase with long range orientational order and no translational order", Phys. Rev. Lett. *53*, 1951-1953.
2. Swartzendruber, L. J., Shechtman, D., Bendersky, L. and Cahn, J. W. (1985), "Nuclear γ-ray resonance observations in an aluminium based icosahedral quasicrystal", Phys. Rev. *32*, 1383-1385.
3. Dunlap, R. A., Dini, K., Stroink, G., Collins, G. S., and Jha, S. (1986), "An Fe Mössbauer effect study of metastable $Al_{86}Fe_{14}$ prepared by rapid solidification", Hyperf. Interact. *28*, 963-966.
4. Dunlap, R. A., Stroink, G., Dini, K. and Jones, D. F. (1986), "Structural electrical and magnetic properties of icosahedral Al-Co alloys", J. Phys. F: Met. Phys. *16*, 1247-1254.
5. Eibschütz, M., Chen, H. S., and Hauser, J. J. (1986), "Study of local environment in quasicrystalline $Al_{86}Mn_{14}$ by Mössbauer effect", Phys. Rev. Lett. *56*, 169-172.
6. Schurer, P. J., Koopmans, B., van der Woude, F., and Bronsveld, P. (1986), "A study of amorphous, quasicrystalline and crystalline Al-Fe alloys by Mössbauer effect and diffraction techniques", Solid State Commun. *59*, 619-623.
7. van der Woude, F., and Schurer, P. J. (1987), "A study of quasicrystalline Al-Fe by Mössbauer effect spectroscopy and diffraction techniques", Can. J. Phys. *65*, 1301-1308.
8. Dunlap, R. A., McHenry, M. E., O'Handley, R. C., Srinivas, V., and Bahadur, D. (1989), "Mössbauer indications for high transition metal site symmetry in quasicrystalline $Ti_{56}Ni_{23}Fe_{5}Si_{6}$", Phys. Rev. B *39*, 1942-1945.
9. Lawther, D. W. and Dunlap,to be published (1991).
10. Janot, C., Audier, M., de Boissieu, M. and Dubois, J. M. (1991), "Al-Cu-Fe quasicrystals: low temperature unstability via a modulation mechanism", Europhys. Lett. *14*, 355-360.
11. Eibschütz, M., Lines, M. E., Chen, H. C., Waszczak, J. V., Papaefthymiou, G., and Frankel, R. B. (1987), "Observation of subsitutional site preference in a quasicrystal and implications on local structure", Phys. Rev. Lett. *59*, 2443-2446.
12. LeCaer, G., Brand, R. A., and Dubois, J. M. (1987), "Sign determination of the ^{57}Fe electric field gradient in Al-based crystals and icosahedral quasicrystals", Phil. Mag. Lett. *56*, 143-151.
13. Dunlap, R. A., Lawther, D. W., and Lloyd, D. J. (1988), "An ^{57}Fe Mössbauer effect study of preferential site occupancy in quasicrystalline $Al_{86}Cr_{14-X}Fe_{X}$ alloys", Phys. Rev. B *38*, 3649-3652.
14. Czjzek, G. (1982), "Distribution of nuclear quadrupole splitting of ^{57}Fe in amorphous ferric fluorides: structural implications", Phys. Rev. B *25*, 4908-4910.
15. Srinivas, V., and Dunlap, R. A. (1991), "Mixed hyperfine interactions in the ferromagnetic quasicrystals $Al_{40}Mn_{25}Cu_{10-X}Fe_{X}Ge_{25}$", J. Phys. Condens. Matter *3*, 2411-2419.
16. Bahadur, D., Srinivas, V., and Dunlap, R. A. (1989), "Structural relaxation in icosahedral Al-Cr-Fe-Ge", J. Phys. Condens. Matter *1*, 2561-2567.
17. Lawther, D. W., Dunlap, R. A., Srinivas, V., Lloyd, D. J., and Jha, S. (1990), "^{57}Fe Mössbauer effect studies of quadrupole splitting distributions in Al-TM-Ge", Hyperf. Interact. *60*, 785-790.
18. Window, B. (1971), "Hyperfine field distributions from Mössbauer spectra", J. Phys. E *4*, 401-402.
19. Srinivas, V., Dunlap, R. A., Bahadur, D., and Dunlap, E. (1990), "Physical properties of icosahedral Al-Cr-Fe-Ge alloys", Phil. Mag. B *61*, 177-188.
20. Srinivas, V., McHenry, M. E., and Dunlap, R. A. (1989), "Magnetic properties of icosahedral Al-Mo-Fe and Al-Ta-Fe alloys", Phys. Rev. B *40*, 9590-9594.

21. Dunlap, R. A., O'Handley, R. C., McHenry, M. E., and Srinivas, V. (1991), "Magnetic properties of quasicrystals: effects of disorder on Fe magnetic moment formation", Struct. Chem. *2*, 501-505.
22. Werkman, R. D., Schurer, P. J., and van der Woude, F. (1990), "Quasicrystals", Hyperf. Interact. *53*, 241-252.
23. Schurer, P. J., Koopmans, B., and van der Woude, F. (1988), "Structure of icosahedral $Al(M_{1-x}Fe_x)$ alloys (M=Cr, Mn, or Fe)", Phys. Rev. B *37*. 507-510.
24. Bancel, P. A., and Heiney, P. A. (1986), ""Icosahedral aluminum transition metal alloys", Phys. Rev. B *33*, 7917-7922.
25. Friedel, J., and Denoyer, F. (1987), "Les quasicristaux de AlLiCu comme alliages de Hume Rothery", CR Acad. Sci. Paris *305*, 171-174
26. Lawther, D. W., Dunlap, R. A., and Srinivas, V.(1989), "On the question of stability and disorder in icosahedral aluminum-transition metal alloys", Can. J. Phys. *67*. 463-466.
27. Brand, R. A., LeCaer, G., and Dubois, J. M. (1990), "Crystalline and quasicrystalline phases of the Al-Si-Mn system: a comparison of the ^{57}Fe electric field gradient properties", J. Phys. Condens. Matter *2*, 6413-6431.
28. Tsai, A.-P., Inoue, A., and Masumoto, T. (1987), ""A stable quasicrystal in Al-Cu-Fe system", Japan. J. Appl. Phys. *26*, L1505-L1507.
29. Kataoka, N., Tsai, A.-P., Inoue, A., Masumoto, T., and Nakamura, Y. (1988), "Mössbauer effect of icosahedral amorphous and crystalline Al-Cu-Fe alloys", Japan J. Appl. Phys. *27*, L1125-L1127.
30. O'Handley, R. C., Dunlap, R. A., and McHenry, M. E., "Magnetism and Quasicrystals", Chapter 5, in *Ferromagnetic Materials, Vol. 6*, K. H. J. Buschow, ed., North Holland: Amsterdam (1991).
31. Fukamichi, H., Goto, T., Komatsu, H., Wakabayashi, H., Tsai, A.-P., Inoue, A., and Masumoto, T. (1988), "Magnetic properties of AlCuTM (TM: transition metal) quasicrystals", J. de Physique Colloq. *49*, C8-239-C8-240.
32. Srinivas, V., and Dunlap, R. A. (1991), "Structural and electrical properties of AlCuFe quasicrystals", Phil. Mag. B, in press.
33. Kuo, K. H. (1986), "From Frank-Kasper phases to the icosahedral quasicrystal", J. de Physique Colloq. *47*, C3-425-C3-436.
34. Zhang, Z., Ye, H. Q., and Kuo, K. H. (1985), "A new icosahedral phase with m35 symmetry", Phil. Mag. A *2*, L49-L52.
35. Kelton, K. F., Gibbons, P. L., and Sabes, P. N. (1988), "New icosahedral phases in Ti-transition metal alloys", Phys. Rev. B *38*, 7810-7813.
36. Jeong, E. K., Holzer, J. C., Carlson, A. E., Conradi, M. S., Fedders, P. A., and Kelton, K. F. (1990), "Highly symmetric Mn site in icosahedral Ti-Mn", Phys. Rev. B *41*, 1695-1698.
37. Dunlap, R. A., O'Handley, R. C., McHenry, M. E., and Chatterjee, R. (1988), "The quasicrystal structure of rapidly solidified Ti-Ni based alloys", Phys. Rev. B *37*, 8484-8487.
38. Dunlap, R. A., McHenry, M. E., O'Handley, R. C., Bahadur, D., and Srinivas, V. (1988), "High symmetry transition metal sites in $Ti_{56}Ni_{28-x}Fe_xSi_{16}$ quasicrystals", J. Appl. Phys. *64*, 5956-5958.
39. Bahadur, D., Srinivas, V., and Dunlap, R. A. (1989), "Studies of hydrogen diffusion in metastable Ti-Ni-Fe-Si alloys", J. Noncrystalline Sol. *109*, 54-58.
40. Bahadur, D., Srinivas, V., Dunlap, R. A., O'Handley, R. C., and McHenry, M. E. (1989), "Properties of metastable Ti-Ni-Fe-Si alloys prepared by rapid solidification", Phil. Mag. B *60*, 871-880.
41. Christie, I. A., Dunlap, R. A., and McHenry, M. E. (1990), "Properties of rapidly quenched Ti(Ni, Fe)Si alloys", J. Appl. Phys. *67*, 5885-5887.

SMALL METAL PARTICLES STUDIED BY NMR

J.J. VAN DER KLINK
Institut de Physique Expérimentale,
Ecole Polytechnique Fédérale de Lausanne,
PHB-Ecublens, CH-1015 Lausanne
Switzerland

ABSTRACT Small metal particles, such as may occur in catalysts, have Pauli-paramagnetic properties that are governed both by the density of energy levels at the Fermi energy and by the corresponding wavefunctions' intensity at each site : the local density of states (LDOS). From NMR data on ^{195}Pt, the variations of the s-like and d-like LDOS in small platinum particles have been deduced. Data from several sources suggest that Pt particles should show a size-dependent metal-nonmetal transition (as measured by the nuclear spin-lattice relaxation time) in the diameter range 0.7 nm < d < 1.2 nm. Since the Pt s-LDOS does not vary much in the particle, it is suggested that the ^{109}Ag NMR linewidth in small silver particles will not be excessive. Preliminary experiments seem to confirm this idea, but also detect an unexpected effect, probably due to atomic motion.

1. Introduction : susceptibility and NMR shift

In spin-singlet molecules, the magnetic susceptibility is usually temperature-independent, and results from a negative Larmor diamagnetic and a positive van Vleck paramagnetic contribution. Under exceptional circumstances, the latter can lead to a Curie-type temperature dependence. The equivalent quantity in nuclear magnetic resonance (NMR) is the chemical shielding. This is a local expression of the susceptibility, that may have different values at different sites in the molecule. Shieldings are expressed on a relative scale, with respect to some reference compound, and counted more positive if the resonance requires a larger magnetic field than the reference.

Molecules of nonzero spin, such as O_2 (spin triplet) have an additional (and dominant) Curie-type paramagnetic susceptibility : the shift of the ^{17}O NMR signal is proportional to the inverse temperature [1]. There is a slight problem of semantics here : this shift is due to electron spin (as the Knight shift is), not to orbital effects (as the chemical shift) : but Knight shifts, by convention, are counted positive if the resonance occurs at low field with respect to the reference.

In a metal, the same shift mechanisms operate as in singlet molecules, but are now usually called the "orbital Knight shift". In addition there is the (spin) Knight shift, related to Pauli paramagnetism. Whereas the chemical shift (orbital Knight shift) is due to all electrons, only those at the Fermi energy contribute to the (spin) Knight shift. No shift is

P. Jena et al. (eds.), Physics and Chemistry of Finite Systems: From Clusters to Crystals, Vol. I, 537–547.

associated with Landau diamagnetism [2]. Band structure effects bring some complication, since different bands may have different effective hyperfine fields. These may even have different signs, so that an accidental zero Knight shift may be associated with nonzero susceptibility : as an example, consider the Knight shifts in CuPt alloys [3], shown in Fig. 1. In some transition metals, there is considerable structure in the density of states near the Fermi level : as a consequence the susceptibility and the Knight shift have an appreciable temperature dependence. A typical case is Pd.

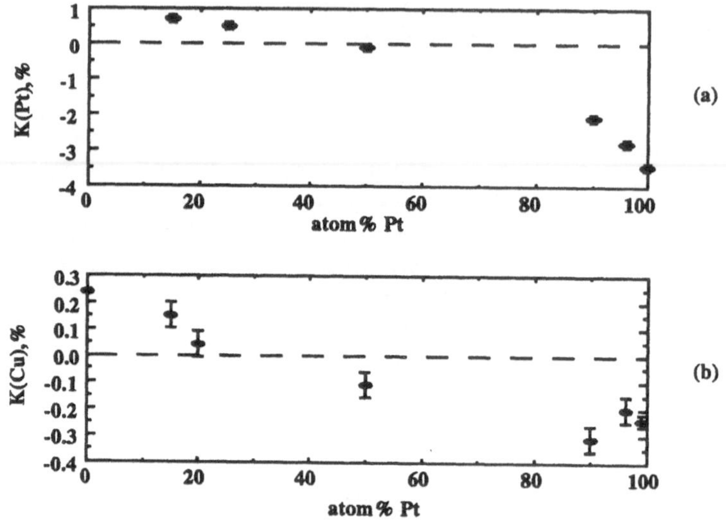

Figure 1. Knight shift of Pt (a) and Cu (b) in Cu-Pt alloy as a function of composition. Note that both change sign, and that zero Knight shift does not mean that the alloys is not metallic.

While the Knight shift is related to the static spin susceptibility, the nuclear spin-lattice relaxation rate T_1^{-1} is connected to the dynamic susceptibility. In usual metals, electron spin fluctuations are rapid on the NMR time scale (this is no longer true for "strongly correlated" metals, such as found near a metal-nonmetal or a superconducting transition) and Fermi-Dirac statistics apply. As a result, the product T_1T is independent of temperature T, and can be related to the value of the Knight shift (the Korringa relation). Whereas the interpretation of shifts ("molecular" or "metallic"; "really zero" or "accidental cancellation") is not always easy, the Korringa product can be seen as the NMR signature of the (normal) metallic state.

A nice study of the susceptibility and Knight shift of an intrinsic semiconductor is the work on tellurium by Kanert's group [4]. At constant number of carriers, the susceptibility (and the shift) should vary inversely with temperature, because Boltzmann statistics apply. At constant temperature, these quantities should be proportional to the number of carriers (that is temperature-dependent because of the gap energy). The result is a very characteristic temperature dependence, of both the shift and the nuclear spin-lattice relaxation time.

2. Spin pairing : superconductors and quantum size effects

In a superconductor, a gap in the density of states starts to open below the transition temperature. Therefore the susceptibility and the Knight shift should fall to zero with decreasing temperature. This is found in bulk specimens, for example on the ^{63}Cu resonance in cuprate superconductors [5]. A different behaviour has been observed in some (but not all) type I metallic superconductors (for a review see [6]). Here the samples are small (compared to the coherence length) in at least one dimension, to avoid problems associated with penetration depth. In the normal state of a grain of such a fine powder, impurities and/or the grain's boundary are supposed to perturb the strict periodicity of the spin-orbit coupling that exists in the pure bulk metal. This has two consequences for the Bloch-like one-electron eigenfunctions : they are no longer strictly periodic, and not eigenfunctions of the z-component of the spin. In the normal state, one expects such a mechanism to give some NMR line broadening (because the nuclear sites are no longer equivalent), but the average resonance frequency and spin-lattice relaxation rate are likely to be comparable to those obtained in a pure bulk sample. However, part of the susceptibility will be due to a second-order, van Vleck-like, spin paramagnetism, that is not impaired by the opening of the superconducting gap. As a consequence, this part of the the susceptibility (and of the Knight shift) persists down to the lowest temperatures in the superconducting region (the Ferrell-Anderson effect).

By analogy, Knight has pointed out [7] that in a system with discrete energy levels of spacing δ, one might say that an "effective gap" opens for $kT < \delta$. Much of the reasoning sketched above is then copied to the case of the even-particle (those with an even number of electrons) quantum size effect. The NMR of the odd particles behaves as that of spin-paramagnetic molecules, but with a size-dependent susceptibility. Whether a residual Knight shift is present or not, the Korringa relaxation mechanism should be suppressed and spin-lattice relaxation times should be long. Kobayashi and coworkers report this effect [8] for copper particles of diameter 10 nm and less (unfortunately, no shift data on these same samples are discussed). They work below 0.5 K, but theoretically the phenomenon should exist at all temperatures. Preliminary results from our laboratory on slightly larger (13 nm) silver particles between 20 K and 300 K show a Korringa relaxation with the same value as the bulk.

To explain metal-nonmetal transitions in bulk samples, several microscopic mechanisms are invoked. The one relevant to our subject is the band-crossing (Wilson) transition, that can occur in divalent metals. Here the metallic character of the bulk is due to the overlap of (typically) the s- and the p-bands. The bandwidths depend on the coordination number Z, and if the latter is lowered, the sp-overlap disappears and a band gap opens. This is thought to cause the transition in liquid Hg, at densities well above critical. It is believed that decreasing density hardly affects the nearest-neighbour distance, but much more the coordination number : the low-density liquid is an "alloy" of Hg atoms and vacancies. This transition has been observed by Warren and El-Hanany monitoring the Knight shift (and thus the spin susceptibility) as a function of density [9] : it falls to zero at $\rho = 9$ g/cm^3, probably corresponding to an average coordination of 6.7. Interesting enough, this is the average coordination number of a 16-atom cluster, reason to call the qualitative change of Hg-cluster properties in the 20-70 atoms range a metal-nonmetal transition [10].

3. Local densities of state : platinum particles

Similar coordination-related NMR effects have been observed in small platinum particles in C.P. Slichter's [11] and in our laboratory [12,13]. In a tight-binding approximation, the d-bandwidth depends on the coordination number as $Z^{1/2}$. Therefore the d-like local density of states at the Fermi energy (d-LDOS) on surface atoms (that have lower Z) is expected to be lower than that in the bulk. This is indeed what is found from analysis of the NMR data, using a model that will be described in some detail.

In the FCC structure, cubo-octahedral particles with triangular (111) and square (100) faces can be formed by adding successive layers of 12, 42, 92, 162 etc. atoms around a central atom to obtain particles containing 1 + 12 + 42 + 92 + 162 ... atoms. A particle of 309 atoms (2.1 nm equivalent diameter, 52% dispersion or fraction of atoms on the surface) will have 162 atoms in the surface layer (that we number layer 0), 92 in the subsurface layer (layer 1), and so on, until 1 atom in layer 4.

Starting from size distributions measured on an electron micrograph, we convert the number-of-particles *vs* diameter relation into number-of-atoms *vs* layer number. In the example of the 309-atom particle, we might further divide the 162 surface atoms into corner-, edge- and face-atoms. Each distinct site would in principle give rise to a distinct NMR line, and the observed spectrum would be the (unresolved) superposition of many such lines. To avoid an excessive number of parameters, we suppose that all the surface sites together give rise to a Gaussian-type signal of a width and a center-frequency that remains to be determined; the subsurface sites give a Gaussian at another frequency and with another width, and so on. The relative areas under the different Gaussians must be proportional to the number of atoms in the corresponding layer, as determined from the electron micrographs. This is the basis for the correlation between catalyst dispersion (fraction of atoms at the surface) and the relative area of the NMR "surface peak", first mentioned by Rhodes et al. [11]. To reduce the number of parameters still further, we impose that the sequence of central frequencies of successive layers must follow a simple law, similar to what has been proposed for successive layers around an impurity [14]. This law has the form

$$K_n - K_o = (K_\infty - K_o)\left(1 - \exp\left(-\frac{n}{m}\right)\right) \qquad (1)$$

As an example, the NMR signals from three neighbouring Pt layers to a Mo impurity have been detected in dilute alloys of Mo in Pt. The value of $K_\infty - K_o$ (the difference in Knight shift between the bulk and the nearest-neighbour to the impurity) is 0.011, and m = 0.97 [14]. To convert m to a distance, multiply by 229 pm, the equivalent thickness of a cubooctahedral layer. A peculiarity of expression (1) is that it decays without oscillation.

For the small particles, we now have one Gaussian-width parameter for each layer, plus the parameters K_o and m, to fit the observed NMR spectrum to the observed layer distribution (size distribution). We now require that spectra on samples that are not too dissimilar (same surface treatment, comparable sizes), but that have measurably different layer distributions, should be fitted by the same set of parameters (only the relative areas of the different Gaussians will be different). This has been done in Fig. 2 [13]. The widths of the surface Gaussian is a significant fraction of the total width, indicating a great variety of

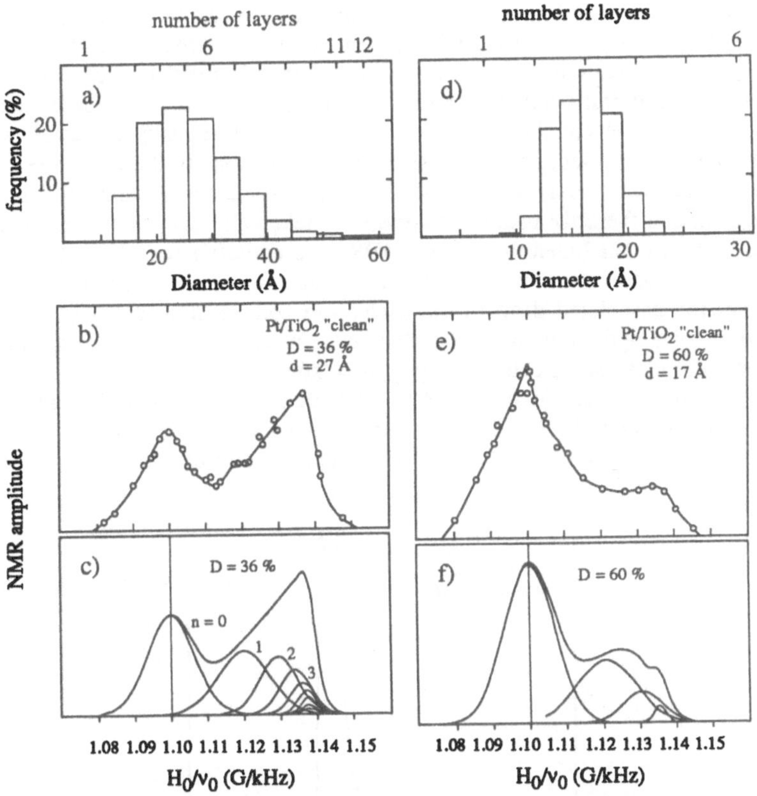

Figure 2. The "layer model" applied to the ^{195}Pt NMR spectrum of two similar catalysts of different particle sizes. Top : the size-distribution measured by TEM. The upper scale indicates the number of layers in a cubo-octahedral particle whose equivalent diameter equals the value on the lower scale. The fraction of atoms in each layer (n = 0 now is the surface, n = 1 the subsurface, and so on) is calculated from this distribution. Bottom : the superposition of Gaussian NMR signals ascribed to each layer using equation (1). The areas of the Gaussians are in proportion to the number of atoms in the corresponding layers, their widths are considered fittable parameters, but are the same at right and at left. Center : observed NMR spectra. The left and the right spectrum have been scaled to the same area.

Knight shifts among different surface sites. Similarly, Zilm et al. [15] have concluded from their magic-angle-spinning (MAS) NMR of ^{13}CO chemisorbed on platinum that a large variety of ^{13}C shifts exists. The parameter m describes how far inside the particle the presence of the surface is felt : for a nominally clean surface its value is 1.35, corresponding to a "healing length" of 0.31 nm. When the analysis is repeated for hydrogen-covered surfaces (supposing that the clean-particle layer distribution stays applicable), m is found to increase to 2.6 (0.59 nm). Perhaps the clearest proof that the effects observed are due to a finite surface/volume ratio (and not to a quantum size effect) is the NMR spectrum of a 0.5 – 1.2 μm commercial platinum powder shown in Fig. 3. These grains have of the order of $10^9 - 10^{10}$ atoms (of the order of a thousand layers) and a dispersion of a few per mille.

The ^{195}Pt nuclear spin-lattice relaxation time T_1 at (or near) the Gaussian positions has also been measured. For all positions (including the clean and H-covered surface positions) Korringa behaviour is found over a factor of 8 in temperature T (10 K – 80 K). From the known relation between Knight shift (position of the Gaussian) and Korringa product T_1T, we have concluded that the (average) local density of states at the Fermi energy (LDOS) in the surface layer is substantially lower than in the bulk (see Fig. 4). That analysis supposed Stoner enhancement for both the s- and the d-band (but no interaction between bands), and the only parameters that were taken site-dependent were the s-like and d-like LDOS; it was concluded that the s-LDOS is essentially site-independent, whereas the d-LDOS drops a factor of two going from the bulk to the surface. Due to the Stoner enhancement, the associated drop in the d-electron contribution to the susceptibility is a factor of four; for a hydrogen-covered surface even a factor of eight. For the clean-surface 36% dispersion Pt catalyst we calculate $\chi/\chi_{bulk} = 0.6$.

In our analysis, the chemical shift (orbital Knight shift) is kept site-independent, and fixed at the (estimated) bulk value. In principle of course this approach is wrong : chemisorption, for example, should make its value site-dependent. Numerically however,

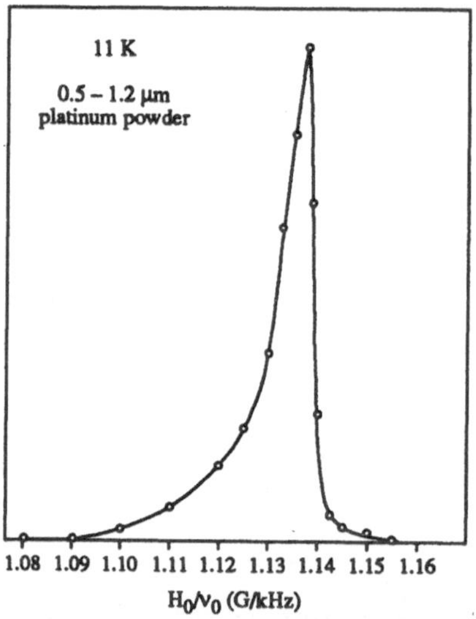

Figure 3 ^{195}Pt NMR spectrum of very large platinum particles. No surface resonance is detected, but the intensity increases rapidly where the "n > = 2 gaussians" start to overlap and pile up.

this effect is overwhelmed by the (spin) Knight shift : as far as ^{195}Pt NMR goes, CO and H_2 chemisorption are undistinguishable. In a qualitative study of the interaction of oxygen with a medium-dispersion (60%) Pt catalyst, we find that the metallic character (as measured by T_1) is completely lost. It is rather fortuitous (a result of an unhappy ratio of s- and d-like effective hyperfine fields) that the "surface peak" of small-particle ^{195}Pt NMR falls at (or very close to) zero shift. This has lead to comparisons between Pt surfaces with CO chemisorbed and Pt-carbonyls that are not very fruitful : the latter are non-metallic and the former metallic, as shown by the absence, resp. presence of a Korringa-type spin-lattice relaxation.

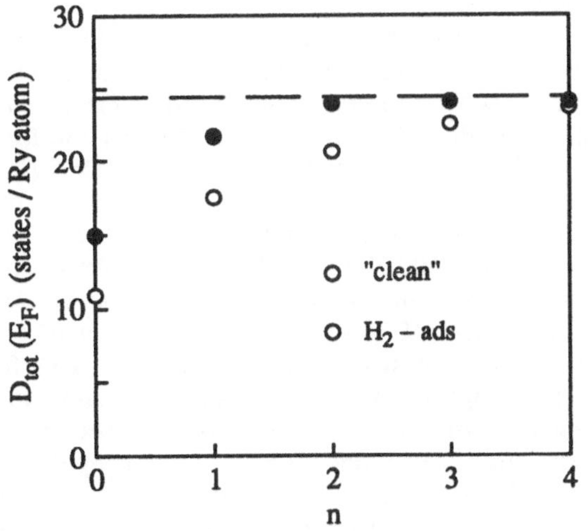

Figure 4. Total local densities of state at the Fermi level, in successive layers of atoms, going from the surface (layer number n = 0) inward. The effect is mainly due to the d-like density of states, that "heals" exponentially from the surface value to the bulk value, with an exponential constant of 1.35 layers for the clean surface and 2.6 layers for the H-covered one.

4. Environment : supported, encaged and ligand-stabilised particles

In a continuing series of experiments, de Jongh, Brom and coworkers study physical properties of high-nuclearity organometallic cluster compounds. One of these has a core consisting of (probably) 561 palladium atoms, forming a cubo-octahedron of five layers. An important number of ligand molecules is coordinated to the surface of the palladium core. (Such systems have also been obtained and studied by Moiseev and coworkers [16]). It is found by SQUID magnetometry that $\chi_{561}/\chi_{bulk} \approx 0.4$ [17]. (There are reasons to believe that the outer 252 Pd atoms are not in a metallic state, and that the value to be considered is $\chi_{309}/\chi_{bulk} \approx 0.7$). Applying the "exponential healing" model to the d-LDOS, keeping the s-LDOS constant, and choosing m = 2.6 (in analogy with the Pt-H surface), we calculate that the d-LDOS of Pd_{561} at the surface equals two-thirds of that in the bulk.

At this conference, ^{195}Pt results by the same group on a Pt_{309} cluster compound are reported [18]. They find a double-peaked spectrum; one of the peaks has a metallic-like T_1, and the other has not, suggesting that the 147 inner atoms (equivalent diameter 1.6 nm) are in a metallic state, but that the outer 162 atoms are not, possibly due to chemical interactions with the ligands (cf. the Pt-O interaction mentioned in the previous paragraph).

This raises the question as to down to what sizes the properties of supported (or encaged) metal particles can be considered as essentially uninfluenced by the support (or embedding medium). As an extreme example, we propose the Na_4^{3+} ion, that has been detected in sodalite cages (using EPR) [19]. According to unpublished calculations [20], that ion is not stable in vacuo, and therefore the presence of the (charged) aluminosilicate framework seems to be essential for its existence. More on the scale of the earlier discussion, we have recently studied two samples of Pt on (in ?) zeolite. If the metal particles are formed inside the undeformed cages of the ideal Y-zeolite structure, there is an upper limit of about 1.3 nm to their diameter (the zeolite is in the form of a powder, and some larger metal particles will form on the surface of the powder), but some deformation and/or damage to the framework seems to occur under certain circumstances. We have not yet obtained a reliable determination of the particle size distribution; for these dimensions, this is quite difficult to do. In one of our samples (labeled "600/300"), a typical particle diameter determined by small-angle X-ray scattering is 1.4 nm; in the other ("300/300") it is 0.9 nm. Some particles definitely larger than these "typical" values have also been seen on electron micrographs. Sample 600/300 has ^{195}Pt NMR characteristics (see Fig. 5) typical of a high-dispersion catalyst : no signal at H/ν = 1.14 G/kHz, and a Korringa T_1 at 1.10 G/kHz. The T_1 measurement of 300/300 shows a fraction of Pt nuclei in a non-metallic environment, as indicated by non-Korringa behaviour. This fraction is between one-half and two-thirds : the signal-to-noise ratio does not permit to say whether this quantity varies with temperature or not. Its relaxation is rapid compared to that in typical diamagnetic platinum compounds (say H_2PtCl_6) at the same temperature, and comparable to what has been found in platinum cluster compounds, in line with the idea that metal clusters in zeolites can be regarded as intermediate between naked and ligand-stabilised clusters.

Although more work is needed (and is actually in progress), I am inclined to speculate that, upon lowering the size of the metal particles, the unavoidable interaction with the matrix or with the ligand shell will suppress the metallic behaviour before the quantum size effect shows up.

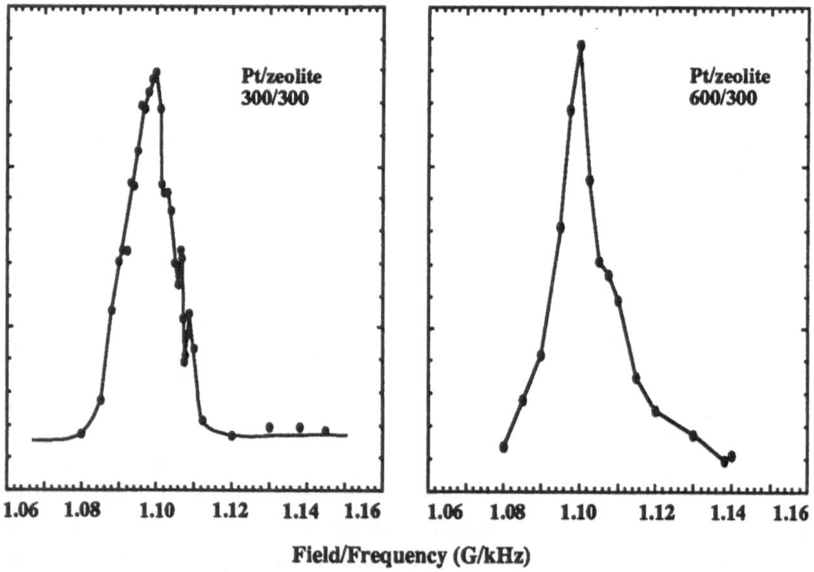

Figure 5. Pt particles can be formed in the large cage of a zeolite. Here the ^{195}Pt NMR spectra are shown of two samples with slightly different particle sizes. From a study of the nuclear spin-lattice relaxation, we find that roughly 60% of the platinum atoms in the smallest-size sample are not in a metallic environment, we speculate that this is due to a size-dependent Pt-zeolite interaction.

5. A simple metal with nuclear spin 1/2 : silver

According to the description of the ^{195}Pt NMR given above, the extreme width of the small particle NMR spectrum is due to a strongly site-dependent d-LDOS. This may also be the reason for the marked size-dependence of the ^{61}Ni signal width in small nickel particles [21] (although domain- and demagnetization effects may play a role as well). By contrast, the non-transition metals (zero d-LDOS by definition) might have relatively narrow small-particle spectra (although the results of Yee and Knight [7] show linewidths of the order of the Knight shift). We are currently testing this idea by studying silver particles. Like platinum, ^{109}Ag (and ^{107}Ag) has spin 1/2, and the bulk metal structure is FCC, both desirable features for NMR. The NMR sensitivity is however a factor of 70 lower. We find that even 13 nm silver particles can be successfully studied, and experiments on smaller sizes are on their way.

At least down to this size, the peak position and relaxation time are like the bulk in the temperature range studied (20 K – 300 K), but the linewidth shows a strong temperature dependence. In the classical experiment of Yee and Knight on small copper particles, the observed linewidth variation was ascribed to (unidentified) localized paramagnetic impurities. In a not less famous experiment on bulk Li [22] the "motional narrowing" (of the nuclear dipole-dipole interaction) due to self-diffusion of the Li atoms was shown. In our silver experiments, the low-temperature linewidth scales with magnetic field, suggesting either the same problem as in Yee and Knight's experiment or a distribution of Knight

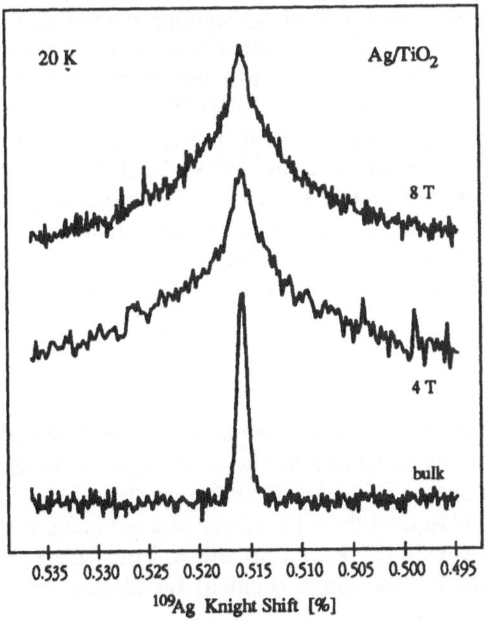

Figure 6. The Knight shift and nuclear spin-lattice relaxation times of ^{109}Ag in silver particles have bulk values, but the NMR line has a temperature- and size-dependent width that scales with magnetic field. Its origin may be a distribution of Knight shifts, accompanied by atomic motion.

shifts, associated with a distribution of s-LDOS. The paramagnetic-impurity linewidth should follow a Curie (or Curie-Weiss) temperature law (as approximation to a Brillouin function, since it scales with magnetic field), whereas motional narrowing (usually) gives a steep change between two plateaus.

We have studied the ^{109}Ag linewidth in a fairly large-particle sample (typical sizes 20 - 100 nm) of silver on TiO_2 between 10 K and 300 K. Below 30 K it is temperature-independent, but it increases linearly in magnetic field (see Fig. 6), as expected for a distribution of Knight shifts (accompanied by motional narrowing at higher T). Analysis of the data with a BPP expression [23], and assuming an Arrhenius behaviour for the motional correlation time leads to unlikely low values (of the order of 10 meV) for the energy barrier to motion. A possible explanation is the existence of a large distribution in rigid-lattice linewidth and in correlation times.

6. Conclusions

The NMR signal of sufficiently small supported, encaged or ligand-stabilised Pt particles has zero Knight shift, even in the metallic regime (T_1T = const). This regime is always found for particles with diameters above 1.4 – 1.6 nm (roughly 150 atoms); and is not qualitatively changed by CO or H-chemisorption. Oxygen however forms oxides, that turn the sample (partly) into an insulator. The transition to the nonmetallic regime (diameters below 1.0 – 1.2 nm, roughly 80 atoms) is more likely due to particle-substrate (or -cage or - ligand) interactions than to the pure quantum size effect (QSE). From calculations on Pt_{13} (d = 0.7 nm) a very large HOMO–LUMO splitting is predicted [24]. Consequently, if the QSE exists in Pt, it must be for particles with 0.7 nm < d < 1.2 nm.

From analysis of spin-lattice relaxation across the NMR spectrum for a 37% and a 60% dispersion catalyst, it is concluded that the s-LDOS is roughly constant throughout the particle (there are a number of assumptions in this analysis, but it is independent of the "layer model"). Consistent with this idea, the ^{109}Ag NMR lines of medium-sized (d > 10 nm) silver particles are only moderately broadened. The broadening is due to a distribution of Knight shifts, size-dependent, and temperature-dependent due to motional narrowing. The latter however cannot be easily described by a single rigid-lattice linewidth and Arrhenius-activated correlation time. Further work on the size-dependence is under way to clarify this question.

Acknowledgments

The work at the EPFL described here was or is part of the Ph.D. research of J.P. Bucher, J.J. Bercier and Y. Tong, partially supported by the CERS and by the Swiss National Science Foundation. Technical assistance in the NMR work is ensured by J. Savoie and A. Vonnez. Samples and advice on how to treat them have been and are being provided by M. Graetzel (EPFL Lausanne), A. Renouprez (CNRS Lyon) and R. Prins (ETH Zürich). I thank D. van der Putten, H.B. Brom and L.J. de Jongh (Leiden) for discussions of their work prior to publication.

References

1. Dundon J.M., (1982) J. Chem. Phys. **76**, 2171.
2. Ebert H., Winter H. and Voitländer J. , (1986) J. Phys. F **16**, 1133.
3. Itoh J. , Asayama K. and Kobayashi S. , (1964) Proc Coll. Ampère **13**, 162.
4. Günther B. and Kanert O. , (1985) Phys. Rev. B **31**, 20.
5. Barrett S.E. , Durand D.J. , Pennington C.H. , Slichter C.P. , Friedmann T.A. , Rice J.P. and Ginsberg D.M. , (1990) Phys. Rev. B **41**, 6283.
6. MacLaughlin D.E. , (1976) Solid State Physics **31**, 1.
7. Yee P. and Knight W.D. , (1975) Phys. Rev. B **11**, 3261.
8. Goto T. , Komori F. and Kobayashi S. , (1989) Z. Phys. D **12**, 415.
9. Warren Jr. W.W. and Hensel F. , (1982) Phys. Rev. B **26**, 5980.
10. Rademann K. , (1990) Ber. Bunsenges. Phys. Chem. **93**, 653.
11. Rhodes H.E. , Wang P.-K. , Makowka C.D. , Rudaz S.L. , Stokes H.T. , Slichter C.P. and Sinfelt J.H. , (1982) Phys. Rev. B **26**, 3569.
12. Bucher J.P. and van der Klink J.J. , (1988) Phys. Rev. B **38**, 11038.
13. Bucher J.P. , Buttet J. , van der Klink J.J. and Graetzel M. , (1989) Surf. Sci. **214**, 347.
14. Inoue N. and Sugawa T. , (1978) J. Phys. Soc. Japan **45**, 450.
15. Zilm K.W. , Bonneviot L. , Hamilton D.M. , Webb G.G. and Haller G.L. , (1990) J. Phys. Chem. **94**, 1463.
16. Vargaftik M.N. , Zagorodnikov V.P. , Stolarov I.P. , Moiseev I.I. , Kochubey D.I. , Likholobov V.A. , Chuvilin A.L. and Zamaraev K.I. , (1989) J. Mol. Cat. **53**, 315.
17. van Ruitenbeek J.M , Jurgens M.J.G.M. , Schmid G. , van Leeuwen D.A. , Zandbergen H.W. and de Jongh L.J. , (1991) Z. Phys. D **19**, 267.
18. van der Putten D. et al., this conference.
19. Edwards P.P. , Harrison M.R. , Klinowski J. , Ramdas S. , Thomas J.M. , Johnson D.C. and Page C.J. , (1984) J. Chem. Soc. Chem. Commun. 982.
20. Reuse F. , private communication.
21. Kohara T. , Yamaguchi M. and Asayama K. , (1985) J. Phys. Soc. Japan **54**, 1537.
22. Gutowsky H.S. and Mc Garvey B.R. , (1952) J. Chem. Phys. **20**, 1472.
23. Bloembergen N. , Purcell E.M. and Pound R.V. , (1948) Phys. Rev. B **73**, 679.
24. Messmer R.P., Knudson S.K. , Johnson K.H. , Diamond J.B. and Yang C.Y. , (1976) Phys. Rev. B **13**, 1396.

SHAPE AND SIZE DEPENDENCE OF ENERGY LEVEL DEGENERACIES

V. SUBRAHMANYAM and MUSTANSIR BARMA
Tata Institute of Fundamental Research
Homi Bhabha Road, Colaba
Bombay 400 005, India.

ABSTRACT. We investigate the occurrence of large degeneracies and their dependence on the size and shape of the confining potential. For a three-dimensional anisotropic harmonic oscillator and a particle in a cuboid, large degeneracies, which grow with size and energy, appear when aspect ratios are commensurable. A crossover from commensurable to incommensurable behaviour can be induced by changing either the size or shape (degree of rationality of aspect ratios). We study this crossover using a combination of analytical and numerical techniques. We find that shape-size scaling is valid, which allows for a quantitative characterization of degeneracies in terms of a single parameter involving both size and shape. The distribution of energy level spacings and scaled degeneracies in the commensurable regime are also found.

1. INTRODUCTION

The electronic properties of nanostructures and clusters depend sensitively on the way in which the energy spectrum evolves as the size is changed. Of particular interest is the occurrence of large degeneracies (growing with size), which go hand in hand with large gaps between distinct energy levels, and thus have a direct bearing on optical and low-temperature properties. It is important to pin down the parameters responsible for large degeneracies, and to quantify the sensitivity of degeneracies to small changes of shape.

While a good deal is known about the general spectral characteristics of integrable systems [1], the shape and size dependence of degeneracies does not seem to have been studied systematically earlier. Accordingly, we have made a detailed study of degeneracies for two families of shapes, *viz.* anisotropic harmonic oscillators (AHO) and cuboids with variable aspect ratios [2]. Large degeneracies, not associated with any obvious symmetry, occur when the aspect ratios are rational (commensurable case) and the size or energy is large enough. By contrast, there are no large degeneracies when aspect ratios are irrational (incommensurable case).

A crossover from commensurable to incommensurable behaviour can be induced by changing either the size or shape (degree of rationality of aspect ratios). An important result is the existence of shape-size scaling, which allows us to quantitatively

P. Jena et al. (eds.), Physics and Chemistry of Finite Systems: From Clusters to Crystals, Vol. I, 549–554.

characterize degeneracies in terms of a single scaling parameter involving both shape and size. The full scaling function is determined numerically. In a practical sense, the result is useful as it allows one to identify optimal small changes of cuboid dimensions which lead to large changes in spectral features (see Fig. 1).

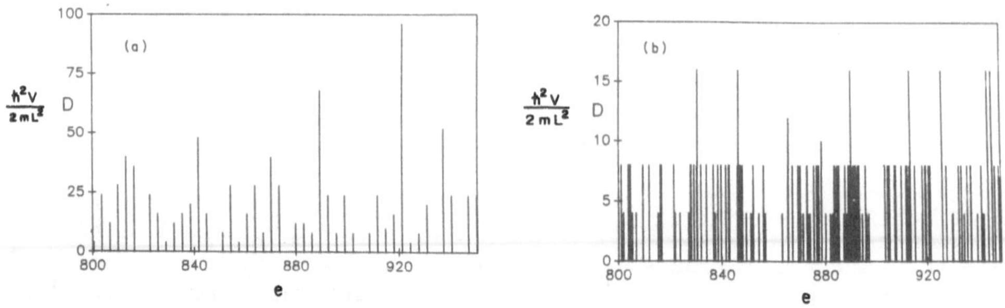

Figure 1 Illustrating large changes in the energy spectrum, particularly degeneracies, brought about by a small change of shape – cuboids with aspect ratios (a) (36,54,90) (b)(34,55,89).

2. INCOMMENSURABLE-COMMENSURABLE CROSSOVER AND SCALING

The total number of states/volume in a given energy range is roughly the same whether or not there are degeneracies, but the number of energy levels (distinct eigenvalues) is sensitive to degeneracies. Accordingly, we examine the total number of levels $N_{level}(e)$ up to e, where e (defined below) is a dimensionless measure of energy ε or size L (e.g. for a cube, $e \sim \varepsilon L^2$). In the incommensurable case, the absence of large degeneracies implies that $N_{level}(e) \sim$ the total number of states, $\sim e^3$ for the AHO, and $\sim e^{3/2}$ for cuboids. In the commensurable case, $N_{level}(e)$ turns out to vary linearly with e if e is large enough, for both AHO and cuboid *i.e.* the function $f(e) \equiv N_{level}(e)/e$ asymptotes to a constant $h(\vec{\alpha})$ as $e \to \infty$, where $\vec{\alpha} \equiv (\alpha_x, \alpha_y, \alpha_z)$ determines the aspect ratios . This saturation property of $f(e)$ implies that for large e, mean degeneracies, given by the ratio of the number of states to the number of levels in an energy interval, are of order e^2 for the AHO and of order $e^{1/2}$ for the cuboid.

If the aspect ratios are rational, but in some sense close to irrational, there is a crossover as function of e. Characteristic rational behaviour sets in only for e much larger than a crossover energy $e^*(\vec{\alpha})$, while for $e < e^*(\vec{\alpha})$ the behaviour is similar to that in the incommensurable case. We study examples of such crossover by examining a sequence of commensurable sets $\vec{\alpha}_m$ which approach an incommensurable set $\vec{\alpha}$ as $m \to \infty$. For each set of aspect ratios $\vec{\alpha}_m$, we study the function $f(e)$ and its asymptotic value $h(\vec{\alpha}_m)$. We are able to find $h(\vec{\alpha}_m)$ for all m for the AHO. For the cuboid, we derive upper bounds on $h(\vec{\alpha}_m)$ for the first few $\vec{\alpha}_m$ and present numerical evidence that in fact the bounds are saturated. $h(\vec{\alpha}_m)$, which characterizes the growth of the mean degeneracy in the commensurable limit $e \to \infty$, provides a very useful parametrization of the shape for scaling purposes.

We investigate the validity of a scaling description of the crossover for asymptotically large e and m, *i.e.* whether we can write

$$f(e, h_m) \approx h_m\, Y(\frac{e}{h_m^{\Delta}}) \tag{1}$$

in the limit $m \to \infty, e \to \infty$,with $y \equiv e/h_m^{\Delta}$ held fixed. Here $h_m \equiv h(\vec{\alpha}_m)$, Y is the scaling function, and Δ is the crossover exponent which depends on the system (AHO or cuboid). Note that $Y(y) \to 1$ as $y \to \infty$ and $Y(y) \sim y^{1/\Delta}$ as $y \to 0$. The full form of the crossover function Y is studied numerically.

2.1. The Three-dimensional Anisotropic Harmonic Oscillator

The eigenvalues of a three-dimensional anisotropic harmonic oscillator (AHO) with frequencies of vibration in the ratios $\alpha_x : \alpha_y : \alpha_z$ are given by $\varepsilon_{\vec{n}} = \hbar\omega_0(\alpha_x n_x + \alpha_y n_y + \alpha_z n_z)$ where $\vec{n} \equiv (n_x, n_y, n_z)$ labels the eigenstates, and $n_{x,y,z} = 0, 1, 2...$ (the zero-point energy has been subtracted off). In terms of the dimensionless, scaled energy $e = (\varepsilon/\hbar\omega_0)(\alpha_x\alpha_y\alpha_z)^{-1/3}$, the number of states with scaled energy $\leq e$ remains roughly the same as the aspect ratios are changed, and varies as e^3.

In the commensurable case, it proves convenient to take α's to be integers with no common divisor. Degeneracies come about from the various ways of representing an integer as a linear form $L_{\vec{\alpha}} = \alpha_x n_x + \alpha_y n_y + \alpha_z n_z$. Now, it can be shown that the asymptotic probability that an integer is of the linear form $L_{\vec{\alpha}}$ is unity if two of the components of $\vec{\alpha}$ do not have a common divisor. This implies that $N_{level}(e)$ grows as e for large e, and $h(\vec{\alpha}) = (\alpha_x\alpha_y\alpha_z)^{\frac{1}{3}}$. With irrationally related aspect ratios, no two sets (n_x, n_y, n_z) can correspond to the same energy, *i.e.* no degeneracies, implying $N_{level} \sim e^3$.

To study the irrational-rational crossover we consider a sequence $\vec{\alpha}_m$ which leads to aspect ratios $1 : \tau : \tau^2$ in the limit $m \to \infty$. Here $\tau = (\sqrt{5}+1)/2$ is the golden mean. We choose $\alpha_x, \alpha_y, \alpha_z$ to be $\beta_m^2, \beta_m\beta_{m+1}, \beta_{m+1}^2$ respectively where β_m is the m'th member of the Fibonacci sequence ($\beta_1 = \beta_2 = 1, \beta_m = \beta_{m-1} + \beta_{m-2}$).

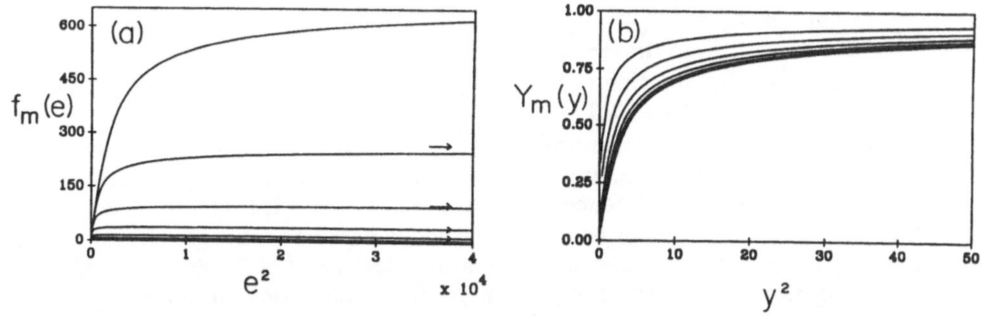

Figure 2 (a) The function $f_m(e)$ is plotted against e^2 for the generations m =3 to 8 (moving upwards) for the AHO. Each of the curves displays incommensurable behaviour at small e and asymptotes, at large e, to a constant $h(\vec{\alpha}_m)$ marked by arrows. (b) $Y \equiv f(e)/h_m$ is plotted against the scaling variable $y \equiv e^2/h_m$ for the AHO for generations $m = 3$ to 8 (moving downwards). The curves seem to approach a limiting curve as the generation index m increases, suggesting the existence of scaling for the incommensurable-commensurable crossover.

Figure 2 shows numerically generated $f(e)$ plotted against e for a few sets $\vec{\alpha}_m$ for the AHO. Each curve shows a crossover from irrational behaviour at low energies $e < e^*(\vec{\alpha}_m)$ to rational behaviour at large energies $e > e^*(\vec{\alpha}_m)$. However, the crossover energy $e^*(\vec{\alpha}_m)$ shifts to larger values as the generation index m gets larger. Scaling would be valid if the asymptotic constant $h(\vec{\alpha})$ essentially determines the behaviour of the whole function $f(e)$. In order to test the scaling form (1), in Fig. 2b we have plotted $f_m(e)/h_m$ against the scaling variable y (with $\Delta = 1/2$ in (1)) for several generations m. The curves for different m seem to approach a limit smoothly and rapidly, suggesting that in fact a scaling form describes the incommensurable - commensurable crossover in the AHO.

2.2. Particle in a Cuboid

For a free particle in a finite cuboid with periodic boundary conditions, with sides in the ratio $\alpha_x : \alpha_y : \alpha_z$, the eigenvalues are $\varepsilon_{\vec{n}} = (\hbar^2/2mL^2)\ (\alpha_x n_x^2 + \alpha_y n_y^2 + \alpha_z n_z^2)$, where $\vec{n}$ labels the eigenstates and $n_{x,y,z} = 0, \pm1, \pm2...$ The lengths of the edges of the cuboid are $2\pi L/\sqrt{\alpha_x}$ *etc.* We define a dimensionless, scaled energy as $e \equiv (2m\pi^2/\hbar^2)\varepsilon L^2(\alpha_x\alpha_y\alpha_z)^{-1/3}$.

In the commensurable case, we choose α's to be integers. Large degeneracies arise from different ways of representing an integer as the quadratic form $Q_{\vec{\alpha}} = \alpha_x n_x^2 + \alpha_y n_y^2 + \alpha_z n_z^2$. Let $p(\vec{\alpha})$ denote the asymptotic probability that an integer is expressible in at least one way as the quadratic form $Q_{\vec{\alpha}}$. Then the asymptotic value of $f(e)$ is given by $h(\vec{\alpha}) = p(\vec{\alpha})(\alpha_x\alpha_y\alpha_z)^{\frac{1}{3}}/\pi^2$. Finding $p(\vec{\alpha})$ for a general set $\vec{\alpha}$ is an unsolved number theoretic problem, but we have estimated it for some cases as explained below.

We consider the sequence $\vec{\alpha}_m$ with the three components chosen as $\beta_m, \beta_{m+1}, \beta_{m+2}$ where β_m is the m'th member of the Fibonacci sequence. As $m \to \infty$, the aspect ratios will approach the ratio $1 : \tau : \tau^2$.

For the simplest case $\vec{\alpha} = (1,1,1)$ (corresponding to $m = 0$ of our sequence), Gauss proved [3] that an integer k cannot be represented as $n_x^2 + n_y^2 + n_z^2$ if and only if k belongs to the set $\mathcal{A}_0 \equiv \{4^l(8t+7);\ l,t = 0,1...\}$. That is, all integers which cannot be written as sums of three squares fall into patterns, each characterized by a period 4^l8 and a phase 4^l7. The fraction of integers contained in $\mathcal{A}_0$ is $\frac{1}{8}(1 + \frac{1}{4} + \frac{1}{4^2}....) = \frac{1}{6}$ which implies that $p(1,1,1) = 5/6$.

For each $\vec{\alpha}$ from the sequence under study, we found patterns in the set $\mathcal{A}_{\vec{\alpha}}$ of integers not representable as $Q_{\vec{\alpha}}$, and thus estimated the asymptotic probability $p(\vec{\alpha})$. There are two important differences from the case $\vec{\alpha} = (1,1,1)$ discussed above. Firstly, not all members of $\mathcal{A}_{\vec{\alpha}}$ belong to periodic patterns, though almost all do. Secondly, the patterns are not necessarily non-intersecting. Let $\mathcal{B}_{\vec{\alpha}}$ denote a subset of $\mathcal{A}_{\vec{\alpha}}$, comprising those members of $\mathcal{A}_{\vec{\alpha}}$ which are fully characterized by patterns. The sets $\mathcal{B}_{\vec{\alpha}}$ can be found analytically and the corresponding densities $1-p(\vec{\alpha})$ estimated; detailed results are given in [2]. Furthermore, by numerical enumeration of $\mathcal{A}_{\vec{\alpha}}$ we found that as $e \to \infty$, the density of $\mathcal{A}_{\vec{\alpha}}$ approaches that of $\mathcal{B}_{\vec{\alpha}}$, for each $\vec{\alpha}$ (see Fig. 2a). This means that the leftovers of $\mathcal{A}_{\vec{\alpha}}$ (those members of $\mathcal{A}_{\vec{\alpha}}$ not characterized by patterns) constitute a vanishing fraction of integers, and do not contribute to $p(\vec{\alpha})$.

The identification of periodic patterns of the form $Pt + a$, and thus $\mathcal{B}_{\vec{\alpha}}$, for a given quadratic form $Q_{\vec{\alpha}}$ is a number-theoretic problem. The method involves examining

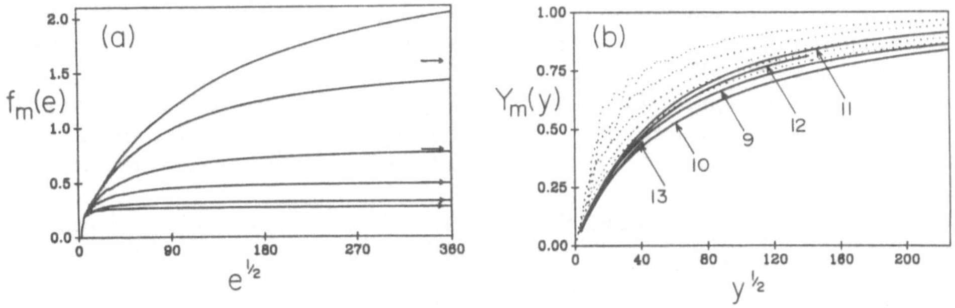

Figure 3 (a) The function $f_m(e)$ is plotted against $e^{1/2}$ for the cuboid for $m = 3$ to 8 (moving upwards). The calculated asymptotic values $h(\vec{\alpha}_m)$ of $f(e)$ are marked by arrows. (b) $Y \equiv f(e)/h_m$ is plotted against the scaling variable $y \equiv \sqrt{e}/h_m$ for the cuboid. The dotted curves are for $m = 4$ to 8 (moving downwards) while the solid curves for $m = 9$ to 13 are marked in the figure. There are oscillations as the generation index m is changed, but their amplitude decreases as m increases, suggesting that they approach an asymptotic scaling function.

the set of remainders generated when $Q_{\vec{\alpha}}$ is divided by P, as n_x, n_y and n_z are allowed to range over all values. A large number of candidate P's was examined on a computer [2]. Our estimate of $p(\vec{\alpha})$ is a definite upper bound as it can be checked that any additional pattern would increase the density of $\mathcal{B}_{\vec{\alpha}}$ and thus decrease $p(\vec{\alpha})$. We tested our estimate of the density of $\mathcal{B}_{\vec{\alpha}}$ against the density of numerically generated $\mathcal{A}_{\vec{\alpha}}$ obtained by examining integers up to 10^8. We found good agreement, suggesting strongly that the bounds are saturated.

Using our estimate of $p(\vec{\alpha}_m)$, we find $h(\vec{\alpha}_m)$, which is the asymptotic value of $f(e)$. Figure 3a shows the numerically generated $f(e)$ plotted against e for a few sets $\vec{\alpha}_m$. The corresponding values of $h(\vec{\alpha}_m)$ are indicated as well. The fact that each of the curves tends to asymptote (as $e \to \infty$) to the calculated value implies that the leftovers (uncharacterized members of $\mathcal{A}_{\vec{\alpha}}$) do not contribute to $h(\vec{\alpha})$, and $f(e)$ exhibits commensurable behaviour asymptotically. However, leftovers abound in the low e regime, and give it its incommensurable character. Each curve displays the incommensurable - commensurable crossover with increasing e. Figure 3b shows a test of the scaling form (1), with $Y_m \equiv f(e)/h_m$ plotted against $y^{1/2} \equiv e^{1/2}/h_m$ (corresponding to $\Delta = 2$). Most curves crowd in patches, but, in contrast to the AHO, the approach to a limit is not monotonic in m. However, oscillations seem to die down with increasing m, suggesting that scaling is valid asymptotically.

3. DISTRIBUTIONS OF ENERGY GAPS AND DEGENERACIES IN THE COMMENSURABLE REGIME

While the mean spacing between successive distinct eigenvalues and mean degeneracy in the commensurable regime are both determinable from the asymptotic constant $h(\alpha_m)$, it is also of interest to ask for the full probability distributions of level spacings and degeneracies.

For cuboids, the distribution of spacings involves knowing the probability of encountering strings of successive integers not expressible as the quadratic form $Q_{\vec{\alpha}}$. Since we know $\mathcal{B}_{\vec{\alpha}}$, which, in the commensurable regime, asymptotes to $\mathcal{A}_{\vec{\alpha}}$, this probability can be figured out, though tediously [2]. The result is that for each m, the probability distribution of spacings consists of a finite sequence of delta functions, in striking contrast to the Poisson distribution, which is supposed to hold in the incommensurable regime.

The probability distribution of degeneracies in cuboids cannot be characterized analytically. Numerical study [4] shows that for each m, there is a nontrivial probability distribution of scaled degeneracies $d \equiv$ degeneracy$/e^{1/2}$. The distribution varies linearly for low d, and decays exponentially for large values of d.

4. CONCLUSION

For both anisotropic harmonic oscillators and cuboids, a crossover from incommensurable to commensurable behaviour – with concommitant strong changes in the spectrum (Fig. 1) – can be induced by changing either e (a measure of size or energy) or aspect ratios $\vec{\alpha}_m$ (which characterize the shape). The conclusion of this study is that a single parameter y, involving a combination of shape and size, in fact determines the full crossover, through the scaling form Eq. 1 (see also Figs. 2 and 3). For instance, for the cuboid, the scaling parameter is $e^{1/2}/h(\vec{\alpha}_m)$ – i.e. the shape enters through the constant h which characterizes the growth of mean degeneracies in the commensurable regime. An analytical demonstration of shape-size scaling, and a derivation of the scaling function $Y(y)$, would be most desirable.

REFERENCES

[1] M. V. Berry, Proc. Roy. Soc. A 413, 183 (1987); V. Subrahmanyam and M. Barma Phase Transitions 24-26, 419 (1991).
[2] V. Subrahmanyam and M. Barma, J. Phys. A 24, 4303 (1991)
[3] E. Landau *Elementary number theory* (Chelsea: New York, 1927) p. 151-164.
[4] C. Itzykson and J. M. Luck, J. Phys. A19, 211 (1986); M. Barma and V. Subrahmanyam, Pramana - J. Phys. 36, 531 (1991) .

MEASUREMENTS OF LITHIUM CLUSTERS IONIZATION POTENTIALS

Ph. DUGOURD[1], D. RAYANE[1], P. LABASTIE[2], B. VEZIN[1], J. CHEVALEYRE[1], M. BROYER[1]

[1] *Laboratoire de Spectrométrie Ionique et Moléculaire (URA CNRS 171) Université Lyon I, Bât. 205, 43 Bd du 11 Novembre 1918 69622 VILLEURBANNE Cedex, France*

[2] *Laboratoire de Physique Quantique, Université Paul Sabatier 118, Route de Narbonne, 31062 TOULOUSE Cedex, France*

ABSTRACT. The ionization potentials of Li_n clusters are obtained by photoionization. Shell effects are observed at n = 8 and n = 20, as well as strong odd/even alternations. The results are compared with the spherical shell model and the ab initio calculations.

1. Introduction

Among metal clusters, Li_n and Na_n clusters are the most accessible to theoretical calculations and experimental measurements. They are then good candidates to check the variation of electronic properties from the atom to the bulk. These properties are usually interpreted in terms of the shell model [1,2] which provides a global qualitative insight of the problem. Quantitative comparison with the experiments are also very useful, especially when high level ab initio calculations are available [3]. From this point of view, the evolution of the ionization potential as a function of the size from the atomic value to the bulk work function is particularly interesting. If results have been already obtained for sodium [4,5] clusters, nothing is known about lithium essentially because of the great difficulty to operate with highly reactive high temperature lithium vapor. We present in this paper the results obtained for lithium clusters and compare them with the best theoretical calculations.

2. Experiment

The lithium clusters are produced in a highly cooled molecular beam, by coexpansion of 10-100 mbars of lithium vapor and 1-5 bars of argon, through a 100 µm nozzle. The dearth of experimental studies on lithium clusters is due to the high temperatures necessary to heat the metal and to the corrosive properties of lithium. Therefore a special oven was built [6]. It is comprised of a cartridge made of titanium-zirconium-molybdenum which is heated by tungsten filaments and can sustain temperatures up to

P. Jena et al. (eds.), Physics and Chemistry of Finite Systems: From Clusters to Crystals, Vol. I, 555–560.

1800 K. The clusters are then ionized in a second chamber by means of a frequency doubled excimer pumped tunable dye laser. The laser energy per pulse was lower than 100 μJ/cm^2. Ions are collected and mass analysed through a time of flight mass spectrometer for a given laser wavelength. By this technique, mass spectra are recorded for longer and longer wavelengths until clusters ions disappear. They are digitalized and analysed in a computer. The photoionization cross section is then deduced as a function of the photon energy, allowing the determination of the ionization threshold by linear extrapolation.

3. Results and discussion

Figure 1 shows typical mass spectra recorded for different laser wavelengths.

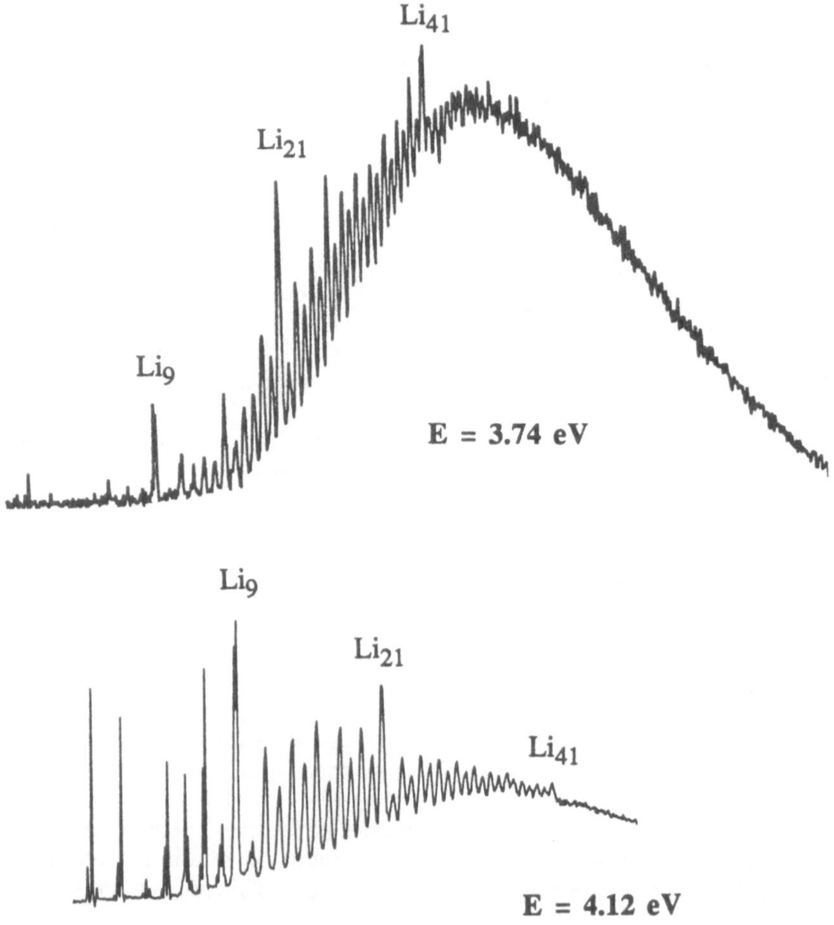

Figure 1. Mass spectra of Li_n clusters as a function of the laser wavelength.

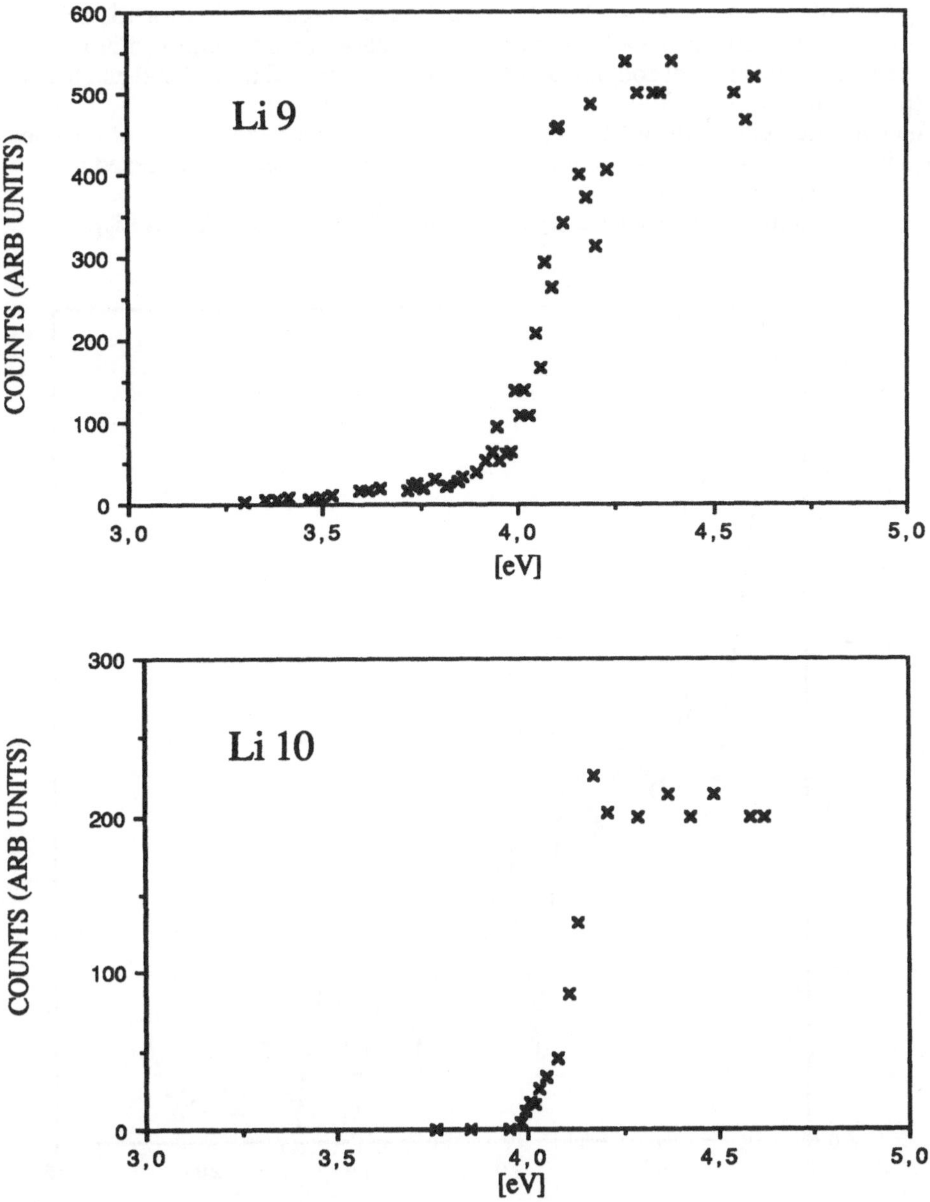

Figure 2. Photoionization efficiency curves for Li_9 and Li_{10}.

Figure 2 illustrates two examples of photoionization cross section for Li_9 and Li_{10}. The Li_9 curve exhibits a characteristic two slopes-shape, already observed in K_9 [7-8]. This behaviour corresponds roughly in terms of the shell model to the band gap between the $1p^6$ and $1d^1$ levels of Li_9.

The ionization potentials of Li_n clusters obtained by linear extrapolation of the photoionization cross section are shown in figure 3. Two main features must be noted :

- a clear even-odd alternation
- an abrupt decrease beyond n = 8 and n = 20 for clusters Li_9 and Li_{21}.

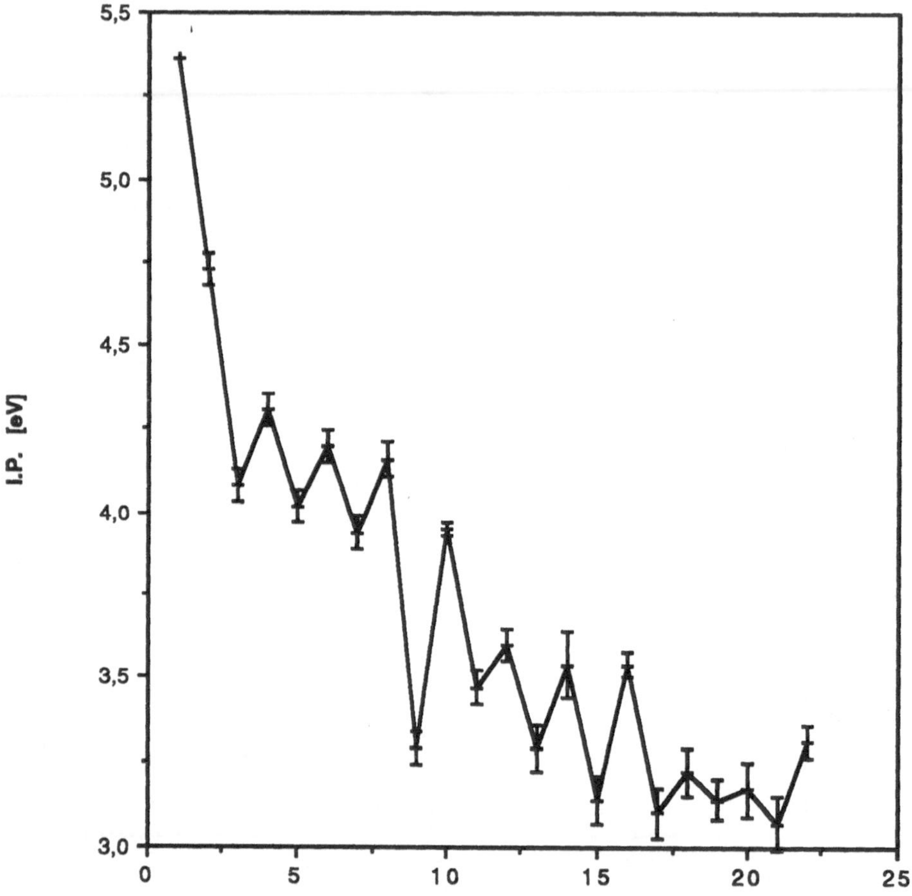

Fig. 3. Ionization potential of Li_n clusters (in eV) as a function of n.

This last effect may be interpreted by the shell model : n = 8 and n = 20 correspond to closed shells, then systems having 8 and 20 electrons (Li_8, Li_{20}, Li_9^+ and Li_{21}^+) are very stable, while the systems having 9 and 21 electrons (Li_9 and Li_{21}) are clearly less stable. However the spherical shell model cannot explain the odd/even alternation which is connected to strong ellipsoïdal deformations [9]. Moreover, for such small systems, the position of the nuclei remains important, and ab initio calculations including geometrical structure [3] are necessary for quantitative comparison with the experiment. This comparison is given in Table 1 for $n \leq 10$. The experimental Ionization Potential obtained by linear extrapolation is expected to be found between the adiabatic and the vertical Ionization Potential. The agreement between theory and experiment remains within ± 0.1 eV and may be considered as very good. It illustrates the high quality of the Koutecky group calculations [3].

TABLE 1. Ionization Potentials (in eV) of Li_n clusters as a function of n. IP (a) and IP (b) are respectively the adiabatic and vertical theoretical Ionization Potentials from ref. 3.

n	IP exp.	Incertitudes	IP (a)	IP (b)
1	5.36	0.01	-	-
2	4.73	0.05	5	-
3	4.08	0.05	3.95	4.18
4	4.31	0.05	4.55	4.61
5	4.02	0.05	4.15	4.48
6	4.20	0.05	4.32	4.53
7	3.94	0.05	3.82	3.91
8	4.16	0.05	4.23	4.57
9	3.29	0.05	3.15	3.70
10	3.95	0.02	3.87	3.91

4. Conclusion

In conclusion, we have measured the Li_n Ionization Potentials. If the results may be interpreted qualitatively by the ellipsoïdal shell model, ab initio calculations are necessary for a satisfactory quantitative interpretation of the experimental results.

References

[1] Knight, W.D., Clemenger, K., de Heer, W.A., Saunders, W.A., Chou, M.Y., Cohen, M.L. (1984) Phys. Rev. Lett. 52, 2414.

[2] Knight, W.D., de Heer, W.A., Saunders, W.A., Clemenger, K. (1987) Chem. Phys. Lett. 134, 1.

[3] Boustani, I., Pervestorf, W., Fantucci, P., Bonacic-Koutecky, V., Koutecky, J (1987) Phys. Rev. B 35, 9437.

[4] Kappes, M.M., Schür, M., Rothlisberger, U.,Yeretzian, C., Schumacher, E. (1988) Chem. Phys. Lett. 143, 251.

[5] Honea, E.C., Horner, M.L., Persson, J.L., and Whetten, R.L. (1990) Chem. Phys. Lett. 171, 148.
[6] Wolf, J.P., Delacretaz, G., and Wöste, L. (1989) Phys. Rev. Lett. 63, 1946.
[7] Knight, W.D., Clemenger, K., de Heer, W.A., Saunders, W.A., Phys. Rev. B 31 (1985) 1359.
[8] Bréchignac, C., Cahuzac, Ph., Z. für Phys. D 3 (1986) 21.
[9] Clemenger, K.(1985) Phys. Rev. B 32, 1953.

DOUBLE ZETA BASIS SETS IN CARBON CLUSTER CALCULATIONS

DAVID W. EWING* AND ISAIAH SHAVITT
Department of Chemistry
Ohio State University
Columbus, Ohio 43210 USA

1. Introduction

Two double zeta plus polarization basis sets are commonly used in ab initio calculations on carbon clusters. The first, Hariharan and Pople's 6-31G* basis set [1], gives a single zeta description of the core electrons and a double zeta description of the valence electrons. The second, the Huzinaga/Dunning DZP basis set [2], gives a double zeta description of core and valence electrons. The 6-31G* basis set has become popular because it is smaller in size, and hence cheaper to use, than DZP, and because of its efficient implementation within the widely used GAUSSIAN system of computer programs.

The 6-31G* basis set does not always give results of double zeta quality, however. Two examples involving carbon clusters are discussed here. First, it is shown that the DZP basis set can produce somewhat better vibrational frequencies and isotopomer frequency shifts. Second, it is shown that the 6-31G* basis set predicts the wrong structures for C_7 and C_9.

2. Methods

The present calculations were performed with the GAUSSIAN 90 [3] and ACES II [4] systems of programs, both running on the CRAY Y-MP8/864 computer at the Ohio Supercomputer Center. Singlet states (all molecules studied except C_4) and triplet states (C_4) were investigated with the restricted and unrestricted Hartree-Fock (HF) methods, respectively. Electron correlation was included via second-order many-body perturbation theory (MP2). Core electrons were included in the correlation scheme except for the largest C_7 calculations.

*Permanent address: Department of Chemistry
John Carroll University
Cleveland, Ohio 44118 USA

P. Jena et al. (eds.), Physics and Chemistry of Finite Systems: From Clusters to Crystals, Vol. I, 561–567.

The carbon basis sets used were the Hariharan and Pople 6-31G* basis set [1], the Huzinaga/Dunning (4s2p) basis set [2] augmented with a set of d functions (DZP), and the Dunning correlation consistent polarized valence triple zeta (4s3p2d1f) basis set (pVTZ) [5]. For the MP2/DZP calculations on C_7 and C_9 a six-component set of d functions, with an exponent of 0.654, was used. For all other MP2/DZP calculations a five-component set of d functions, with an exponent of 0.75, was used. Only linear and bent (C_7 and C_9) structures were studied. These structures were optimized, except for the C_7 calculations utilizing the pVTZ basis set, using analytical energy gradients. C_7 was optimized at the MP2/pVTZ level of theory via parabolic interpolation. The harmonic frequency calculations employed second derivatives computed numerically using analytically calculated first derivatives.

3. Results and Discussion

3.1. FREQUENCIES AND ISOTOPOMER FREQUENCY SHIFTS

Table 1 compares the MP2/6-31G* and MP2/DZP calculated vibrational frequencies to experiment for the vibrational modes of C_n, n=2-5, for which experimental frequencies are known. The MP2/DZP values are often somewhat better, with the notable exception of the Σ_g mode of C_3. There are two modes listed in Table 1 for which MP2/6-31G* and MP2/DZP levels of theory both give poor results. These are the Π_u mode of C_3 and the Σ_u mode of C_5. HF/6-31G* frequencies are also given in Table 1 in order to help explain this.

HF/6-31G* frequencies are systematically too high for three reasons: lack of electron corrleation effects, incomplete basis set, and lack of anharmonicity. As electron correlation effects are included the calculated frequencies should decrease toward experimental values. This is seen to occur in Table 1 for all modes except the C_3 Π_u and the C_5 Σ_u. The Π_u mode of C3 is known to be quite anharmonic [14]; the Σ_u mode of C5 is known to require higher order anharmonic effects for proper treatment [15]. Hence, at the levels of theory used in the present study, the frequencies of these two modes cannot be well predicted.

This trend, the decrease in calculated harmonic frequency with the inclusion of electron correlaton effects, appears to be an indicator of the accuracy of MP2/DZP frequencies. If the calculated frequencies decrease upon inclusion of electron correlation effects, the MP2/DZP frequency is within, and usually higher than, a few percent of the experimental value. This has been more thoroughly discussed in a recent paper by the present authors [16].

If frequencies are well predicted at the MP2/DZP level of theory then isotopomer frequency shifts will also be well predicted. This is illustrated for the Σ_u mode of C_4 in Table 2. Whereas MP2/6-31G* frequency shifts require scaling for good agreement with experiment, MP2/DZP shifts do not. More examples can be found in Reference 16.

TABLE 1. Comparison of vibrational frequencies (cm^{-1}) of linear carbon molecules calculated at various levels of theory. All values are unscaled.

Molecule	Mode	HF/6-31G*[a]	MP2/6-31G*[b]	MP2/DZP	Experiment[c]
C_2	Σ_g	1940	1910	1867	1855[d]
C_3	Π_u	154	190	198	70[e]
C_3	Σ_g	1367	1220	1190	1235[e]
C_3	Σ_u	2311	2183	2128	2040[e]
C_4	Π_u	210	187	175	172[f]
C_4	Σ_u	1740	1586	1550	1543[g]
C_5	Σ_u	2344	2358	2320	2164[h]

[a]For C3 and C5: ref. 6.
[b]For C4: ref. 7; for C5: ref. 8.
[c]Matrix values.
[d]Reference 9.
[e]Reference 10.
[f]Reference 11.
[g]Reference 12.
[h]Reference 13.

TABLE 2. Frequency shifts (cm^{-1}) for the Σ_u band of C_4. Values in parentheses are the frequencies (cm^{-1}) of the unsubstituted isotopomer.

Isotopomer	MP2/6-31G*[a,b] scaled shifts	MP2/DZP unscaled shifts	Experiment[c]
12-12-12-12	(1587.6)	(1550.2)	(1543.4)[d]
13-12-12-12	14.8	14.9	14.6
12-13-12-12	16.1	16.1	15.9
13-13-12-12	32.2	32.4	31.8
13-12-13-12	29.9	30.1	29.6
13-12-12-13	30.1	30.2	29.6[e]
12-13-13-12	30.1	30.2	29.6
13-13-13-12	45.3	45.4	44.6
13-13-12-13	46.4	46.7	45.7
13-13-13-13	60.7	61.0	59.8[f]

[a]Reference 17.
[b]Reference 18.
[c]Reference 12 unless otherwise noted. Measured in argon matrices.
[d]The gas phase frequency is 1548.9 cm^{-1} (Ref. 19).
[e]Reference 20.
[f]Reference 11.

TABLE 3. Structures (bond lengths in Å, angles in degrees) and total energies (a.u.) for C_7 and C_9. The numbering of the carbon chains is from the outside in.

Property		MP2/6-31G*[a]	MP2/DZP	MP2/pVTZ
C_7	C_1-C_2	1.301	1.314	1.297
	C_2-C_3	1.295	1.306	1.292
	C_3-C_4	1.282	1.291	1.278
	$<C_1C_2C_3$	178.5	180.0	180.0
	$<C_2C_3C_4$	178.3	180.1	180.0
	$<C_3C_4C_5$	176.7	179.9	180.0
Total energy		-265.5735	-265.65719	-265.77525
C_9	C_1-C_2	1.302	1.315	---
	C_2-C_3	1.298	1.309	---
	C_3-C_4	1.281	1.290	---
	C_4-C_5	1.286	1.294	---
	$<C_1C_2C_3$	178.7	180.1	---
	$<C_2C_3C_4$	178.5	180.0	---
	$<C_3C_4C_5$	177.4	180.0	---
	$<C_4C_5C_6$	177.9	180.1	---
Total energy		-341.5106	-341.61623	---

[a]Reference 18.

TABLE 4. MP2/DZP vibrational frequencies (cm^{-1}) and relative IR intensities for C_7.

Symmetry	Frequency	Rel. int.	Symmetry	Frequency	Rel. int.
Π_u	84	0.01	Σ_g	561	a
Π_g	184	a	Σ_u	1068	0.01
Π_u	274	b	Σ_g	1543	a
Π_u	467	b	Σ_u	1927	0.26
Π_g	471	a	Σ_g	2183	a
			Σ_u	2390	1.00

[a]Not IR active.
[b]Less than 0.01.

TABLE 5. MP2/DZP vibrational frequencies (cm^{-1}) and relative IR intensities for C_9.

Symmetry	Frequency	Rel. int.	Symmetry	Frequency	Rel. int.
Π_u	53	b	Π_g	464	a
Π_g	130	a	Σ_u	854	b
Π_u	215	b	Σ_g	1237	a
Π_g	289	a	Σ_u	1604	0.09
Π_u	411	b	Σ_g	1898	a
Σ_g	442	a	Σ_u	2102	0.38
Π_u	452	b	Σ_g	2260	a
			Σ_u	2422	1.00

[a]Not IR active.
[b]Less than 0.01.

3.2. STRUCTURES OF C7 AND C9

Kurtz and Adamowicz recently published a MP2/6-31G* study of several carbon clusters [18]. Among their conclusions is that C_7 and C_9 are somewhat bent, rather than perfectly linear. Table 3 gives these results, along with the present MP2/DZP structures. It is seen that the MP2/DZP calculations produce the qualitatively different result that these clusters are linear, not bent. To further explore this question, a MP2/pVTZ optimization was performed on C_7. This structure is also given in Table 3, where it is seen that the use of this more complete basis set predicts C_7 to be linear. The MP2/pVTZ level of theory is expected to produce excellent structural results.

Table 4 gives MP2/DZP vibrational frequencies for linear C_7. The lack of imaginary frequencies confirms that the linear structure is a minimum on the potential energy surface. Experimental gas phase frequencies for C_7 have been measured at 2138 cm^{-1} [21] and 1898 cm^{-1} [22]. Table 5 gives MP2/DZP vibrational frequencies for linear C_9. Again, no imaginary frequencies are calculated. An experimental gas phase frequency for C9 has been measured at 2014 cm^{-1} [23].

4. Conclusions

MP2/DZP vibrational frequencies and isotopomer frequency shifts for small linear carbon clusters are generally significantly closer to experiment than the corresponding MP2/6-31G* values. Whereas the MP2/6-31G* level of theory predicts C_7 and C_9 to be bent, MP2/DZP correctly predicts these molecules to be linear.

5. Acknowledgements

This work was supported by a grant of Cray time from the Ohio Supercomputer Center. D.W.E. thanks John Carroll University for a George E. Grauel Faculty Fellowship and the Ohio State University for supporting his stay.

6. References

[1] P.C. Hariahran and J.A. Pople, Chem. Phys. Lett. **66**, 217 (1972).
[2] S. Huzinaga, J. Chem. Phys. 42, 1293 (1965); T.H. Dunning, Jr., J. Chem. Phys. 53, 2830 (1970).
[3] GAUSSIAN 90, M.J. Frisch, M. Head-Gordon, G.W. Trucks, J.B. Foresman, H.B. Schlegel, K. Raghavachari, M.A. Robb, J.S. Binkley, C. Gonzalez, D.J. Defrees, D.J. Fox, R.A. Whiteside, R. Seeger, C.F. Melius, J. Baker, R.L. Martin, L.R. Kahn, J.J.P. Stewart, S. Topiol, and J.A. Pople, Gaussain, Inc., Pittsburgh, PA, 1990.

[4] ACES II is a new program package from the Quantum Theory Project of the University of Florida. The SCF, integral transformation, and correlation energy codes in this package were written J.F. Stanton, J. Gauss, J.D. Watts, W.J. Lauderdale, and R.J. Bartlett. See, e.g., J.F. Stanton, J. Gauss, J.D. Watts, and R.J. Bartlett, J. Chem. Phys. **94**, 4334 (1991). The package also includes the VMOL integral program written by P.R. Taylor and J. Almlof.
[5] T.H. Dunning, Jr., J. Chem. Phys. **90**, 1007 (1989).
[6] K. Raghavachari and J.S. Binkley, J. Chem. Phys. **87**, 2191 (1987).
[7] D. Michalska, H. Chojnacki, B.A. Hess, Jr., and L.J. Schaad, Chem. Phys. Lett. **141**, 376 (1987).
[8] J.M.L. Martin, J.P. Francois, and R. Gijbels, J. Chem. Phys. **90**, 3403 (1989).
[9] B. Rosen, Spectroscopic Data Relative to Diatomic Molecules (Pergamon, New York, 1970).
[10] W. Weltner, Jr. and D. McLeod, Jr., J. Chem. Phys. **40**, 1305 (1964).
[11] P.A. Withey, L.N. Shen, and W.R.M. Graham, J. Chem. Phys. **95**, 820 (1991).
[12] L.N. Shen and W.R.M. Graham, J. Chem. Phys. **91**, 5115 (1989).
[13] M. Vala, T.M. Chandrasekhar, J. Szczepanski, R. Van Zee, and W. Weltner, Jr., J. Chem. Phys. **90**, 595 (1989).
[14] See, e.g., F.J. Northrup, T.J. Sears, and E.A. Rohlfing, J. Mol. Spec. **145**, 74 (1991).
[15] P. Botschwina and P. Sebald, Chem. Phys. Lett. **160**, 485 (1989).
[16] D.W. Ewing and I. Shavitt, manuscript in preparation.
[17] J.M.L. Martin, J.P. Francois, and R. Gijbels, J. Chem. Phys. **93**, 8850 (1990).
[18] J. Kurtz and L. Adamowicz, Astrophys. J. **370**, 784 (1991).
[19] J.R. Heath and R.J. Saykally, J. Chem. Phys. **94**, 3271 (1991).
[20] L.N. Shen, P.A. Withey, and W.R.M. Graham, J. Chem. Phys. **94**, 2395 (1991).
[21] J.R. Heath, R.A. Sheeks, A.L. Cooksy, and R.J. Saykally, Science **249**, 895 (1990).
[22] J.R. Heath, A. Van Orden, E. Kuo, and R.J. Saykally, Chem. Phys. Lett. **182**, 17 (1991).
[23] J.R. Heath and R.J. Saykally, J. Chem. Phys. **93**, 8392 (1990).

ELECTRONIC STRUCTURE OF DEPOSITED MASS-SELECTED Pt CLUSTERS ON SILVER FILMS

P. FAYET, F. PATTHEY, H.-V. ROY, Th. DETZEL and W.-D. SCHNEIDER

Université de Lausanne
Institut de physique expérimentale
CH-1015 Lausanne
Switzerland

The knowledge of electronic structure as a function of cluster size is crucial for the understanding of many properties of condensed matter at the nanometer scale such as the metal-insulator transition, the behaviour of the melting temperature and the critical size for heterogeneous catalysis. Up until now, most photoemission studies of the electronic structure of supported small particles have been carried out on evaporated or sputtered films of metallic elements deposited on amorphous carbon or on silica [1-5]. It has been only recently that data has been obtained from deposited monosize mass-selected clusters [6,7]. Here we report on the evolution of the 5d states and 4f core levels of Pt with cluster size using photoemission spectroscopy (XPS, UPS) on submonolayer quantities of mass-selected Pt clusters (Pt_1, Pt_2 and Pt_3) deposited *in situ* on cold condensed Ag films at 110 K [8].

Platinum clusters were produced by bombarding a platinum target with a Xe-ion beam provided by a cold reflex discharge ion source (CORDIS) with typical current densities of 10 mA/cm^2 at 30 keV kinetic energy. Ionized clusters were then mass-selected by a quadrupole mass spectrometer and transported to the substrate surface for deposition by an RF-mode only quadrupole. The substrates were cold-condensed films of silver held at 110 K in the center of a custom-built (VSW) electron spectrometer at 2×10^{-10} mbar, which intercepted the monodispersed cluster beam perpendicularly. Pt coverages, corresponding to 0.1 monolayer (ML) as determined by the cluster beam density and the deposition time have been prepared. During deposition, the kinetic energy of the beam was 10±5 eV. Subsequently X-ray (MgK_α) and UV (HeI) photoelectron spectra were recorded with an instrumental resolution of 1.0 eV and 0.1 eV respectively (see ref. [8] for more details).

P. Jena et al. (eds.), Physics and Chemistry of Finite Systems: From Clusters to Crystals, Vol. I, 569–573.

The excitation spectra of clean silver films, measured with HeI exhibit a wide (4 eV) and flat s-p band extending from the Fermi edge to the onset of the intense 4d band. With the presence of Pt atoms or clusters additional spectral features appear on the top of the silver s-p band, centered at about 2 eV below the Fermi energy.

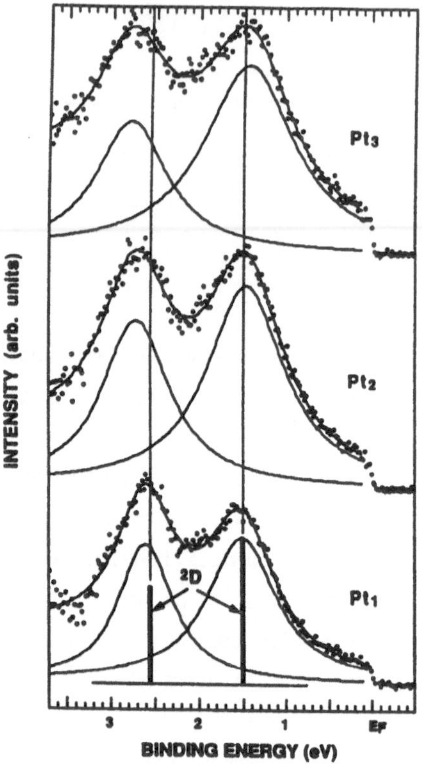

Figure 1. Pt 5d emission (dots) obtained after subtraction of the silver substrate signal from the valence spectra of supported Pt clusters on silver at 110 K. The solid curves are the result of a fit where the data were approximated by two Lorentzians convoluted with Gaussians of 0.1 eV FWHM. The vertical bars represent the spin-orbit splitting of 1.05 eV for Pt^+ in the $5d^9$ (^2D) configuration [9].

Figure 1 shows the cluster-substrate difference spectra for Pt_1, Pt_2 and Pt_3 from the Fermi level to about 4 eV below the Fermi energy. For each size we observe a two-peak emission structure positioned at about 1.5 eV and 2.6 eV binding energy. The interpretation of this doublet structure in the case of the Pt monomers (bottom spectrum) is rather straightforward when compared with the energy levels from free Pt-ions in their $5d^9$ (^2D) electronic configuration. From optical data the Pt^+ in the ^2D state shows a spin-orbit doublet with a splitting of 1.05 eV [9]. In Figure 1, this splitting is represented by the vertical bars and it is seen that the experimentally derived splitting for Pt_1 of 1.1±0.1 eV is in good agreement with these data and with earlier photoemission results [3,10]. In addition, there is a broadening of these peaks due to solid-state effects which is discussed below. Photoemission from the 5d shell creates a screened Pt-ion and a photoelectron. Therefore, the observation of spin-orbit split states in the photoemission spectra indicates that the Pt initial state in the present adsorbate system is characterized by a full $5d^{10}$ shell. These observations suggest that a strong interaction takes place between the Pt atoms and the metallic substrate which changes the $5d^96s^1$ electronic configuration of Pt atoms into a $5d^{10}$ one of the adsorbate system. Although the broadening of the split states could be due to different Pt adsorption sites on the cold-condensed silver film, we suggest that the atomic-like d levels have been mainly broadened by hybridization with the silver conduction electrons. From this point of view the Pt on Ag represents a quasi-two dimensional system of a localized level embedded in a conduction electron sea with the formation of virtual bound states [11,12].

With increasing cluster size, both spin-orbit components broaden and the higher $5d_{3/2}$ component shifts slightly towards higher binding energy. These observation can be quantified by fitting the doublet with two Lorentzians each convoluted with a Gaussian of 0.1 eV FWHM simulating the experimental resolution function of the electron energy analyser. The evolution of the binding energy position and linewidth (FWHM) of the two Pt 5d split states given by the fitting procedure are displayed in Figure 2. Although the same remark concerning the peak enlargement could be made here as well, the observed broadening of the components with increasing cluster size is believed to be caused mainly by an increase of the 5d-5d overlap within the cluster as a function of the number of atoms per cluster [13].

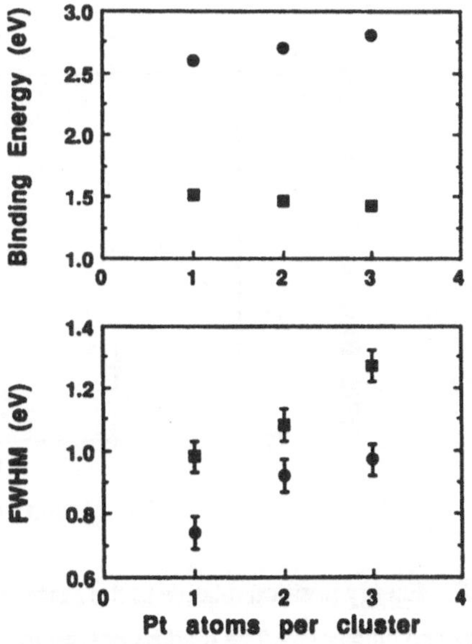

Figure 2. As function of the cluster size, the upper panel displays the energy position of the Lorentzian components for the Pt $5d_{5/2}$ (square) and Pt $5d_{3/2}$ (circles) peaks. The lower panel displays the corresponding peak width (FWHM). For the upper panel data the statistical error is within the size of the circles and squares.

The Pt 4f core level spectra have been analysed by a set of Doniach-Sunjic line shapes [14] convoluted with Gaussians of 1.0 eV FWHM representing the overall resolution of the spectrometer. The observed asymmetry ($0.05 \leq \alpha \leq 0.1$) points to the presence of metallic screening of the Pt core hole by the Ag conduction electrons. This fitting procedure allows us to plot the binding energy of the 4f levels as a function of the cluster size. Since the spin-orbit splitting between the $4f_{7/2}$ and $4f_{5/2}$ levels does not depend on the cluster size, it is sufficient to display in Figure 3 only the binding energy change of the $4f_{7/2}$ component relative to the Fermi level. The important observation here is that the binding energy shows negative shifts, i.e. the Pt-cluster core level binding energies are found to be lower than for bulk Pt-metal (71.2 eV [15]). On SiO_2 [7] and on amorphous carbon [16] substrates, Pt core levels which shift in the opposite direction (positive shifts) have been reported. The reason for negative shifts for Pt clusters on metallic substrates is twofold: i) a surface core level shift induced by the low coordination of Pt surface atoms, ii) an increase in the Pt 5d occupation in the small cluster leads to an increase of the repulsive Coulomb interaction between Pt 4f and Pt 5d electrons, hence the core level binding energy should decrease with increasing d occupation [17]. This is exactly what we are observing here: the core level binding energy for Pt atoms is about 0.9 eV lower than for the bulk Pt-metal; this difference could be directly correlated to i) the difference in cohesive energy between a Pt atom in a dilute bulk Pt-Ag alloy system and a Pt atom on the surface of such system [1 and ref. therein] and ii) the change in the 5d occupation between the adsorbed Pt-atoms ($5d^{10}$) and the Pt metallic bulk ($5d^{9+x}6s^{1-x}$, where $x \approx 0.4$).

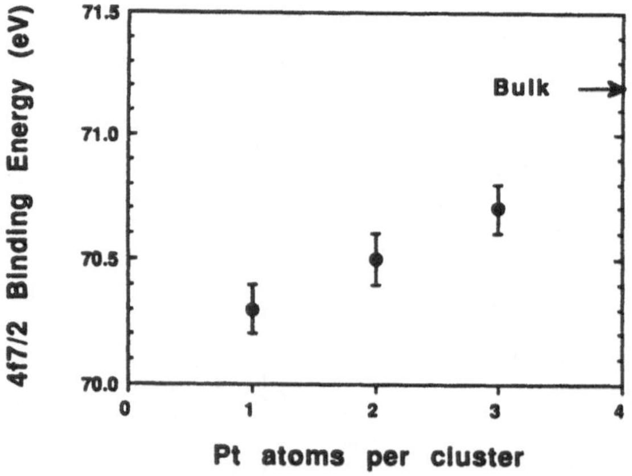

Figure 3. Energy position relative to the Fermi energy of the Pt $4f_{7/2}$ core levels as a function of the deposited cluster size.The binding energy for the bulk Pt core level (71.2 eV) is indicated by the arrow.

In conclusion, the negative core level shifts for supported Pt-clusters on metallic silver films confirm the findings of Mason [1], who attributes all observed shifts for valence bands and for core levels mainly to initial state effects. The broadening of the 5d peaks for Pt-monomers points to a considerable adsorbate-substrate hybridization. The 5d broadening and the energy shift with increasing cluster size are interpreted as being due to 5d-5d overlap between the atoms within the cluster.

Acknowledgements

The authors wish to thank E. Bullock for stimulating discussions. This work is supported by the Swiss National Science Foundation.

References :

1. M. G. Mason, Phys. Rev. B 27, 748 (1983) .
2. G. K. Wertheim, Z. Phys. D-Atoms, Molecules and Clusters 12, 319 (1989); and references therein.
3. S. Raaen and M. Strongin, Phys. Rev. B 32, 4289 (1985).
4. M. Cini, M. De Crescenzi, F. Patella, N. Motta, M. Sastri, F. Rochet, R. Pasquali, A. Balzarotti and C. Verdozzi, Phys. Rev. B 41, 5685 (1990).
5. Ch. Kuhrt and M. Harsdorff, Surf. Sci. 245, 173 (1991).
6. S. B. DiCenzo, S. D. Berry and E. H. Hartford Jr., Phys. Rev. B 38, 8465 (1988).
7. W. Eberhart, P. Fayet, D. M. Cox, Z. Fu, A. Kaldor, R. Sherwood and D. Sondericker, Phys. Rev. Lett. 64, 780 (1990).
8. P. Fayet, F. Patthey, H.-V. Roy, Th. Detzel and W.-D. Schneider, to be published in Surf. Sci.
9. C. E. Moore : Atomic Energy Levels, U. S. National Bureau of Standards, National Standard Reference Data Serie - 35 (U. S. GPO, Washington, D. C., 1971), Vol. 3.
10. S. Hüfner, in : Photoemission in Solids, eds. L. Ley and M. Cardona, Topics in Applied Physics 27, 173 (1979).
11. J. Friedel, Nuovo Cimento (Suppl.) 7, 41 (1961).
12. P. W. Anderson, Phys. Rev. 124, 41 (1961).
13. N. J. Shevchik, J. Phys. F 5, 1860 (1976).
14. S. Doniach and M. Sunjic, J. Phys. C3, 285 (1970).
15. G. P. Williams, X-Rays Data Booklet (1985).
16. W. Eberhart, P. Fayet, D. M. Cox, Z. Fu, A. Kaldor, R. Sherwood and D. Sondericker, Physics Scripta 41, 892 (1990).
17. E. E. Watson and M. L. Perlman, Struct. Bonding (Berlin) 24, 83 (1975).

HOMONUCLEAR METALLIC MICROCLUSTERS: STRUCTURE AND ENERGETICS

CLARENCE W. FINLEY
Chemistry Department, Pennsylvania State University, New Kensington, PA 15608

JOHN FERRANTE and PHILLIP ABEL
NASA Lewis Research Center, Cleveland, OH 44135

ABSTRACT. We have calculated the energy and equilibrium geometry of Ni, Pd and Pt microclusters (N ≤ 10) using the Universal Binding Energy version of the Embedded Atom Method. Agreement with available experimental data is discussed. A systematic trend in the average binding energy per atom was found for each element studied. Parameterizations of the trend are discussed.

1 Introduction

The energy and equilibrium geometry of microclusters is important in understanding vapor phase nucleation, the properties of thin films, and the reactivity of dispersed metal catalyst. Because the surface to volume ratio in microclusters is large their energies and structures are expected to be dissimilar from those of the bulk crystalline metals. However, as the size of the cluster grows its properties are expected to become those of the bulk metals. How the properties of microclusters differ from bulk is becoming understood.

The minimum energy of metal clusters, M_N, up to N = 10 and M = Ni, Pd and Pt have been calculated using the Universal Binding Energy version of the Embedded Atom Method (EAM). Agreement with experimental data is discussed, especially for the dimers. Trends in the average binding energy as a function of N are fit to a model function whose form was first reported by R. C. Tolman[1] to explain how the surface tension of a droplet depends on its size. A similar functional from was recently reported by J. P. Perdew, Y. Wang and E. Engel[2] to show how atomic corrugation varies with the Wigner-Seitz radius.

2 Method of Calculation

The Embedded Atom Method of Daw and Baskes[3, 4] is used to calculate the minimum energy of metal clusters. EAM is a semi-empirical method based on effective-medium theory. It was developed to study the solid state properties of several fcc metals. Previously we[5] applied the method to the calculation of surface energy as a function of plane separation, slip barrier heights and Poisson's ratio. These are types calculations for which EAM was designed. We also applied the method to the calculation of nickel cluster energies with up to six atoms with encouraging results. The version of EAM used in the earlier work has the embedding function and the pair interaction energy expressed as a table

P. Jena et al. (eds.), Physics and Chemistry of Finite Systems: From Clusters to Crystals, Vol. I, 575–580.

of cubic spline knots. In the present work we have used the version of EAM[6] that is based on the universal binding energy relation of Rose, Ferrante and Smith[7,8].

The Morse potential

$$V(R) = D_e [(1 - e^{-\beta(R - R_e)})^2 - 1] \quad (1)$$

has many advantages. One of these is that it contains only two adjustable parameters, D_e, the depth of the potential, which is know from our EAM calculation, and β where

$$\beta = 1/2\ \omega_e (2\mu / D_e)^{1/2} \quad (2)$$

and ω_e is the equilibrium vibrational frequency and μ is the reduced mass. Another advantage of this potential is that the Schrodinger equation can be solved analytically and the vibrational energy levels are

$$E_n = (n + 1/2)\ (h/2\pi)\ \omega_e - (n + 1/2)\chi_e\omega_e \quad (3)$$

where $\chi_e = (h/2\pi)\beta^2/(2\mu\omega_e)$. A further advantage is that it can be linearized as

$$\ln[1 +/- (V(R)/D_e + 1)^{1/2}] = -\beta(R - R_e). \quad (4)$$

The plus sign is taken for $R \leq R_e$. A least-squares fit to equation four can be accomplished and β is the slope of the best-fit line.

3. Results

We have used EAM to calculate the potential function for the dimers at 301 points along the dimer axis with $\Delta R = 0.01$ angstroms. At this scale the difference between the EAM curve and the Morse potential can not be seen. That is the correlation coefficient is 1.00. Once the slope, β, is known then ω_e and χ_e can be easily calculated. These results are shown in Table 1. As can be seen from Table 1, EAM predicts the experimentally available quantities very well and compares favorably with calculations requiring significantly more computational complexity.

The N = 3 configuration space was searched exhaustively. For the clusters larger than 3 the 'downhill simplex method' of Nelder and Mead, presented by Press *et al.*[10], was modified as a technique for discovering the minimum energy configuration for these larger clusters. The problem with steepest descent minimization is that of finding the global minimum., i.e. it can not be identified with absolute certainty. The procedure then is to search the configuration space so that in all probability the lowest value found can be identified , with reasonable certainty, as the global minimum. These calculations were done without symmetry constraints and each minimum was relaxed until a difference in energy of 10^{-6} ev was obtained, which assures the accuracy coordinates to 10^{-4} angstroms. The positions of the atoms in the N = 3 through 7 clusters correspond to the the regular

Table 1. Equilibrium Parameters for the Morse Potential[a]

		r_e (A)	D_e (ev)	ω_e (cm^{-1})	χ_e (10^{-3})
Ni_2	Exp.	2.24	1.92		
	Ex. Huck.	2.2	2.71	395	
	CNDO	2.75	1.96		
	SCF-CI	2.20	1.42	289	
	HF-GVB-CI	2.26	1.88		
	POLSDCI	2.39	1.30	190	
	DFT	2.18	2.70	320	
	This Work (EAM)	2.13	1.92	292	4.71
Pd_2	Exp.	2.65	1.04	133	
	Exp.	2.57	1.71	344	
	Rel. Eff. C.	2.81	0.76	216	
	Rel. Model	2.30	1.10	320	
	This Work (EAM)	1.77	1.51	178	3.65
Pt_2	Exp.	2.34	3.72	259	
	Rel. Eff. C.	2.58	0.93	267	
	This Work (EAM)	1.83	4.19	278	2.06

[a] Experimental values and other calculated values where taken from reference 9.

polyhedra. The N = 8 cluster is a slightly distorted octahedron with the extra atoms over the center of two of the faces. The N = 9 and the N = 10 clusters are the lowest N clusters where an atom is inside the cluster. The N = 9 cluster is cubical with a atom in the body center and the N = 10 cluster is a distorted cube with the an atom near the body center and and extra atom over the center of one of the faces.

4. Discussion

As small clusters grow the atoms are in an environment most similar that of surface atoms in a crystal. Hence, we expect the average binding energy in small clusters to approach the surface energy before any bulk-like atoms begin to appear in these small clusters. Both the Tolman and Perdew functions have parameters that relate to the size of a transition region between surface-like atoms and bulk-like atoms.

The function chosen to model the average binding energy (AB) as a function of N to is

$$E_{AB}(N) = A_1 / (1 + A_2 / (N-1)^{1/3}) \tag{5}$$

This is a function of N-1 because the binding energy of one atom as referenced in EAM is zero. A_1 and A_2 are determined by a best fit to equation 4. The results of the fitting are shown in Table 2.

Table 2. Parameters for the function $E_{AB}(N) = A_1 / (1 + A_2 / (N-1)^{1/3})$

Ni Clusters			Pd Clusters			Pt Clusters		
A1	A2 (ev)	MSqD (10^{-3})	A1	A2 (ev)	MSqD (10^{-4})	A1	A2 (ev)	MSqD (10^{-4})
9.86	3.98	5.41	15.8	9.89	9.00	6.11	0.479	2.03

In Figures 1 and 2 the results of the fit to the model (equation 5) vs the average binding energy as calculated by EAM have been plotted. If the model exactly described the average binding as a function of cluster size the points in Figures 1 and 2 would lie on the line y = x. As can be seen from the figures and Table 2, equation 5 describes very well the increase in the average binding energy as the number of atoms in the cluster grows with a mean square deviation on the order of 10^{-3} ev for Ni and 10^{-4} ev for Pd and Pt. The fact that the fits are excellent is encouraging for two reason: 1) we can be confident that the global minimum energy structures have been identified, and 2) as the cluster size grows the average binding energy varies in a way that is to be expected from references 1 and 2.

5. Summary

The Embedded Atom Method has been applied to the computation of the energy of some transition metal microclusters. This application EAM is far from the original intent of the authors who parameterized it to describe bulk properties. The advantage of this approach is in its simplicity and speed, compared to first principle level calculations. Furthermore, many of these calculations have been done on 22 IBM model 50z PCs available to CWF at PSU.

This work was supported by Summer Faculty Fellowships at the NASA Lewis Research Center and by a half-year sabbatical leave from The Pennsylvania State University.

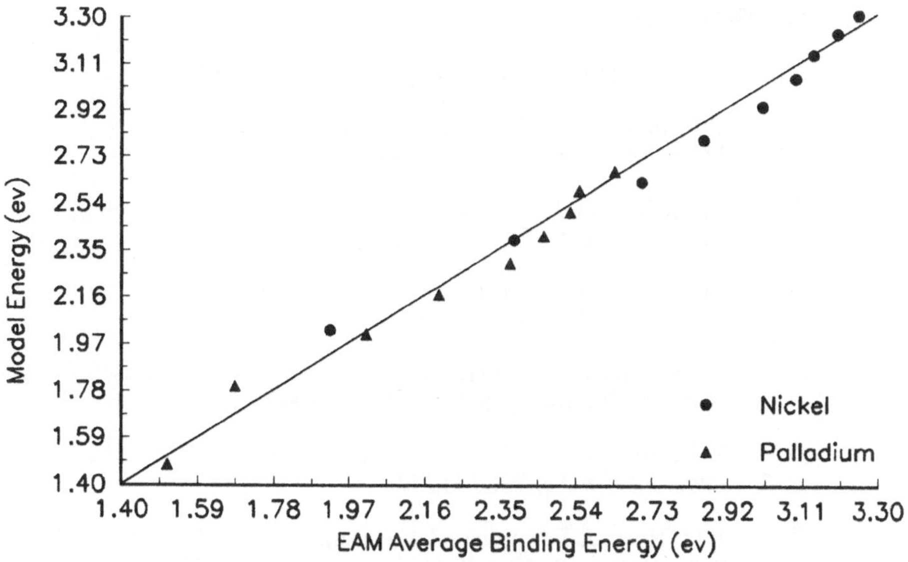

Figure 1. Model Energy vs Average Binding Energy

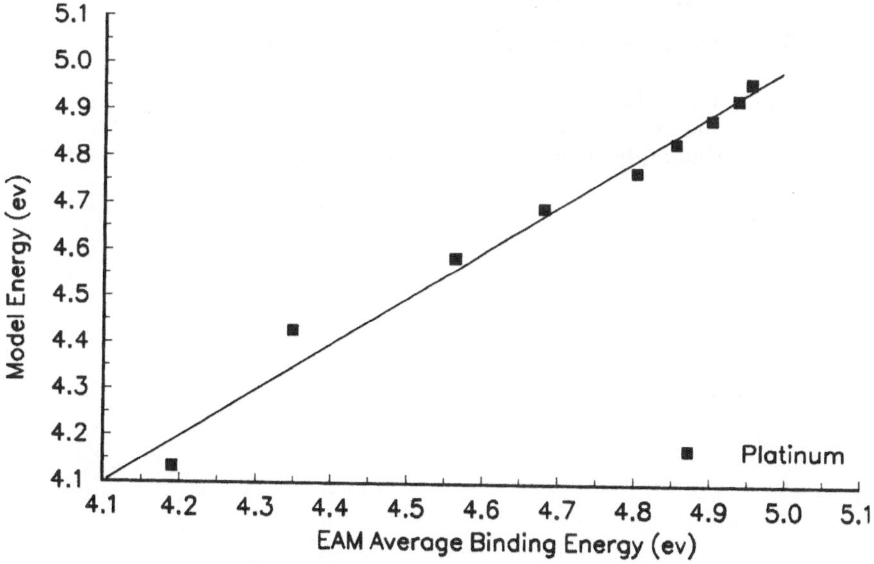

Figure 2. Model Energy vs Average Binding Energy

6. References

1. R. C. Tolman, J. Chem. Phys., **17,** 333 (1949)

2. J. P. Perdew, Y, Wang , and E. Engel, Phys. Rev. Lett., **66,** 508 (1991)

3. M. S. Daw and M. I. Baskes, Phys. Rev. Lett. **50**, 1285 (1983).

4. M. S. Daw and M. I. Baskes, Phys. Rev. B **29**, 6443 (1984).

5. C. W. Finley and J. Ferrante, Surf. and Interface Analysis **15**, 73 (1990).

6. S. M. Foiles, M. I. Baskes, and M. S. Daw, Phys. Rev. B **33**, 7983 (1987).

7 J. H. Rose, J. Ferrante and J. R. Smith, Phys. Rev. Lett. **47**, 675 (1981).

8. J. H. Rose. J. R. Smith and J. Ferrante, Phys. Rev. B **28**, 1835 (1983).

9. M. D. Morse, Chem. Rev., **86**, 1049 (1986).

10. W. H. Press, B. P. Flannery, S. A. Teukolsky, and W. T. Vetterling, Numerical Recipes. Cambridge University Press, Cambridge (1987)

Ionization Energies of Large Sodium Clusters: Direct Evidence for Atomic Shell Structure

H. Göhlich, T. Lange, T. Bergmann, U. Näher and T.P. Martin
Max-Planck-Institut für Festkörperforschung
Heisenbergstr. 1, 7000 Stuttgart 80, FRG

ABSTRACT. Recently it has been shown that the mass spectra of sodium clusters show evidence for magic numbers up to 22000 atoms. In the present investigation we have measured the ionization potential of large sodium clusters. The results indicate that magic numbers observed earlier are caused by an increase in the ionization energy each time a geometric shell of atoms is completed.

Introduction:

The quasi-free electrons in sodium clusters have been shown to arrange themselves into subshells and shells in a manner similar to that found in atoms [1–4]. If a cluster contains exactly the right number of electrons to fill one of these shells, it has a high ionization energy, just as for atoms. The analogy goes even further. It has recently been shown that clusters with closed electronic shells are also relatively inert chemically [5].

Another type of shell structure can modulate the properties of metal clusters, namely, shells of atoms. Shells of atoms were first observed in rare gas clusters in 1981 [6]. Peaks in the mass spectra of Xe-clusters could be correlated to the number of atoms needed to construct perfect icosahedra and cuboctahedra. Similar mass spectra were later obtained for magnesium and calcium clusters [7,8]. In all three cases a close examination of the mass spectra between shell closings indicate that the clusters grow by adding onion-like layers on tiny but perfect icosahedra.

Recently, we found that these same icosahedral (or cuboctahedral) magic numbers show up as minima in the mass spectra of sodium clusters obtained using ionizing photon energies near the ionization threshold [9]. At that time we suggested that these minima might reflect a size dependence of the ionization energies. In this paper we show that this is indeed the case. We have measured the photoionization threshold curves for sodium clusters containing up to 22000 atoms. It was found that geometric shell structure modulates the ionization threshold in a manner similar to that of electronic shells.

P. Jena et al. (eds.), Physics and Chemistry of Finite Systems: From Clusters to Crystals, Vol. I, 581–586.

Experimental Section:

The experimental setup we used is described in detail elsewhere [10–12]. The cluster source is a low pressure condensation cell [10]. Sodium vapor was quenched in cold He gas having a pressure of about 1 mbar. Na clusters condensed out of the quenched vapor and were transported by the gas stream through two chambers of intermediate pressure into a high vacuum chamber where they were photoionized with a $5\mu J/cm^2$, $2\times$ 1mm, 15 nsec dye laser pulse. For mass selection we used a high resolution time-of-flight mass spectrometer. It has a mass resolution of better than 20000 and a mass range of over 600000 amu.

The mass spectra showed only pure clusters without impurities. The ion signals were linearly dependent on the dye laser power. The laser photon energy was scanned over three laser dye ranges from 450 to 367 nm with mostly 1 nm step size. A complete photoionization efficiency curve consists of 103 wavelength steps showing normalized cluster intensities from prethreshold to postthreshold photon energies.

Discussion and Results:

Ionization efficiency curves have previously been measured for clusters containing up to 200 atoms [3,4,13–17]. Here we extend this type of experiment to larger clusters containing up to 22000 atoms. It would not have been practical to attempt to obtain a photoionization spectrum for each of the 22000 clusters observed in the mass spectrum. Using our source conditions it would have been necessary to measure several hours at each wavelength in order to obtain a significant number of events in narrow mass intervals which encompass just one cluster size. Even if such data were gathered it is questionable if such an enormous number of spectra could be usefully analyzed. Therefore, ionization efficiency curves were obtained for not each cluster size but rather for large groups of clusters having nearly the same mass. The 480 mass intervals used were chosen to be equally spaced on an $n^{1/3}$ scale. Ionization spectra were obtained for each mass interval in three overlapping wavelength regions.

There exist in the literature [16,18–21] several procedures for extracting ionization energies from photoionization spectra . We have chosen an error function fit because it considers the total shape of the spectrum and no information is lost [21,22]. According to this procedure each photoionization spectrum must be fit (using a least squares method) with the error function

$$f(\epsilon) = \frac{1}{\sqrt{\pi}} \int_{b(E-\epsilon)}^{\infty} e^{-t^2}\, dt \, .$$

E and b are the fit parameters. The maximum of the first derivative of $f(\epsilon)$ occurs at E, b is proportional to the inverse of the half width of the first derivative. If the width of the ionization threshold were due only to electron-vibration interaction, the characteristic energy, E, could be interpreted as a vertical ionization energy and with some further assumptions the adiabatic ionization energy could be extracted from the data. However, we believe the width is probably due to a multitude of isomers, each isomer having its own distinct ionization energy. For this reason the energy E might be better characterized as

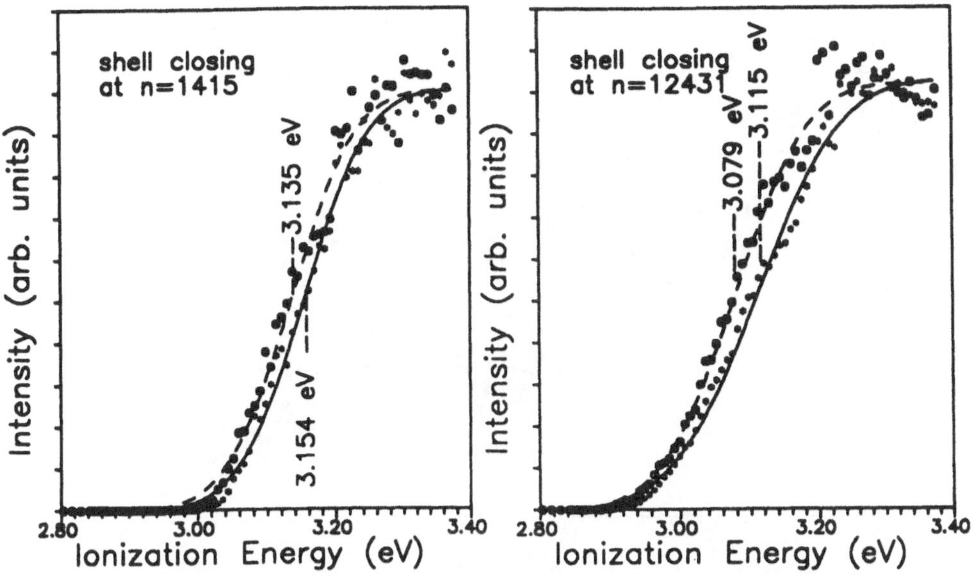

Figure 1: Photoionization efficiency curves for selected clusters. We compare closed shell clusters (asterisks and solid line) with nonclosed (circles and dashed line). The numbers indicate the ionization energy determined from an error function fit. The right plot shows data for the smallest observed atomic shell closing. Data for the 15th shell with n = 12431 are plotted in the left curve.

the average ionization energy of an ensemble of isomers. The values of E obtained using this procedure are indicated in Fig. 1. The lines represent best fits by the error function. The dashed line and the circles correspond to mass intervals between closed shells, the solid line and asterisks to closed shells. The right graph shows the smallest observable atom shell closing. In the left of Fig. 1 we have plotted the ionization behaviour of large clusters. All ionization energy data are summarized in Fig. 2 where E is plotted versus cluster size on an $n^{1/3}$ scale. The maxima of the periodic structure coincide very well with the shell closings expected for icosahedra and cuboctahedra (also indicated in Fig. 2). The modulation of the ionization energies is about 20 ± 10 meV. Any procedure for determining the ionization energy which does not include the total shape of the curve, e.g. a Watanabe plot, pseudo Watanabe plot or a linearization, smears out the oscillating structure.

The fast modulation of the ionization energy due to electronic shell structure at small masses could not be extracted from this data. The signal to noise ratio was too low. The slow atomic shell modulations, however, can be seen quite well at the expected masses.

Why do closed shell clusters have high ionization energies? Studies in surface science offer us a plausible explanation. The work function of a metal is lowered by the presence of steps on its surface. The degree of lowering is linearly proportional to the step surface density [23]. It is thought that edges or steps decrease the ionization energy by creating an electric dipole moment [24]. Since steps can be expected to appear on the surface of nonclosed shells in clusters, this effect could explain the modulation of the photoionization energy.

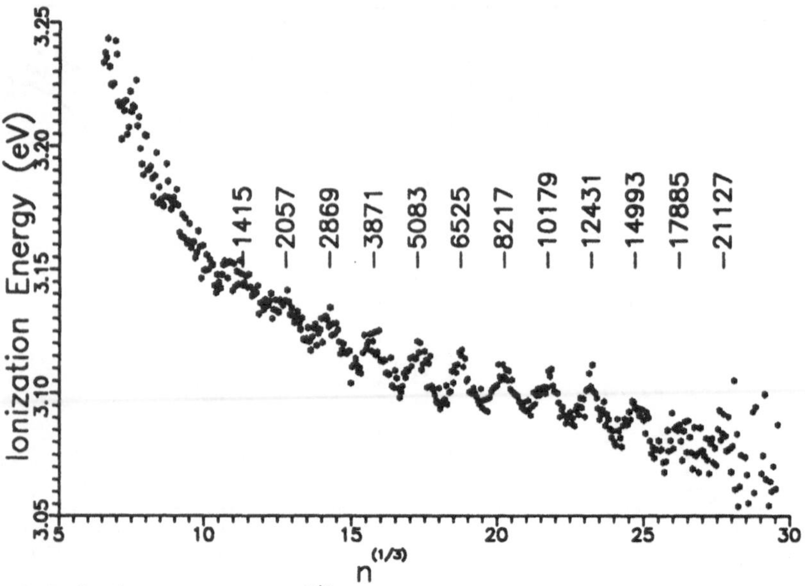

Figure 2: Ionization energy versus $n^{1/3}$ for Na_n clusters. Notice that the modulations occur at equal intervals. The theoretical values of n for icosahedral (cuboctahedral) shell closings are indicated.

Another point of interest concerns the asymptotic size dependence of the ionization potential. The classical law predicts in a first order approximation a linear behaviour of the ionization potential when plotted against $n^{-1/3}$ [25,26]:

$$E(n) = a\,n^{-1/3} + b,$$

where a is $\frac{1}{2}$ or $\frac{3}{8}$ (a recent discussion concerning the appropriate value of the parameter a can be found in [27]) and b is the bulk work function value. Our results permit a comparison with this law for very large cluster sizes. In Fig. 3 we have plotted a linear regression of the error function data on an $n^{-1/3}$ scale. The measured values follow quite well the linear behaviour, except for the nonclassical modulations due to the shell closings. But the slope of the line is smaller than $\frac{3}{8}$ and the value b (intersection with the y-axis) too large when compared to the measured bulk work function [28]. Our definition of the ionization energy is not an 'adiabatic value'. So in a next step we performed a so-called pseudo Watanabe analysis of the data [16] which yields ionization potentials closer to the adiabatic values (Fig. 3, dashed line). The straight line fit to this data yields a value of b close to the bulk work function (for polycrystalline sodium it is 2.75 eV [28]) and a slope (0.39) close to the classical value $\frac{3}{8}$.

Concluding Remarks

Our experiments show clear evidence for a periodic variation of the ionization energy for large sodium clusters. We interpret this as an effect caused by the formation of shells of

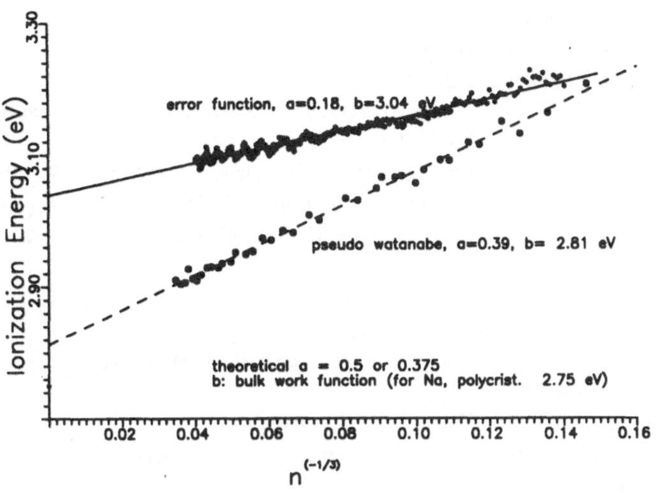

Figure 3: Ionization energy versus $n^{-1/3}$ for Na_n clusters. The asterisks represent data from the error function fit, the circles from pseudo Watanabe plots. The lines were obtained by a linear regression of the respective data points. Factor a is the slope of the line, b represents the intersection with the y-axis.

atoms. We cannot determine whether these shells have icosahedral or cuboctahedral symmetry because shell closings require the same number of atoms for both structures. The resolution of the measurements was too low to observe subshells, even if they should occur. Only with information concerning subshells is it possible to determine the symmetry.

References

[1] W.D. Knight, K. Clemenger, W.A. de Heer, W. A. Saunders, M.Y. Chou, and M.L. Cohen, *Phys. Rev. Lett.* **52**, 2141 (1984).

[2] W. Ekardt, *Ber. Bunsenges. Phys. Chem.* **88**, 289 (1984).

[3] A. Herrmann, E. Schumacher, and L. Wöste, *J. Chem. Phys.* **68**, 2327 (1978).

[4] A. Herrmann, S. Leutwyler, E. Schumacher, and L. Wöste, *Hel. Chim. Acta* **61**, 453 (1978).

[5] R.E. Leuchtner, A.C. Harms, and A.W. Castleman, Jr., *J. Chem. Phys.* **94**, 1093 (1991).

[6] O. Echt, K. Sattler, and E. Recknagel, *Phys. Rev. Lett.* **47**, 1121 (1981).

[7] T.P. Martin, T.Bergmann, H. Göhlich, and T. Lange, *Chem. Phys. Lett.* **176**, 343 (1991).

[8] T.P. Martin, U. Näher, T. Bergmann, H. Göhlich, and T. Lange, submitted to *Chem. Phys. Lett.*

[9] T.P. Martin, T. Bergmann, H. Göhlich, and T. Lange, *Chem. Phys. Lett* **172**, 209 (1990); T.P. Martin, T. Bergmann, H. Göhlich, and T. Lange, *Z. Phys. D* **19**, 25 (1991).

[10] T.P. Martin, *J. Chem. Phys.* **81**, 4426 (1984).
[11] T. Bergmann, T.P. Martin, and H. Schaber, *Rev. Sci. Instrum.* **60**, 347 (1989); **60**, 792 (1989); **61**, 2592 (1990).
[12] T. Bergmann, H. Göhlich, T.P. Martin, H. Schaber, and G. Malegiannakis, *Rev. Sci. Instrum.* **61**, 2585 (1990).
[13] E.J. Robbins, R.E. Leckenby, and P. Willis, *Adv. Phys.* **16**, 739 (1967).
[14] P.J. Foster, R.E. Leckenby, and E.J. Robbins, *J. Phys. B* **2**, 478 (1969).
[15] K.I. Peterson, P.D. Dao, R.W. Farley, and A.W. Castleman, Jr, *J. Chem. Phys.* **80**, 1780 (1984).
[16] M.M. Kappes, M. Schär, U. Röthlisberger, C. Yeretzian, and E. Schumacher, *Chem. Phys. Lett.* **143**, 251 (1988).
[17] E.C. Honea, M.L. Homer, J.L. Persson, and R.L. Whetten, *Chem. Phys. Lett.* **171**, 147 (1990).
[18] K. Watanabe, *J. Chem. Phys.* **26**, 542 (1957).
[19] P.M. Guyon and J. Berkowitz, *J. Chem. Phys.* **54**, 1814 (1971).
[20] W.A. Saunders, K. Clemenger, W.A. de Heer, and W.D. Knight, *Phys. Rev. B* **32**, 1366 (1985).
[21] H. Limberger and T.P. Martin, *J. Chem. Phys* **90**, 2979 (1989).
[22] T. Bergmann and T.P.Martin, *J. Chem. Phys.* **90**, 2848 (1989).
[23] K. Besocke and H. Wagner, *Phys. Rev. B* **8**, 4597 (1973).
[24] K. Besocke, B. Krahl-Urban, and H. Wagner, *Surf. Sci.* **68**, 39 (1977).
[25] J.M. Smith, *Amer. Inst. Aeronaut. Astronaut. J.* **3**, 648 (1965).
[26] D.M. Wood, *Phys. Rev. Lett.* **46**, 749 (1981).
[27] M. Seidl, K.-H. Meiwes-Broer, and M. Brack, to be published.
[28] R.J. Whitefield and J.J. Brady, *Phys. Rev. Lett.* **26**, 380 (1971).

STRUCTURAL AND ELECTRONIC PROPERTIES OF ICOSAHEDRAL QUASICRYSTALS

J. HAFNER AND M. KRAJČÍ
Institut für Theoretische Physik, Technische Universität
Wiedner Hauptstrasse 8-10, A-1040 Wien, Austria

ABSTRACT. We have studied a series of rational approximants to icosahedral alloys by molecular dynamics annealing using realistic, interatomic forces and calculations of the electronic density of states. Both investigations suggest an important electronic contribution to the stability of the quasicrystal via a Hume–Rothery or Peierls mechanism.

1. Introduction

A central problem in the study of quasicrystalline materials is to understand why nature should prefer quasiperiodic to periodic long–range order [1-3]. It is now reasonably certain that a major contribution to the stability of quasicrystalline alloys comes from the electronic ground state energy [4,5]. The composition of many quasiperiodic phases appears to be governed by a Hume–Rothery rule that states that if the Fermi level falls into a minimum in the electronic density of states induced by a strong diffraction peak of the quasicrystalline lattice, the structure is energetically favourable. The Hume–Rothery rule seems to hold not only for the quasicrystalline alloys formed by simple and noble metals (typefied by icosahedral (i-) Al-Mg-Zn and i–Al-Li-Cu), but also for the icosahedral and decagonal phases formed by Al with noble and transition metals [6]. Evidence for the existence of Hume–Rothery–like pseudogaps comes mainly from electronic–structure calculations for crystalline approximants to the quasicrystalline phases [4,5]. To date, such studies have been performed only for the lowest order approximants. Alternatively, the Hume–Rothery or Peierls argument may also be formulated in real space: the stabilization arises from a constructive interference between the interatomic distances and the oscillatory effective pair interaction. The Friedel wavelength of these oscillations is determined by the Fermi momentum. This description has proved to be very fruitful for understanding the structures of the liquid elements [7] and the structure and stability of liquid and amorphous alloys [8]. In this paper we discuss both aspects of the Hume–Rothery mechanism for stabilizing a quasicrystalline phase.

2. Interatomic forces and structural stability

2.1 STABLE ONE–COMPONENT QUASICRYSTALS

Some years ago, Hafner and Heine [9] have shown that the trends in the crystal structures may be understood in terms of the variation of the effective interatomic forces with

P. Jena et al. (eds.), Physics and Chemistry of Finite Systems: From Clusters to Crystals, Vol. I, 587–592.

electron density and the electron–ion pseudopotential. E.g. the trend from close–packed metallic to open covalent structures in the series Na-Mg-Al-Si-P arises from the fact that with increasing electron density, the Friedel oscillations in the effective pair interactions are shifted relative to the repulsive core in such a way that at high electron densities the nearest–neighbour distance in a close–packed structure falls on a maximum of the potential. Hence the lattice is unstable against a Peierls distortion which places the atoms of the split nearest–neighbour shell into the minima of the pair potential. This real–space explanation for the occurrence of a Peierls distortion does not require periodicity, so it applies to non-periodic and quasiperiodic phases as well. These results can be expressed in the form of a phase diagram showing the crystal lattices minimizing the energy against three parameters: valence, electron density, and a core radius characterizing the pseudopotential. Very recently Smith [10] has extended this analysis to quasicrystals. He could show that stable one–component quasicrystals exist in a region (in parameter space) separating the close–packed and open crystalline lattices. No element falls into this region, but it corresponds rather closely to the virtual–crystal parameters of the observed quasicrystals. Rational approximants to the quasicrystals are also stable, but only in a narrow region bordering the open structures.

2.2 COMPUTER SIMULATION OF QUASICRYSTALS

To perform a similar investigation for the known ternary quasicrystals would be impossible. However, we can study the stability of an assumed quasicrystalline lattice as a function of alloy composition by molecular dynamics simulation [11]. The system chosen for study are the icosahedral Al-Zn-Mg quasicrystals. In this case the lowest–order crystalline approximant is the Frank–Kasper phase $(Al_xZn_{1-x})_{49}Mg_{32}$, and it has been shown that the icosahedral phase is well described by a three-dimensional Penrose lattice with Henley–Elser decoration of the basic structural units [12]. Computer simulations may be performed on a hierarchy of rational approximants [13] obtained by replacing the Golden Mean τ by a quotient of Fibonacci numbers, $\tau \sim \tau_n = F_{n+1}/F_n$. It has been shown [14] that the pair distribution function calculated for the 5/3–approximant (12380 atoms in the periodically repeated cell) reproduces the experimental pair distribution for distances up to 50 Å.

Observed quasicrystals have electron–per–atom ratios between e/A $\simeq$ 2.15 and e/A $\simeq$ 2.40. Fig. 1 demonstrates the variation of the interatomic potentials with electron density for $(Al_xZn_{1-x})_{0.56}Mg_{0.64}$ alloys for $0 \leq x \leq 1$. The potentials have been generated using standard pseudopotential perturbation theory based on nonlocal orthogonalized plane–wave pseudopotentials [15]. The stability of the quasicrystalline phase under these pair forces has been tested by isothermal molecular–dynamics annealing of the 3/2–approximant with 2920 atoms in the periodically repeated cell. We find that for alloys with $x \geq 0.5$ (i.e. for e/A $\geq$ 2.3) the quasicrystalline state is unstable, it transforms to an amorphous state (Fig. 2(a)). The QC phase is stable at lower Al content ($0.2 \leq x \leq 0.4$ corresponding to $2.10 \leq e/A \leq 2.25$), in agreement with experimental observation. The comparison of Figs. 2(b) and 1(b) shows that for the stable phase there is a good correspondence between the minima in the pair interactions and the maxima of the pair correlation functions. Evidently this requires that the Friedel wave length of the oscillations $\lambda_F = \pi/k_F$ (k_F is the Fermi momentum) is equal to the distances between the first-, second- etc coordination shells. It also requires a strong electron–lattice interaction. At constant density λ_F scales according to $\lambda_F \approx (e/A)^{-13}$.

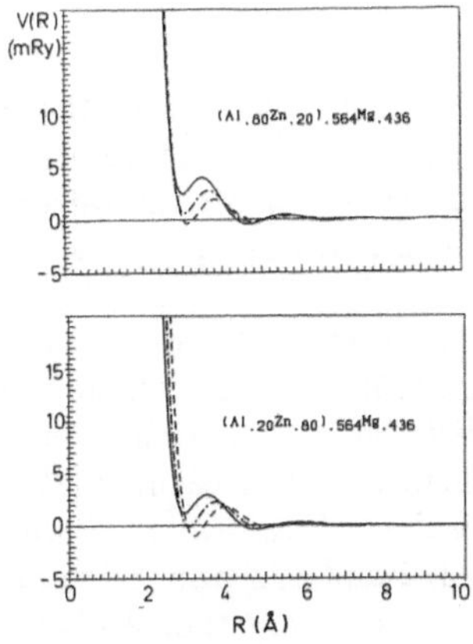

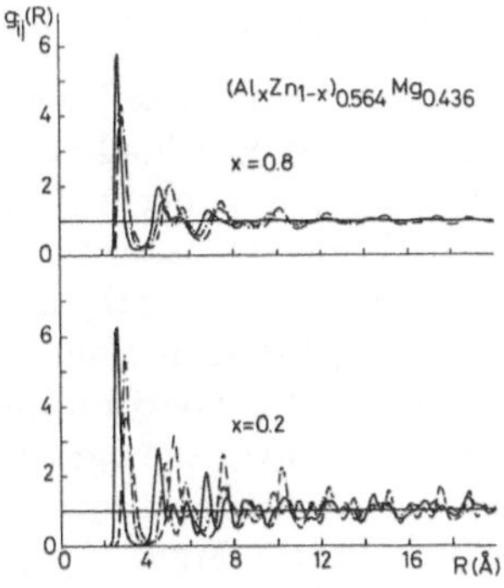

Fig.1: Effective interatomic potentials in the quasibinary alloy $(Al_xZn_{1-x})_{0.564}Mg_{0.436}$ with x = 0.8 (a) and x = 0.2 (b).

Fig.2 Partial pair correlation functions for the $(Al_xZn_{1-x})_{0.564}Mg_{0.436}$ alloy, obtained by molecular dynamics relaxation of the 3/2–quasicrystalline approximant (key see Fig.1). The system with x = 0.8 transforms to an amorphous state, while the alloy with x = 0.2 retains the initial symmetry.

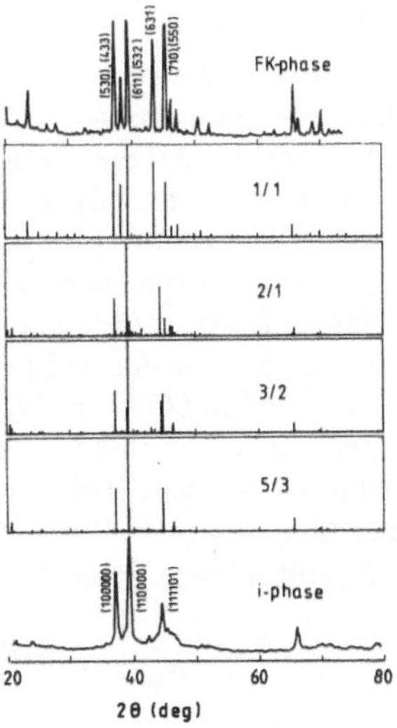

Fig.3: Powder diffraction patterns calculated for the 1/1, 2/1, 3/2 and 5/3 rational approximants of icosahedral Al-Zn-Mg, compared with the measured scattering intensities of the Frank–Kasper– and the icosahedral phases.

2.3 MODULATED QUASICRYSTALS

The relaxation of the idealized periodic Penrose lattice to a minimum of the free energy leads to a structure in which all atoms are displaced. This is a consequence of the fact that the basic structural units occur in many different local environments. However, when one computes the Fourier transform of the relaxed configuration, one obtains a pattern with the same point group symmetry and sharp diffraction peaks. The reason is that the displacement field has the same symmetry as the idealized structure. Therefore, the structure of a real quasicrystal is always displacively modulated [16]. We have already shown that the modulation improves the agreement of the pair distribution function with experiment [11,14]. This is also confirmed by the comparison of calculated and measured [17] diffraction intensities, both for the Frank–Kasper and for the icosahedral phases. Fig.3 demonstrates that the powder–diffraction patterns of the 1/1, 2/1, 3/2 and 5/3–approximants converge very nicely toward the observed diffraction pattern of the quasicrystalline phase.

3. Electronic structure of quasicrystals

For the lowest–order crystalline approximants, the electronic structure may be calculated using modern band–structure techniques, such as the linear–muffin–tin–orbital (LMTO) method [20]. Calculations for the 1/1–approximants of i-Al-Mg-Zn (FK-phase), i-Al-Li-Cu (R-phase), and α-AlFe ($Al_{13}Fe_4$) all show a distinct pseudogap induced by a strong scattering peak of the quasicrystal, and a Fermi level falling into the pseudogap [4,5]. However, to date it is an assumption that the result obtained for the crystalline approximant holds also for the quasicrystal. This assumption is not contradicted by the available data on the electronic specific heat [18,19], but it appears to be highly desirable to verify it by calculations for the higher–order approximants.

Standard band–structure calculations are feasible for elementary cells up to about 150 atoms, hence even the 2/1 approximant cannot be treated using $\vec{k}$–space techniques. Electronic density-of-states determinations for larger systems can be performed in $\vec{r}$–space, based on a tight–binding Hamiltonian and the recursion technique [21]. Here we present calculations for the 1/1, 2/1, 3/2, and 5/3 approximants of i-Al-Mg-Zn. The starting point is a selfconsistent LMTO–calculation for the 1/1–crystal. It has been shown [22,23] that the long–ranged basis of the MTO's may be transformed to an orthonormal tight–binding (TB) basis formed by the most localized TB-LMTO's. In the LMTO description the Hamiltonian yielding the eigenvalues and eigenvectors may be factorized into a canonical TB–structure constant matrix S^α characterizing the geometrical distribution of the atoms and potential parameter–matrices d^α, c^α describing the one–electron potentials at the atomic positions. To first order in (E - E_ν), the Hamiltonian is given by

$$H^{(1)} = E_\nu + h^\alpha = c^\alpha + (d^\alpha)^{1/2} S^\alpha (d^\alpha)^{1/2} \tag{1}$$

where E_ν is some reference energy and the TB structure matrix S^α may be expressed as

$$S^\alpha = S^0(\alpha - \alpha S^0)^{-1} \tag{2}$$

in terms of the conventional structure constants S^0 and the screening matrix α. The screening matrix α is diagonal and independent of the atomic position to a good approximation [22,23]. Higher order Hamiltonians may be constructed as well.

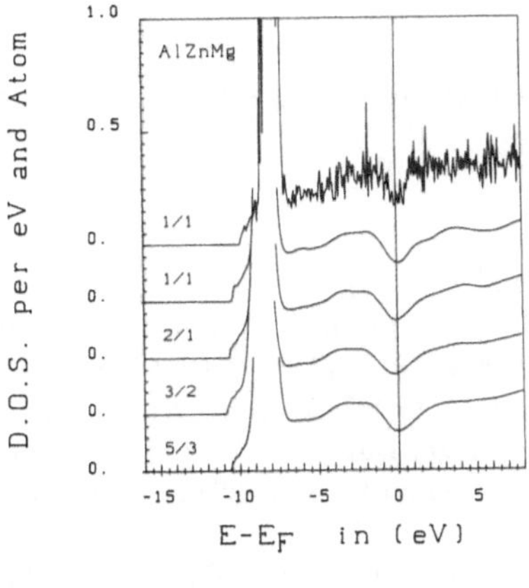

Fig.4: Electronic densities of states for the 1/1, 2/1, 3/2 and 5/3 rational approximants of icosahedral Al-Zn-Mg. For the 1/1 approximant we show the result of the self consistent LMTO calculation and a Brillouinzone integration with 55 $\vec{k}$- points in the irreducible part and the TB-LMTO recursion calculations, for the higher approximants only the recursion result is shown. Energy scales match at the bottom of the band, arrows mark the Fermi level.

Fig. 4 shows the electronic density of states for the first four approximants of i-Al-Mg-Zn containing 162, 694, 2920, and 12380 atoms in the elementary cell. The calculations have been performed using a second-order Hamiltonian, fifty recursion levels and a Gaussian-quadrature [24] terminator for the continued fraction (details will be published elsewhere). Even with fifty exact recursion levels, the DOS has a finite resolution so that the recursion calculation cannot resolve the 'fine-structure' in the exact DOS. Note, however, that although the number of reciprocal lattice points increases with the number of atoms in the elementary cell (and the quasicrystalline reciprocal lattice is dense every where), the diffraction pattern shows that the intensity is gradually concentrated in the peaks corresponding to the low-order reflections of the quasicrystal lattice. The gap at the Fermi level is induced by the (631) and (710)+(550) reciprocal lattice vectors of the Frank-Kasper phase, corresponding to the (111101) scattering peak of the icosahedral phase (see Fig. 3). Our results show that the pseudogap at the Fermi level is a generic property of all rational approximants to quasicrystals, and the small variation of the DOS along the hierarchy of the rational approximants suggests that this holds for the quasicrystalline phase as well.

4. Conclusions

To conclude: the computer simulation studies of the stability and the calculation of the electronic DOS of higher-order rational approximants to icosahedral alloys suggest an important electronic contribution to the stability of the quasicrystalline phase via a Hume-Rothery or Peierls mechanism.

Acknowledgements

This work has been supported by the Austrian Science Foundation under project no. 8148.

References:

1 Bancel P.A. and Heney P.A. (1986) Phys.Rev. B33, 7917
2 Friedel J. and Denoyer F. (1987) C.R.Acad.Sci. (Paris) 305, 171
3 Smith A.P. and Ashcroft N.W. (1987) Phys.Rev.Lett. 59, 1365
4 Fujiwara T. and Yokokawa T. (1991) Phys.Rev.Lett. 66, 333
5 Hafner J. and Krajčí M. (1991) Europhys.Lett. (in print)
6 Tsai A.P., Inoue A. and Masumoto T. (1991) Sci.Rep.Res.Inst.Tohoku University A, 36, 112
7 Jank W. and Hafner J. (1990) Phys.Rev. B41, 1497; B42, 6926; B42, 11530; and in print
8 Hafner J. (1980) Phys.Rev. B21, 604
9 Hafner J. and Heine V. (1983) J.Phys. F13, 2479
10 Smith A.P. (1991) Phys.Rev. B43, 11635
11 Hafner J. and Krajčí M. (1990) Europhys.Lett. 13, 335
12 Henley C.L. and Elser V. (1986) Philos.Mag.Lett. 53, 115
13 Krajčí M. and Fujiwara T. (1988) Phys.Rev. B38, 12903
14 Hafner J. and Krajčí M. (1991) J.Non-cryst. Solids (in print)
15 Hafner J. (1987) "From Hamiltonians to Phase Diagrams", Springer, Berlin 1987
16 Jansen T. (1989) in Ch. Jannot and J.M. Dubois (eds.), "Quasicrystalline Materials", World Scientific, Singapore
17 Rajasekharan, Akhitar D., Gopalan R. and Muraleedharan K. (1986) Nature 322, 528
18 Wagner J.L., Biggs B.D., Wong K.M. and Poon S.J. (1988) Phys.Rev. B38, 7436
19 Mizutani V., Sakabe Y. and Matsuda T. (1990) J.Phys.: Condens.Matter 2, 6153
20 Andersen O.K. (1975) Phys.Rev. B12, 3060
21 Haydock R., Heine V. and M.J. Kelly (1975) J.Phys.C 8, 2591
22 Andersen O.K. and Jepsen O. (1984) Phys.Rev.Lett. 53, 2571
23 Šob M., Jepsen O. and Andersen O.K. (1988) Z.Phys.Chemie 157, 515
24 Nex C.M.M. (1985) in Pettifor D.G. and Weaire D.L. (eds.) "The Recursion Method and its Applications", Springer, Berlin, p. 52

HARTREE-FOCK INVESTIGATIONS OF PURE ANTIMONY AND ALKALI-ANTIMONY CLUSTERS.

F.HAGELBERG, S.NEESER, N.SAHOO and T.P.DAS
Department of Physics
State University of New York at Albany, Albany, NY, 12222

K.G.WEIL
Institut für Physikalische Chemie
D-6100 Darmstadt, Germany

ABSTRACT. The structures of the SB_4 cluster as well as of various compounds of the type X_nSb_4 (X = Li, Na, K, Cs) have been explored by Hartree-Fock calculation. From total energy optimization using the pseudopotential approach, we propose geometries and equilibrium bond lengths for the Alkali-Antimony clusters X_2Sb_4, K_4Sb_4 and Na_6Sb_4.

I. Introduction.

The present investigation is directed at Alkali-Antimony clusters of the form X_mSb_4 (X: Li, Na, K, Cs) detected by Knudsen Effusion Mass Spectrometry /1/. A theoretical understanding of the stability and electronic aspects of these clusters, we believe, should be preceded by a first principle investigation of the structures of the most abundant species observed so far.

The focus of this work is on clusters containing the Sb_4 unit. The pure Antimony cluster Sb_4 is found to be substantially more abundant than Sb_2 and Sb_3 from mass spectroscopic measurements /1/ and thus exhibits higher stability than these species. In preparation of our work on the Alkali-Antimony clusters we first carried out a structure investigation for the Sb_4 unit. This first step of the research presented here was also motivated by the need of testing current theoretical methods on a simpler system than the relatively complex Alkali-Antimony clusters. Whereas most of these compounds require the use of the pseudopotential approach /2/ the pure Antimony cluster Sb_4 can be investigated using both all-electron and pseudopotential procedures. Thus,a comparison between the results

P. Jena et al. (eds.), Physics and Chemistry of Finite Systems: From Clusters to Crystals, Vol. I, 593–598.

of Sb_4 structure calculations employing the first principle Hartree-Fock method and the pseudopotential treatment could provide a justification for the use of the latter in the investigation of Alkali-Antimony clusters. Although there is substantial experimental information currently available on the Sb_4 cluster which is known to have tetrahedral geometry, no direct observation of its bond length exists so far.

Following the presentation of our work on the Sb_4 unit we will discuss clusters of the form X_2Sb_4 (X=Li,Na, K,Cs) and some species of the form X_4Sb_4 and X_6Sb_4. Here, our interest will be focused on the changes of both the tetrahedral geometry and equilibrium bond lengths of the Sb_4 cluster induced by the addition of Alkali atoms.

II. Procedure.

To demonstrate theoretically from first principles that the tetrahedral structure of Sb_4 is the stablest one, it is necessary to show that it has the least total energy compared to other possible structures. Keeping this in mind we have investigated both the regular tetrahedron and several planar structures. The four geometries considered in this work: regular tetrahedron, square, rectangle and rhombus are summarized in Fig.1. In all cases, a singlet

Fig.1: Geometrical arrangements examined for for the Sb_4 cluster.

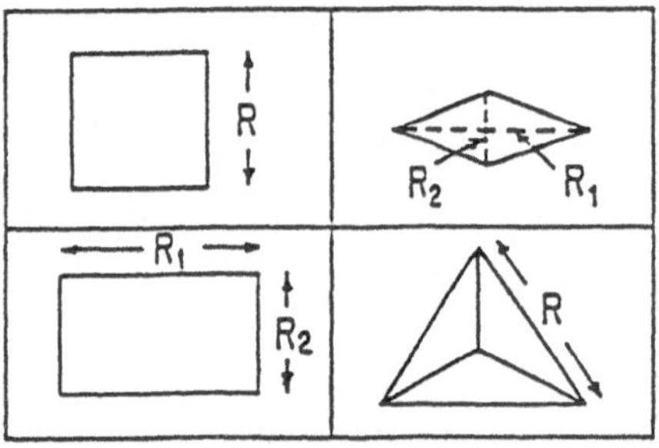

spin configuration was examined, except for the square arrangement, which was proved to be unstable from arguments based on the Jahn-Teller theorem (as is further explained in Sec.III) in the singlet spin state.

The stabilities of the various cluster geometries and configurations studied were compared for their total ground state energies as obtained from self consistent field Hartree-Fock calculations. In all cases, Gaussian functions were used as basis state primitives /3,4/. To examimine the possible sensitivity of our results on the basis set chosen, we studied the tetrahedral arrangement of Sb_4 not only using STO-3G /4/ but also a more extensive basis set of the form (43222/4222/33) /5/, hereafter referred to as "basis set II".

In view of the good agreement between our pseudopotential result for the Sb_4 tetrahedron and the result obtained by a calculation using basis set II - see Sec.III - , all structure investigations for Alkali-Antimony clusters were based on the pseudopotential approach.

For all calculations presented in this work we have employed the program "Gaussian 88" on the IBM ES/3090 600J Supercomputer at the Cornell National Supercomputing Facility.

III. Results and Discussion.

In this section, we will present the results for the pure Sb_4 cluster before turning to Alkali-Antimony clusters of the form X_nSb_4 (X=Li, Na, K, Cs).

a. Structure of the Sb_4 cluster.

To prove conclusively by Hartree-Fock theory that the tetrahedral structure of the Sb_4 cluster is the actual one it is necessary to compare the total energy of a Sb_4 tetrahedron with those for possible planar structures. The most obvious structure among these is the square. However, no stability can be found for the Sb_4 square arrangement in spin singlet configuration (S=0). As is evident from the scheme of the highest occupied levels for the Sb_4 square in both, singlet and triplet configuration (see Fig.2a.and b.), one of the two degenerate E_g levels will be left empty

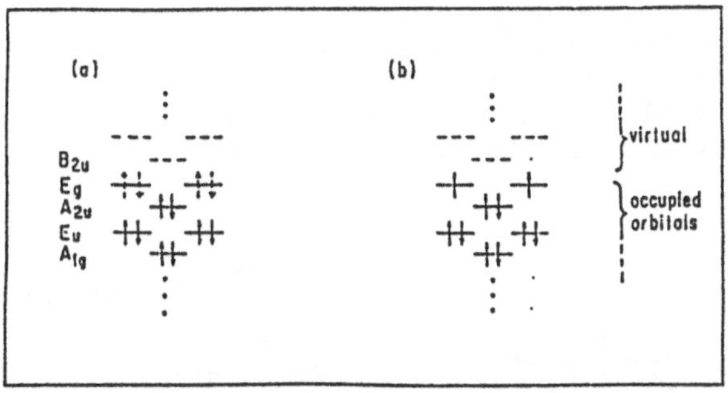

Fig.2: The highest occupied electronic levels of the Sb_4 square cluster in (a) spin singlet (b) spin triplet configuration. The dashed arrows in (a) indicate alternate possibilities of occupation for the two E_g states.

in the case S=0 (2a.) so that Jahn Teller distortion will occur resulting in a structure of lower symmetry. Angular distortion leads to rhombus-, linear distortion to rectangular geometry. For the triplet spin configuration (S=1, see 2b.), by the Pauli principle both E_g levels have to be occupied so that the square geometry is expected to be stable for S=1.

Table 1. Geometries investigated for Sb_4 using different basis sets.

Structure	Minimum energy [a.u.]	R1[Å]	R2[Å]
Rhombus	-25005.513[a]	4.00	3.80
Rectangle	-25005.517[a]	2.95	2.80
Square (S = 1)	-25005.566[a]	2.70	2.70
Tetrahedron	-25005.646[a]	2.80	2.80
	-25227.647[b]	2.95	2.95
	-20.930[c]	3.05	3.05

a : STO-3G basis set; b : (43222/4222/33) basis set;
c : Pseudopotential STO-3G basis set.

Table 1 contains. the minimal energies and optimized bond lengths for the four structures investigated by all-electron calculation employing an STO-3G basis set. For the tetrahedral structure which turns out to be the most stable one we give the equilibrium bond lengths obtained for the three different choices of basis functions mentioned in Sec.II. These three results are seen to agree within 10% with each other as well as with a value for the Sb_4 bond length of 2.77Å estimated by Capwell and Rosenblatt /6/. From this agreement the use of the pseudopotential basis set appears justifiable for the exploration of Alkali-Antimony clusters too.

b. Structure of Alkali-Antimony clusters.

In continuation of our structure research we studied first compounds of the type X_2Sb_4 (X=Alkali atom). From symmetry considerations, a very plausible geometry for the X_2Sb_4 cluster seems to be a square arrangement of 4 Sb atoms with one Alkali atom above and another below their plane (see Fig.3). As is readily concluded from the nature of the highest occupied levels - see Fig.4 - of the X_2Sb_4 system,

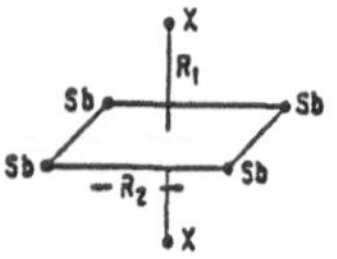

Fig.3: The geometry proposed for X_2Sb_4 clusters.

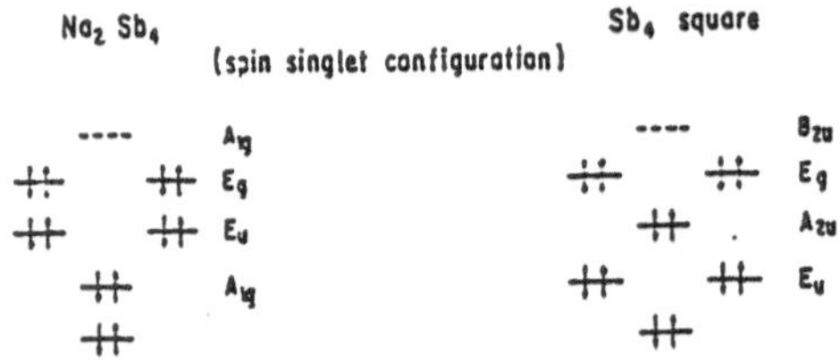

Fig.4: The highest occupied electronic levels of Na_2Sb_4 as compared to those of the Sb_4 square.

the Jahn Teller instability of the Sb_4 square cluster in spin singlet configuration is removed by the addition of the 2 Alkali atoms which thus have a stabilizing effect on the Sb_4 square arrangement. Table 2 gives the geometrical parameters for the four X_2Sb_4 systems under examination.

Table 2. Geometrical parameters for the X_2Sb_4 systems Na_6Sb_4 clusters.

System	R1[Å]	R2[Å]
Li_2Sb_4	2.11	2.97
Na_2Sb_4	2.79	2.95
K_2Sb_4	3.26	2.93
Cs_2Sb_4	3.62	2.92

Another conceivable structure of the X_2Sb_4 compounds that could preserve the tetrahedral geometry of the pure Sb_4 cluster is an X_2-dumbbell intersecting one edge of the Sb_4 tetrahedron, as shown in Fig.5. From a consideration of the

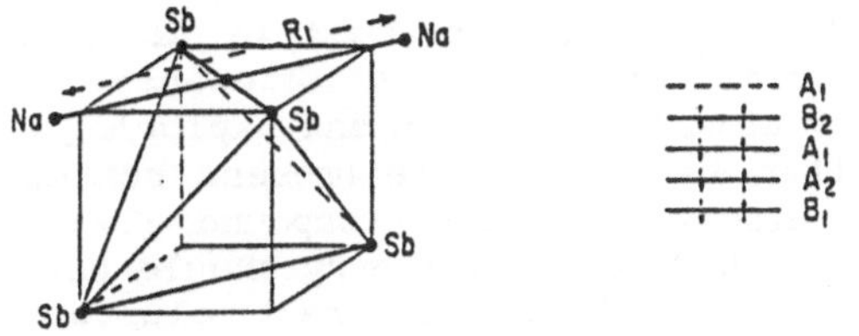

Fig.5: An alternative stable structure of the Na_2Sb_4 cluster shown together with its highest occupied electronic levels.

symmetries of the highest occupied electronic states, this structure is expected to be stable too. However, energy optimization for Na_2Sb_4 assuming a Sb-Sb bond length of R=2.7Å and and using R_1 as a free variation parameter resulted in an energy minimum that is by 6.7 eV higher than the minimum obtained for the geometry displayed in Fig.3. This finding favours the latter as the actual structure for compounds of the type X_2Sb_4.

Additionally, we investigated the more complex clusters K_4Sb_4 and Na_6Sb_4. For these compounds, we propose the highly symmetric structures shown in Fig.6. Pseudopotential optimization using the geometrical parameters indicated in Fig.6 as variables yielded stability for both of these structures with a Sb-Sb equilibrium bond length of R_1 = 3.58Å for K_4Sb_4 and of R_1 = 4.55Å for Na_6Sb_4.

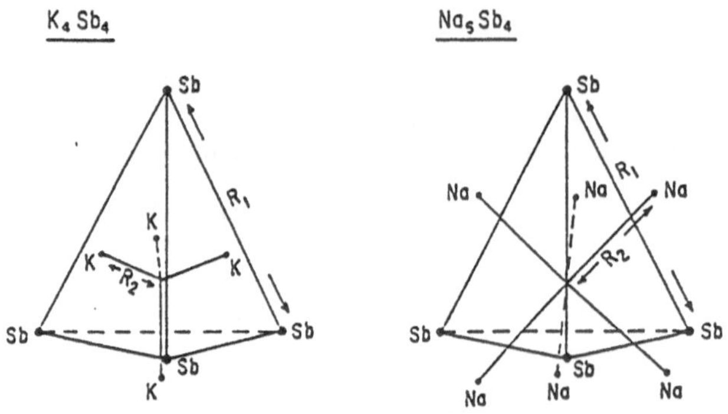

Fig.6: The geometries proposed for the K_4Sb_4 and the Na_6Sb_4 clusters.

IV. Conclusion and Prospects.

In the present work, we have discussed possible structures of the pure Antimony Sb_4 clusters as well as of some Alkali-Antimony clusters of the form X_2Sb_4, X_4Sb_4 and X_6Sb_4. This research is to be continued by more comprehensive structure investigations making allowance for a higher number of variational parameters and exploring a wider range of clusters than is done in the present context, including the recently discovered Cs_6Sb_4 compound /7/. Moreover, an explanation from Hartree-Fock theory should be given for the observed abundances of the different cluster species /1/ which requires a theoretical understanding of the processes involved in the formation and dissociation of these clusters. Lastly, their binding characteristics and the systematics of their ionization potentials both vertical and adiabatic should be explored from first principle theory.

The support of the Alexander von Humboldt Foundation given to one of us (F.H.) is gratefully acknowledged.

References.

/1/ T.Scheuring and K.G.Weil, Surf.Sci.156, 457 (1985).
/2/ W.R.Wadt and P.J.Hay, J.Chem.Phys., 82, 284 (1985).
/3/ W.J.Hehre, R.F.Stewart, and J.A.Pople, J Chem.Phys., 51, 2657 (1969).
/4/ J.B.Collins, P.v.R.Schleyer, J.S.Binkley, and J.A.Pople, J.Chem.Phys.64, 5142 (1976).
/5/ S.Huzinaga et al.,"Gaussian Basis Sets for Molecular Calculations", Elsevier, New York (1984).
/6/ R.J.Capwell,G.M.Rosenblatt, J.Phys.Chem.,72,4327 (1968).
/7/ A.Hartmann, K.G.Weil, High Temp.Sci., Vol.27, 31 (1990).

ELECTRONIC STRUCTURES OF METAL CLUSTERS ON SURFACE

T.IKEDA* and S.OHNISHI
NEC Scientific Information System Development, Ltd.*,
34, Miyukigaoka, Tsukuba, Ibaraki, 305, Japan ;
Fundamental Research Labs., NEC Corp.,
34, Miyukigaoka, Tsukuba, Ibaraki, 305, Japan

ABSTRACT. The interaction of Au_{13} and Al_{13} clusters of icosahedral symmetry with the Si surface is discussed based on the Harris functional scheme.

1. Introduction

We discuss the electronic structures of Au_{13} and Al_{13} clusters of icosahedral symmetry and their interaction with the Si surface. Instead of applying the Kohn-Sham scheme for the large scale complex materials like cluster molecular and the cluster on surface, we introduced the Harris functional scheme to study their interaction and binding mechanism [1][2]. Electronic structures of individual clusters are self-consistently determined by the Kohn-Sham scheme based on the local density functional scheme based on the linear combination of atomic orbitals (LCAO) method using the norm-conserving pseudopotentials [3][4]. The nodeless feature of atomic radial wave functions makes it feasible to perform numerical integrations for Hamiltonian and overlap matrix elements and expansions for augmentation of wave function tail and charge density in the LCAO scheme. Using these results, we applied the Harris functional method to calculate the interaction energy between clusters and the surface.

2. Method

In the Harris functional method, the potential energy of the coupled system is approximated by superimposing self-consistently determined charge density of separated clusters as

$$n(x) = \sum_k n_k(x) \tag{1}$$

The kinetic energy is given by the eigenvalue of the coupled system. The

P. Jena et al. (eds.), Physics and Chemistry of Finite Systems: From Clusters to Crystals, Vol. I, 599–604.

density functional total energy is represented as

$$E = \sum_i f_i \varepsilon_i - \int dx\ n(x)\,[\tfrac{1}{2}\phi(x) + \mu_{xc}(x)] + E_{xc}[n] + E_N \qquad (2)$$

where f_i and ε_i are the occupation number and eigenvalue determined by solving the one-electron Schrodinger equation of the coupled system. $\phi(x)$ and $\mu_{xc}(x)$ are the Coulomb, and the exchange-correlation potential. $E_{xc}[n]$ and E_N are the exchange-correlation, and the internuclear repulsion energy.

Fig.1 shows binding energies for uniform scaling of Au_{13} and Al_{13} clusters vs bond distance calculated by both the Harris functional and Kohn-Sham scheme. In the Harris functional case, binding energy is given by superimposing atomic charge density. Fig.2 shows the density of states (DOS) of these clusters by both methods. The highest and lowest occupied molecular orbitals (HOMO,LUMO) for Au_{13} and Al_{13} comprises 6s and 3p orbitals, respectively.

Since determined equilibrium distances and electronic structures are almost same in both cases, the Harris functional method is found to be very reliable in predicting complicated systems like cluster-cluster and cluster-surface interaction.

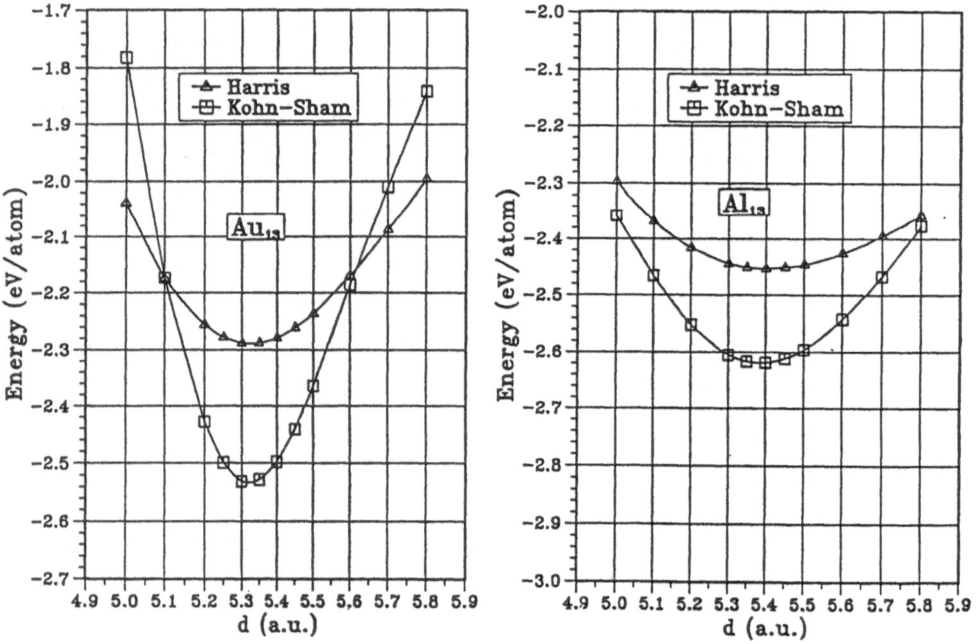

Figure 1. Binding energy curves of Au_{13} and Al_{13} clusters. The triangle is Harris method. The cube is Kohn-Sham scheme.

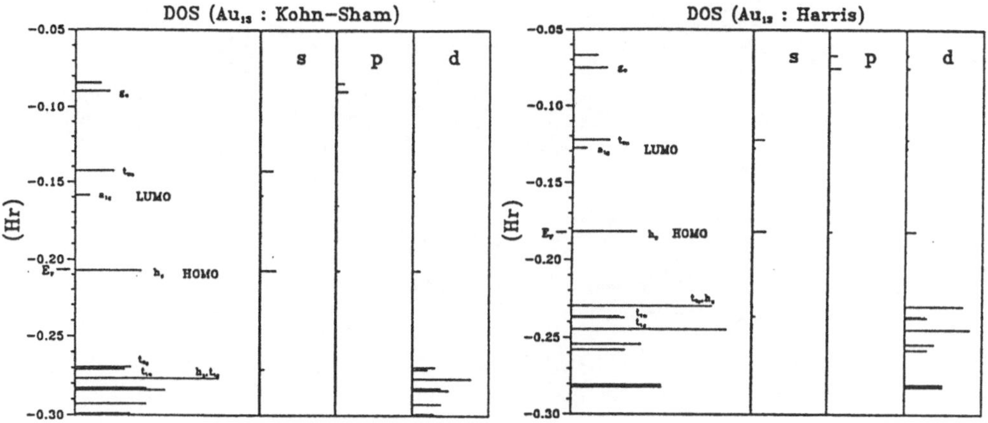

Figure 2. The DOS of Au_{13} cluster by the Kohn-Sham scheme and the Harrias scheme.

3. Results

3.1. Cluster dimers

Fig.3 shows binding energy curves of $(Au_{13})_2$ and $(Al_{13})_2$ vs distance between cluster centers. Atomic distances and charge densities of Au_{13} and Al_{13} are calculated by the Kohn-Sham scheme as shown in Fig.2. The HOMO and LUMO for Au_{13} and Al_{13} of equilibrium distances comprises these 6s and 3p orbitals, respectively.

3.2. Cluster on Si(111) surface

Fig.4 shows the model of M_{13}-$Si_{28}V_{58}$ for the M(=Au,Al)$_{13}$ cluster on Si(111) surface, where V_{58} represents surrounding 58 muffin-tin type potentials given by

$$V_M = \sum_l \frac{V_{loc}\,|\,\psi_l\rangle\langle\psi_l\,|\,V_{loc}}{\langle\psi_l\,|\,V_{loc}\,|\,\psi_l\rangle} \qquad (3)$$

V_M enforces sp_3 type bonding between nearest-neighbor atoms and the easy convergence of self-consistent calculations. Fig.6 shows the binding energy curves vs distance between the first layer Si and the cluster center. Charge densities of Au_{13}, Al_{13} and $Si_{28}V_{58}$ are given by the Kohn-Sham scheme. Au_{13} on Si(111) found to be stable , while Al_{13} is unstable. In Au_{13} case, strong binding is considered to be due to the 5d states.

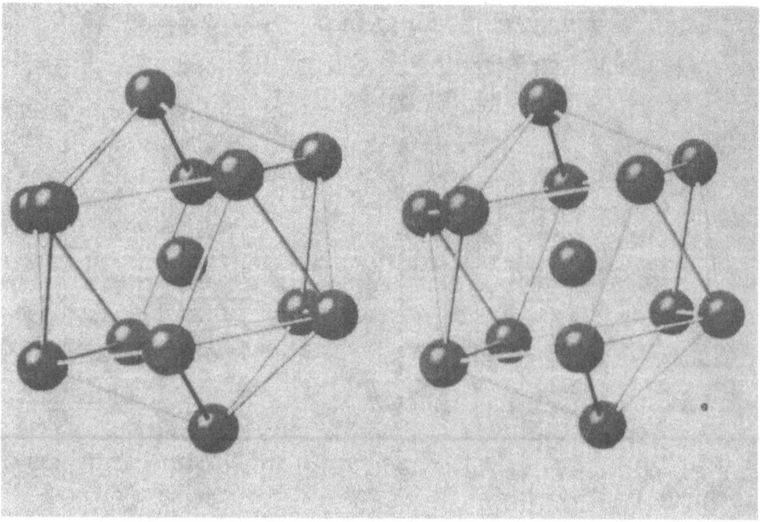

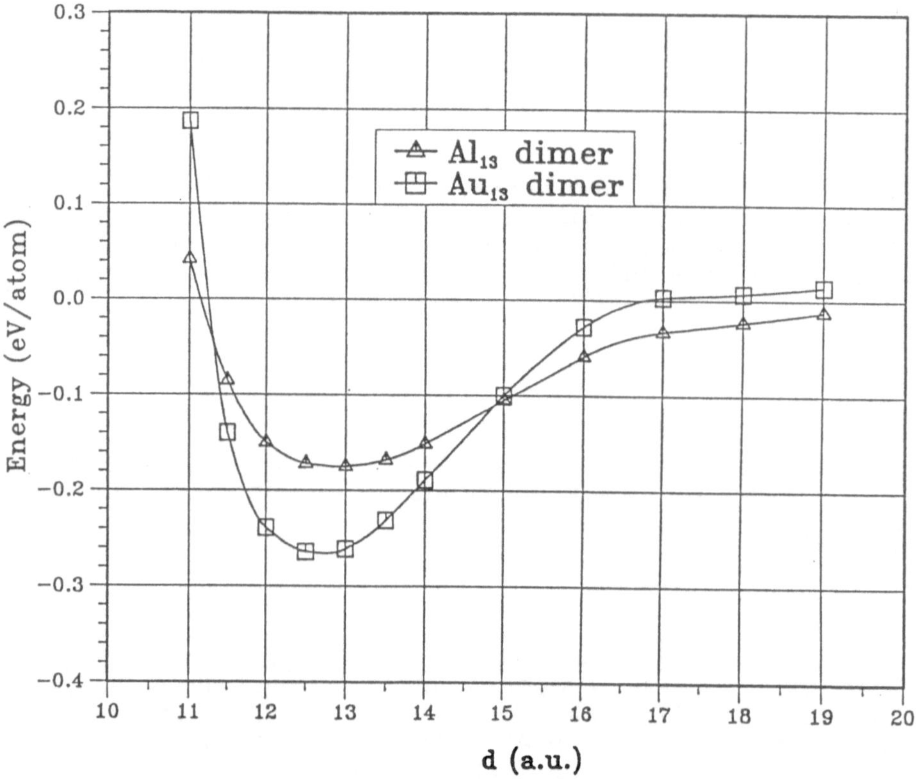

Figure 3. Binding energy curves of $(Au_{13})_2$ and $(Al_{13})_2$ clusters

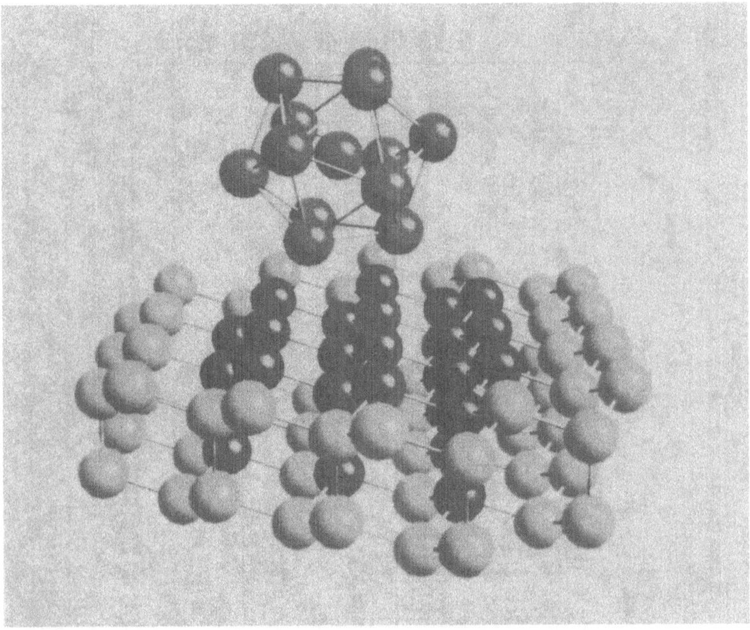

(a)

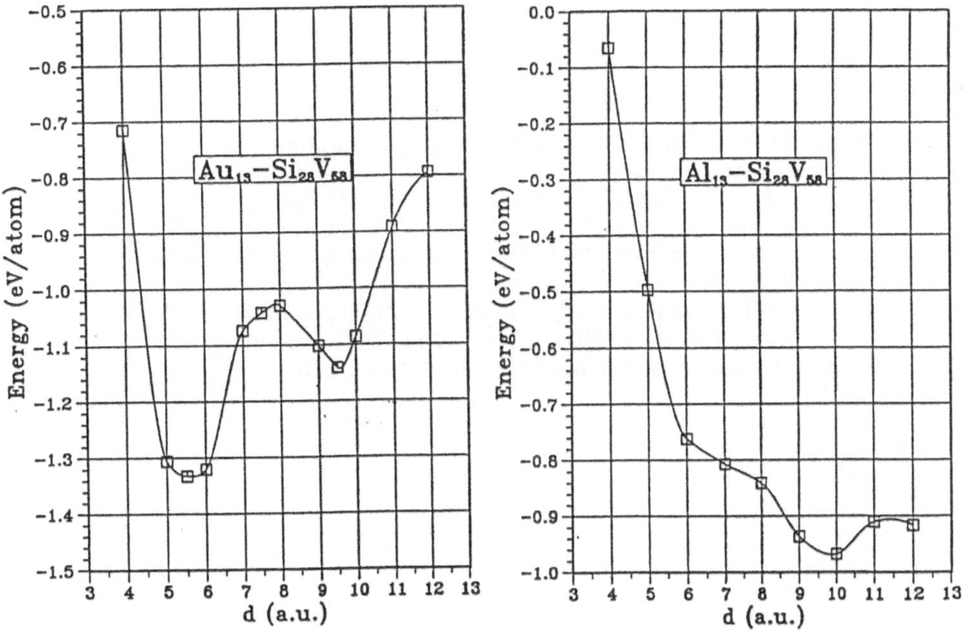

Figure 4. Binding energy curves of Au_{13},Al_{13}-$Si_{28}V_{58}$ (a), and the DOS of Au_{13}-$Si_{28}V_{58}$ (b).

(b)

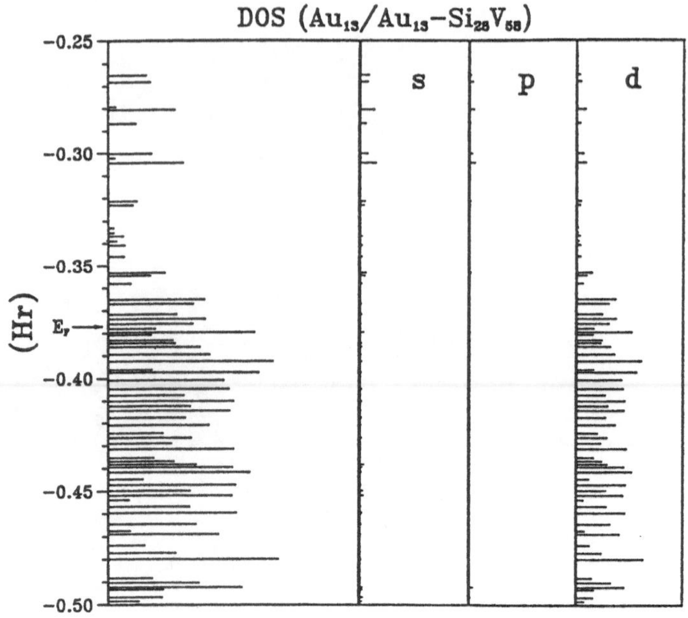

Figure 4. Binding energy curves of Au_{13},Al_{13}-$Si_{28}V_{58}$ (a), and the DOS of Au_{13}-$Si_{28}V_{58}$ (b).

References

[1]W.Kohn,L.J.Sham, (1965), 'Self-consistent equations including exchange and correlation effects', Physical Review, 140, A1133
[2]J.Harris, (1985), 'Simplified method for Calculating the energy of weakly interacting fragments', Physical Review, B31, 1770
[3]G.B.Bachelet,D.R.Hamann,M.Schluter, (1982), 'Pseudopotentials that work: From H to Pu', Physical Review, 26, 4199
[4]S.Ohnishi,T.Ikeda,N.Watari, (1991), 'Effective potential method for electronic structure calculations in materials design' Computer Aided Innovation of New Materials, 175

A STUDY OF QUANTUM MONTE CARLO SIMULATION FOR THE HYDROGEN CLUSTER IONS

Satoshi Itoh and Sigeo Ihara
Central Research Laboratory, Hitachi Ltd.
1-280 Higashi-Koigakubo, Kokubunji
Tokyo 185 Japan

ABSTRACT. The total energies and the equilibrium configurations of H_2, H_3^+ and H_5^+ are calculated by using the Quantum Monte Carlo (QMC) method which can fully take into account electron-correlation. The total energy of these clusters obtained by the QMC method was found to be lower than those obtained by the Configuration Interaction (CI) method. For H_5^+, C_{2v} symmetry is found to be more energetically favorable than the D_{2d} symmetry. Furthermore, we found a new energy minimum structure for the C_{2v} symmetry.

1. Introduction

For the hydrogen cluster ions, a lot of experimental (Clampit and Gowland, 1969; Deursen and Reuss, 1973; Hiraoka, 1987) and theoretical analyses (Kolos et al., 1968; Carney, 1980; Yamaguchi et al., 1987) have been performed. The hydrogen cluster ions are suitable for verifying the accuracy of the calculation methods. Up to now, molecular orbital methods are popular methods to analyze these clusters. Among these methods, the Configuration Interaction (CI) method appears to be the most accurate. On the other hand, Quantum Monte Carlo (QMC) methods (Mentch and Anderson, 1981; Reynolds et al., 1982; Ceperley et al., 1984) have become established in the last 10 years. Both the CI method and the QMC method can fully take into account electron-correlation. The CI method can calculate excited states; however, CPU time and memory for the CI method are proportional to the fourth power of the number of base functions. If the scale of the system becomes large, it is very difficult to apply the CI method to a large system. On the other hand, CPU time and memory for the QMC method is proportional to the second power of the number of electrons. In calculating the ground state for large scale systems, it is expected that the QMC method is more effective than the CI method. In this study, a comparison of the accuracy between the QMC method and the CI method is performed. And new stable structure for H_5^+ which is obtained by the QMC method is reported.

2. Calculation Methodology

Our QMC method is similar to previous work (Reynolds et al., 1982). We start with the Hamiltonian for the hydrogen cluster ions is given by the following form using the Born-

P. Jena et al. (eds.), Physics and Chemistry of Finite Systems: From Clusters to Crystals, Vol. I, 605–610.

Oppenhimer approximation,

$$H = \sum_{i=1}^{N} \{ -\frac{1}{2}\nabla_i^2 - \sum_{\alpha=1}^{M} \frac{1}{r_{\alpha i}} + \frac{1}{2}\sum_{j \neq i}^{N} \frac{1}{r_{ij}} \} + \sum_{\alpha > \beta}^{M} \frac{1}{r_{\alpha\beta}}, \tag{1}$$

where, N and M are the number of electrons and hydrogen atoms respectively, and $r_{\alpha\beta}$, $r_{\alpha i}$, r_{ij} are the distances between two specified particles. For Coulomb interactions, we use an exact formulation and no approximations such as that introduced by model potentials are necessary. In this paper, the atomic unit is mainly used.

For Hamiltonian (1) the imaginary time-dependent Schrödinger equation is given by

$$-\frac{\partial}{\partial \tau} \Phi(\boldsymbol{R},\tau) = (H - E_T)\, \Phi(\boldsymbol{R},\tau), \tag{2}$$

where, $\boldsymbol{R}$ is a $3N$ dimensional vector which represents all N electrons. The imaginary time-dependent Schrödinger equation is analytically solved. As the limit of τ goes to infinity, only the ground state is left. The QMC method is based on this principle.

The imaginary time-dependent Schrödinger equation (2) is translated to the diffusion equation with a probability distribution f,

$$-\frac{\partial}{\partial \tau} f(\boldsymbol{R},\tau) = -\frac{1}{2}\nabla^2 f + (E_L(\boldsymbol{R}) - E_T) f + \frac{1}{2}\nabla\cdot(f\, \boldsymbol{F}_q(\boldsymbol{R})), \tag{3}$$

where, $f(\boldsymbol{R},\tau) = \Psi_T^*(\boldsymbol{R})\, \Phi(\boldsymbol{R},\tau)$ is the product of the trial wave function Ψ_T and true wave function Φ. $E_L(\boldsymbol{R}) = H\Psi_T(\boldsymbol{R}) / \Psi_T(\boldsymbol{R})$ is the local energy. Total energy is evaluated by the expectation value of this function. E_T is the trial energy which must coincide with true energy. $\boldsymbol{F}_q(\boldsymbol{R}) = 2\nabla\Psi_T(\boldsymbol{R}) / \Psi_T(\boldsymbol{R})$ is called the quantum force using classical analogy. This diffusion equation for f is solved by Green's function G,

$$f(\boldsymbol{R}',\tau+\Delta\tau) = \int d\boldsymbol{R}\, f(\boldsymbol{R},\tau)\, G(\boldsymbol{R}',\boldsymbol{R},\Delta\tau). \tag{4}$$

The evaluation of this integral is carried out by the Monte Carlo method. Starting from an arbitrary initial distribution $f(\boldsymbol{R},0)$, imaginary time increases by $\Delta\tau$. After a sufficient amount of imaginary time τ the ground state will be obtained. The number of sampling for the Monte Carlo method is 10,000. The discretized imaginary time $\Delta\tau$ is chosen to be 0.005 so that the statistical error is comparable to the systematic error occurring in finite $\Delta\tau$.

3. Results and Discussion

3.1. HYDROGEN MOLECULE AND H_3^+ CLUSTER ION

As a first step, the hydrogen molecule and H_3^+ are investigated as test cases. The total energies and the equilibrium distances obtained here are summarized in table 1 as compared with previous works. The total energies of these clusters for several internuclear distances R are calculated. For the hydrogen molecule, the exact solution was obtained (Kolos et al., 1968). In a wide range of R, the results obtained by our QMC calculations coincide with the exact solution within the error bars. The equilibrium distance R_e is determined by the least square fitting of the total energy to 4th order polynomial functions. For the case of H_3^+, it is confirmed that the energy minimum configuration is D_{3h} symmetry (Christoffersen, 1964). Therefore, only the restricted configuration of an equilateral triangle is calculated. Equilibrium distance is determined in the same way as the hydrogen molecule.

In the case of the QMC method, since the total energy contains numerical error, it is difficult to decide the equilibrium distance accurately. However, the total energies obtained by the QMC method is lower than standard Full-CI calculations except for the Hylleraas-CI method, in which the results are comparable.

TABLE 1. A comparison of the total energy for the hydrogen molecule and H_3^+.

Method	R_e	Energy	Ref.
Exact	1.401078..	-1.174475	Kolos, et al. (1968)
Full-CI	1.4	-1.173387	Dykstra (1978)
Full-CI	1.4046	-1.171430	Yamaguchi, et al. (1987)
Hyllerass-CI	1.4011	-1.174467	Urdaneta, et al. (1988)
QMC	1.40(2)	-1.1744(3)	This work
Singer	1.6500	-1.34335	Salmon and Poshusta (1973)
Full-CI	1.6504	-1.34278	Dykstra (1978)
Full-CI	1.6500	-1.33936	Carney (1980)
QMC	1.65	-1.3439(2)	Mentch and Anderson (1981)
Full-CI	1.6538	-1.34029	Yamaguchi, et al. (1987)
Hyllerass-CI	1.6405	-1.34350	Urdaneta, et al. (1988)
QMC	1.65	-1.34407(18)	This work

3.2. H_5^+ CLUSTER ION

The candidates for the most stable configuration of the H_5^+ structure are the D_{2d} and C_{2v}symmetries (see Fig. 1). D_{2d} and C_{2v} are fixed by two and four parameters respectively. Much research has been done for these two structures. Some researchers concluded that the D_{2d} structure is more stable than the C_{2v} structure (Hirao and Yamabe, 1983; Ahlrichs, 1975; Kaplan and Radimova, 1971). Others concluded that the C_{2v} is more stable(Yamaguchi et al., 1987; Salmon and Poshusta, 1973).

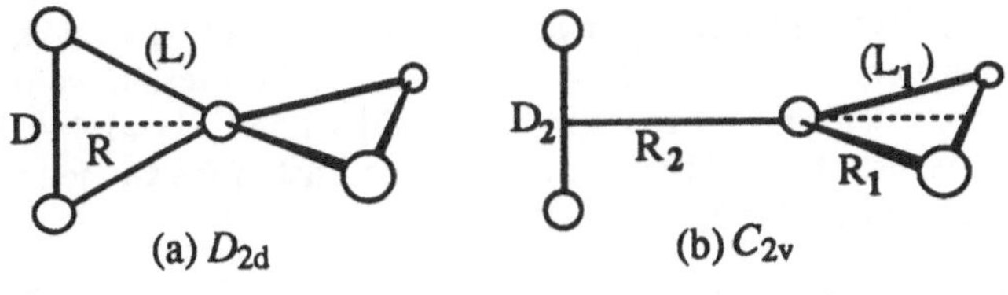

Figure 1. Candidates for the H_5^+ structure; (a) D_{2d} symmetry, (b) C_{2v} symmetry.

In our simulation, the H5+ structure is optimized using the following sequence. 1) optimization within the D2d symmetry, 2) optimization within the intermediate structures between D_{2d} and C_{2v}, and 3) optimization within the C_{2v} symmetry. At first, the H_5^+ structure is optimized within the D_{2d} symmetry. Fig. 2 shows the potential energy surface in the D and R parameter space. The optimized configuration (D=1.48, R=2.00) is illustrated by a closed circle, which is very close to the results (D=1.48 R=1.99) obtained by the CI method (Yamaguchi et al., 1987). However, the total energy of -2.52839 obtained by our QMC simulation is lower than the results of -2.52258 obtained by the CI method (Yamaguchi et al., 1987).

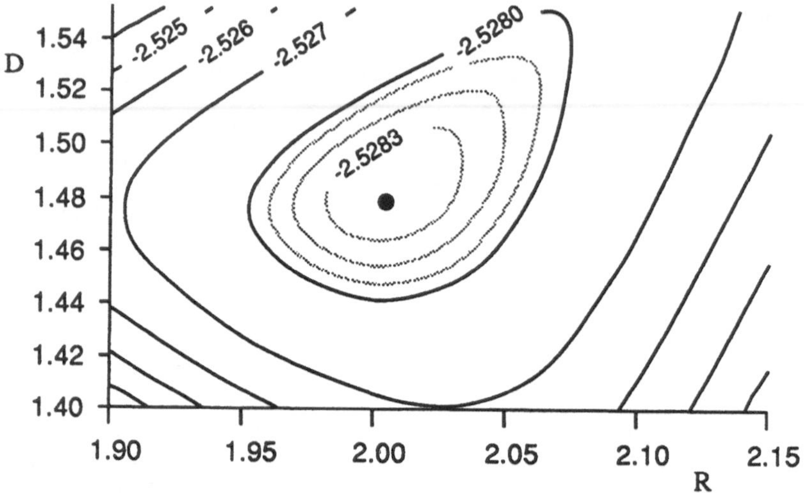

Figure 2. The potential energy surface of H_5^+ within D_{2d} symmetry.

Next, the intermediate structures between D_{2d} and C_{2v} are considered. These structures correspond to the C_{2v} structure with D_1=D_2=D. D is fixed to 1.48 which is optimized for D_{2d} symmetry. Fig 3 shows the potential energy surface in the R_1 and R_2 parameter space. The dotted line L corresponds to D_{2d} symmetry, and the energy minimum structure in the D_{2d} symmetry is indicated by P. However in Fig. 3, structure P is a saddle point, and there are more stable asymmetric configurations C_1 and C_2. The point C_1 is specified using three parameters, R_1=1.68, R_2=2.44, D= 1.48, and the total energy of the point C_1 is -2.5291. If the restriction of D_1=D_2=D is removed, it is possible to obtain a lower energy structure than point C_1, since points C_1 and C_2 will become saddle points in the full 4 parameters space. Searching for the minimum energy configuration using 4 parameters, however, is a formidable task. Therefore we use the parameters D1 and D2 obtained by the CI method where the 4 parameters were varied independently (Yamaguchi et al., 1987). Using these two parameters D_1 =1.544, and D_2 =1.444, we optimized the remaining two parameters R_1 and R_2 for the C_{2v} structure. As a result, we obtained a new stable structure, which is specified using four parameters, D_1 =1.544, D_2 =1.444, R_1=1.66, and R_2=2.64. The stable structure obtained here is different from that of the CI method, and is more energetically stable, as shown in table 2.

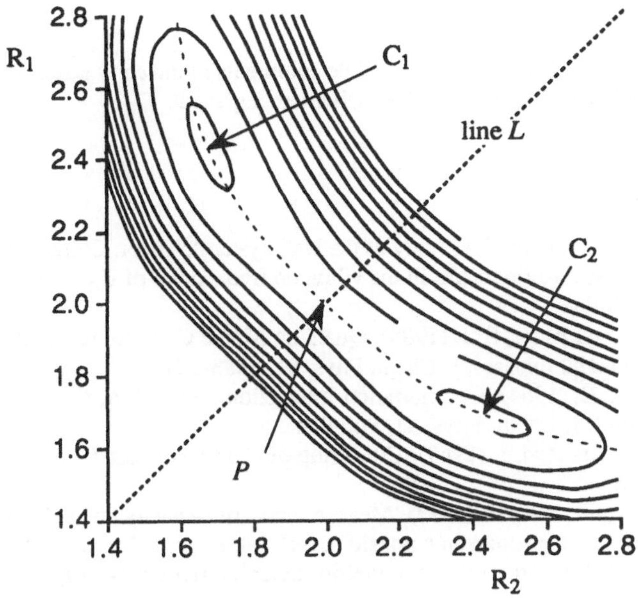

Figure 3. The potential energy surface of H_5^+ within C_{2v} symmetry with D_1=D_2

TABLE 2. A comparison of the minimum energy of H_5^+ within C_{2v}.

Method	Workers	D_1	R_1	D_2	R_2	Energy
CI	Hirao, Yamabe (1983)	1.584	1.684	1.412	3.014	-2.508200
CI	Yamaguchi et al. (1987)	1.544	1.644	1.444	2.530	-2.523150
QMC	This work	1.544	1.66	1.444	2.64	-2.5293

4. Conclusion

For the hydrogen molecule, the total energy is in agreement with the exact solution (Kolos et al., 1968). For H_5^+ and H_3^+ total energies obtained by our QMC simulation are lower than that obtained by the CI method. Therefore, we believe that the QMC method is a more accurate method to calculate the ground state than the CI method. In computing the D_{2d} and the C_{2v} symmetry, it is found that the C_{2v} symmetry is energetically more favorable than the D_{2d} symmetry. Furthermore, we found a new energy minimum structure for H_5^+, which is different from previous works (Yamaguchi et al., 1987).

5. Acknowledgements

We would like to thank Dr. T. Uda for his valuable comments in the early stages of this work. We also thank Drs. A. Yajima and H. Johnishi for their encouragement.

6. References

Ahlrichs, R. (1975) 'Theoretical study of the H_5^+ system', Theor. Chim. Acta 39, 149-160.

Carney, G.D. (1980) 'Refinements in the vibration frequencies of H_3^+ and D_3^+', Mol. Phys. 39, 923-933.

Ceperley, D. M. and Alder, B. J. (1984) 'Quantum Monte Carlo for molecules: Green's function and nodal release', J. Chem. Phys. 81, 5883-5844.

Christoffersen, R. E. (1964),'Configuration-Interaction Study of the Ground State of the H3+ Molecule', J. Chem. Phys. 41, 960-971.

Clampit, R. and Gowland, L.(1969),'Clustering of Cold Hydrogen Gas on Protons', Nature 223, 815-816.

Deursen, A. van and Reuss, J. (1973),'Measurements of Intensity and velocity distribution of clusters from a H_2 supersonic nozzle beam', Int. J. Mass Spect. 11, 483-489.

Dykstra, C.(1978) 'The uncoupled symmetric stretching frequency of H_3^+', J. Chem. Phys. 68, 3951-3952.

Hirao, K. and Yamabe, S. (1983) 'Theoretical study on the structure and stability of hydrogen-ion clusters H_n^+ and H_n^- (n=3,5,7,9,11,13)', Chem. Phys. 80, 237-243.

Hiraoka, K. (1987),'A determination of the stabilities of $H_3^+(H_2)_n$ with n =1-9 from measurements of the gas-phase ion equilibria $H_3^+(H_2)_{n-1} + H_2 = H_3^+(H_2)_n$ ', J. Chem. Phys. 87, 4048-4055.

Kaplan, I. G. and Rodimova, O. B. (1971) 'Ab initio calculation of electronic states of the ions H_4^+ and H_5^+', Int. J. Quantum Chem. 5, 669-682.

Kolos, W. and L. Wolniewicz, 'Improved Theoretical Ground-State Energy of the Hydrogen Molecule', J. Chem. Phys. 49, 404-410 (1968).

Mentch, F. and Anderson, J. B. (1981) 'Quantum chemistry by random walk: Importance sampling for H_3^+', J. Chem. Phys. 74, 6307-6311.

Reynolds, P. J., Ceperley, D.M., Alder, B. J. and Lester, Jr.W.A. (1982) 'Fixed-node quantum Monte Carlo for molecules', J.Chem. Phys. 77, 5593-5603.

Salmon, L. and Poshusta, R. D. (1973) 'Correlated gaussian wavefunctions for H_3^+', J. Chem. Phys. 59, 3497-3503.

Salmon, L. and Poshusta, R. D. (1973) 'Polarized-orbital valence-bond calculations on the ground state properties of H_5^+', J. Chem. Phys. 59, 4867-4870.

Urdaneta, C., Largo-Cabrerizo, A., Lievin, J. and Clementi, E. (1988) 'Gaussian functions in hylleraas-CI calculations. I. Ground state energies for H_2, HeH^+, and H_3^+', J.Chem. Phys. 88, 2091-2093.

Yamaguchi, Y., Gaw, J. F. and Schaefer III,H. F. (1987) 'The H_5^+ potential energy hypersurface: Characterization of ten distinct energetically low-lying stationary points', J. Chem. Phys. 86, 5072-5081.

A Molecular Orbital Calculation Method Considering Crystal Fields

Hiroki Kumahora, Kazutami Tago, Kinya Kobayashi, Noriyuki Kurita
Energy Research Laboratory, Hitachi Ltd.
1168 Moriyama-cho, Hitachi-shi, Ibaraki-ken, 316, JAPAN

ABSTRACT. An evaluation method for crystal fields in the Discrete Variational (DV)-Xα molecular orbital calculation is presented. Its main features are as follows. First, molecular orbitals for a cluster are calculated. Then the electrostatic potential of a unit cell is added to the hamiltonian of the cluster. This electrostatic potential is obtained by using the Coulomb potential of the electron density. The electron density is estimated by the square of the molecular orbitals of the cluster so that the potential of the unit cell is consistent with that of the cluster. Finally the potential of the unit cell is summed over the infinite crystal by using the Ewald sum method. The approach is applied to the MgO crystal.

1. Introduction

Molecular orbital calculations have been performed for the electronic structure analysis of various atomic clusters [1-8]. These molecular orbital calculations treat a cluster which consists of a finite number of atoms, so that the effects of infinite fields such as electrostatic fields of a crystal are ignored. Molecular orbital calculations for a large number of atoms are needed to estimate the infinite fields. But these calculations for large clusters are usually impossible because of insufficient memory size or long calculation time of the computer systems.

Some ideas for estimation of the crystal fields have been proposed [2,6,8] by the calculation of the electrostatic potential using the Coulomb potential from point charges. In these methods , fixed values of the point charges are spread outside of the cluster. Then the Coulomb potential from these point charges is added to the hamiltonian of the cluster as an external field. But these point charges are not consistent with charges in the cluster.

In the present calculation the electron density obtained by the molecular orbital of the cluster is used for the source term of the Coulomb potential. The entire calculation consists of three steps. The first is the molecular orbital calculation for the cluster. This uses the ordinary discrete variational method described previously [9]. Then the electron density is obtained for the individual atoms in the cluster. In the second part of the calculation, the potential of a unit cell of

P. Jena et al. (eds.), Physics and Chemistry of Finite Systems: From Clusters to Crystals, Vol. I, 611–616.

the crystal is obtained. This potential consists of the hamiltonian of the cluster and the Coulomb potential from the electron densities outside the cluster. In the third step, the potential of the unit cell is summed over the infinite crystal by using the Ewald sum method. In the present approach, the Coulomb potential from the electron densities outside the cluster is obtained by the molecular orbitals in the cluster. So that the electron densities in the cluster and those outside it are consistent.

In section 2, the evaluation method is presented. An example application for the MgO crystal is given in section 3. Comparison of the source term of the unit cell is also discussed there.

2. Molecular orbital calculation with crystal field

The hamiltonian consists of the usual one for the cluster and the additive potential of the crystal fields. The latter is estimated by the electrostatic potential of the cluster. This section outlines the hamiltonian, additive potential of the unit cell and the potential of the infinite crystal.

One electron Shrodinger equation for the cluster is written as

$$H(\vec{r})\ \psi_i(\vec{r}) = E_i\ \psi_i(\vec{r}) \tag{1}$$

where H(r) is a one electron hamiltonian, $\psi(r)$ is a basis function and E is a level energy, respectively. The hamiltonian H(r) of the cluster, in which the crystal potential is not included, is the sum of an electron kinetic energy T(r) and potential energy $V_{in}(r)$.

$$H(\vec{r}) = T(\vec{r}) + V_{in}(\vec{r}) \tag{2}$$

where V_{in} (r) is the sum of the nuclear attractive potential $U_c(r)$, the electron repulsive potential $U_r(r)$, and exchange correlation potential $U_e(r)$.

$$V_{in}(\vec{r}) = U_c(\vec{r}) + U_r(\vec{r}) + U_e(\vec{r}) \tag{3}$$

The nuclear attractive potential $U_c(r)$ is given by

$$U_c(\vec{r}) = -\sum_{a}^{M_{in}} \frac{Z_a}{|\vec{r} - \vec{R}_a|} \tag{4}$$

where M_{in} is the total number of atoms in the cluster and Z_a and R_a is the atomic number and coordinates of individual atoms in the cluster, respectively.

The electron repulsive potential $U_r(r)$ is given by

$$U_r(\vec{r}) = \int \frac{\rho(\vec{x})}{|\vec{r} - \vec{x}|} d\vec{x} \tag{5}$$

The electron density $\rho(r)$ is written as

$$\rho(\vec{r}) = \sum_{i}^{N} | \psi_i(\vec{r}) |^2 \tag{6}$$

where N is the total number of electrons in the cluster. The exchange correlation potential $U_e(r)$ is given by

$$U_e(\vec{r}) = c\rho(\vec{r})^{1/3} \tag{7}$$

where the constant c is set to 0.75.

Then the crystal potentials, $V_{term}(r)$ and $V_{cryst}(r)$, are added to the hamiltonian H(r) of the cluster as follows;

$$H(\vec{r}) = T(\vec{r}) + V_{in}(\vec{r}) + V_{term}(\vec{r}) + V_{cryst}(\vec{r}) \tag{8}$$

The potential of the unit cell is the sum of T(r), $V_{in}(r)$ and $V_{term}(r)$. The electrostatic potential of the unit cell is extended over the infinite crystal by using the Ewald sum method. The potential for the infinite crystal is $V_{cryst}(r)$ in equation (8). The potential $V_{term}(r)$ consists of the two parts of the source terms as follows:

$$V_{term}(\vec{r}) = U'_c(\vec{r}) + U'_r(\vec{r}) \tag{9}$$

where the source term of the potential is given by

$$U'_c(\vec{r}) = - \sum_{b}^{M_{term}} \frac{Z_b}{|\vec{r} - \vec{R}_b|} \tag{10}$$

and M_{term} is the total number of sites around the cluster. Z_b and R_b are the atomic number and coordinates of individual atoms around the cluster, respectively.

$$U'_r(\vec{r}) = \sum_{b}^{M_{term}} \int \frac{\rho'_b(\vec{x})}{|\vec{r} - \vec{x} - \vec{R}_b|} d\vec{x} \tag{11}$$

The source term of equation (11) is given by

$$\rho'_b(\vec{r}) = \sum_{i \in b}^{N} | \psi_i(\vec{r}) |^2 \tag{12}$$

where the summation of equation (12) is done over the sites around the cluster. The potential $V_{cryst}(r)$ is the infinite crystal potential estimated by using the

Ewald sum method.

$$V_{cryst}(\vec{r}) = V_1(\vec{r}) + V_2(\vec{r}) \tag{13}$$

The first term in equation (13) is given by

$$V_1(\vec{r}) = \sum_j{}' Q_j \sum_k \frac{\mathrm{erfc}(\alpha R_{jk})}{R_{jk}} \tag{14}$$

where the first summation denotes the atomic charge of the j-th site in the unit cell, which is defined as follows:

$$Q_j = - \int \rho'_j(\vec{x})\, d\vec{x} + Z_j \tag{15}$$

where Zj is the atomic number of the j-th site in the unit cell having coordinates defined as follows :

$$R_{jk} = \left| \vec{r} - \vec{r}_j - \vec{L}\cdot\vec{k} \right| \tag{16}$$

and erfc in equation (14) means the usual error function.

$$\mathrm{erfc}(x) = \frac{2}{\sqrt{\pi}} \int_x^{\infty} \exp(-u^2)\, du \tag{17}$$

The second term of equation (13) is given by

$$V_2(\vec{r}) = \frac{1}{\pi L} \sum_j{}' Q_j \sum_k \frac{\exp(-\pi^2 |\vec{k}|^2 / \alpha^2 L^2)}{|\vec{k}|^2} \cos\left(\frac{2\pi R_{jk}}{L}\right) \tag{18}$$

$$L = \sqrt{L_x^2 + L_y^2 + L_z^2} \tag{19}$$

$$k = \sqrt{k_x^2 + k_y^2 + k_z^2} \tag{20}$$

where L in equation (19) is the lattice size of the unit cell in the x-, y- or z-direction. The integer k in equation (20) is the number of unit cells. The calculation of the infinite crystal potential $V_{cryst}(r)$ is performed for a large value of the integer k while adjusting the convergence parameter α.

The entire calculation consists of an initial step and three calculation steps. In the initial step, input data for the calculation are provided. The main information included in the input data is basis functions, coordinates of atoms in the cluster and coordinates of the unit cell. The first calculation step is the molecular orbital calculation for the cluster. The second step is the Coulomb potential calculation for the unit cell. The third step is the calculation of the infinite crystal potential by using the Ewald sum method. These three steps are iterated with self-consistent field (SCF) until the molecular orbitals converge.

3. Results and discussion

This method is adopted for an electronic structure analysis of the MgO crystal. An MgO_6 cluster, in which the Mg atom is surrounded by O atoms at octahedral sites, is chosen for an electronic structure analysis. The bond length between the Mg atom and an O atom is taken to be 0.21nm [10,11]. A geometric structure of $Mg_{13}O_{14}$ is chosen for the unit cell through calculations. This unit cell is the cubic cell for which bond length between the Mg atom and an O atom is also taken to be 0.21nm. In the present calculation, the unit cell consists of the MgO_6 cluster and additive sites of $Mg_{12}O_8$.

The extended Slater type orbitals [14] are used for the basis sets in the calculations. For the Mg ion the basis sets from the 1s to 3s orbitals and for the O ion the basis sets from the 1s to 2p orbitals are used for the calculations.

Five calculation cases are made to compare the effect of the source term of the Coulomb potential for the unit cell. No source terms are added in case 1. Both cases 2 and 3 are calculations for the unit cell without the infinite crystal potential. The source terms of the Coulomb potential for the additive sites of $Mg_{12}O_8$ are point charges for case 2 and electron densities for case 3. Both cases 4 and 5 are calculations with infinite crystal potential. The source terms of the Coulomb potential for the additive sites of $Mg_{12}O_8$ are point charges for case 4 and electron densities for case 5. These source terms are summarized in table 1.

Results of the calculations are listed in table 1. The energy gaps, denoted by Eg, between the highest occupied molecular orbital (HOMO) and the lowest unoccupied molecular orbital (LUMO) are listed in table 1. The value of Eg from previous works [10,11] was 7.8eV. The energy width of the valence level, denoted by Ew is also listed there. The value of Ew from previous works [12,13] was 3.0eV.

Table 1. Summary of calculation.

Case.	Adding Potentials		Eg	Ew	CPU Time	Memory
	Unit Cell	Infinite Crystal	(eV)	(eV)	(s)	(MB)
1	none	none	8.1	2.4	2.1	3.3
2	point charge	none	6.4	4.9	2.2	3.3
3	electron density	none	7.9	4.0	19.1	3.3
4	none	Ewald sum	7.0	3.4	26.2	3.3
5	electron density	Ewald sum	7.2	3.4	33.6	3.3

The energy width Ew in case 1 is smaller than those of the other cases. This is because case 1 ignores both the potential of the unit cell and the infinite crystal potential.

Comparing Eg of previous works [10-13], cases 2 and 3, the value of case 3 agrees with the previous works [10-13] than case 2 does. So the electron densities are a better source term of the unit cell than point charges.

Comparing Ew of previous works [10-13], cases 4 and 5, their values are almost the same, but Eg of case 5 agrees with the previous works [10-13] better than case 4 does. Then the effect of the electron densities in the unit cell is important.

In conclusion the best calculation to estimate the electrostatic potential of the crystal is case 5, in which electron densities in the cluster are used for the Coulomb potential calculation. The maximum errors from the previous works [10-13] are 0.6eV for Eg and 0.4eV for Ew, respectively.

For the calculations, a supercomputer, the HITAC S-820/60 was used. The cpu times for every SCF iteration and memory sizes with the computer required for each calculation case are listed in table 1. The estimated values of cpu time and memory size for the unit cell cluster $Mg_{13}O_{14}$ are 104 (s) and 28 (MB), respectively. Comparing these estimated values and those of case 5, the cpu time is reduced to about 1/3 and the memory size is reduced to about 1/8.

4. Conclusion

The DV-Xα molecular orbital calculations considering crystal fields were performed. The main feature of this method was the evaluation of the crystal fields by using electron densities of the cluster. The electron densities of the cluster was shown to be a better source term than the point charges for calculation of the Coulomb potential. This method was adopted to the MgO crystal. The maximum errors from previous works [10-13] are 0.6eV for the energy gap between the HOMO and the LUMO and 0.4eV for the energy width of valence electrons, respectively. Comparing this method and a cluster calculation for the unit cell cluster $Mg_{13}O_{14}$, the cpu time is reduced to about 1/3 and the memory size is reduced to about 1/8.

References

[1] H. Adachi et al., (1978),J.Phys. Soc. Japan 45, 875
[2] C. Satoko et al., (1978), J.Phys. Soc. Japan 45, 1333
[3] H. Adachi et al., (1979), J.Phys. Soc. Japan 47, 1528
[4] M. Tsukada et al., (1979), J.Phys. Soc. Japan 47, 1610
[5] M. Tsukada et al., (1980), J.Phys. Soc. Japan 48, 200
[6] J. Kondou et al., (1988), J.Phys. Soc. Japan 57, 4334
[7] P. F. Walch and D. E. Ellis, (1973), Phys. Rev. B8, 5920
[8] J. P .Hansen, (1973), Phys. Rev. A8, 3096
[9] K. Tago et al., (1988), Int. J. of Supercomputer Application 2, 58
[10] R. C. Whited and W. C. Walker, (1966), Phys. Rev. Lett. 17, 1106
[11] M. W. Williams and E. T. Arakawa, (1967), J. Appl. Phys. 38, 5272
[12] L. F. Mattheiss, (1972), Phys. Rev. B5, 290
[13] L. F. Mattheiss, (1972), Phys. Rev. B5, 306
[14] E. Clementi and C. Roetti, (1974), Atomic Data and Nuclear Data Tables 14, 177

CLUSTER RESONANCES AND BOUND STATES AND CRYSTAL BAND SPECTRUM

P. KURASOV
Institute for Physics, St. Petersburg University, 198904 St. Petersburg, RUSSIA, USSR
and
Manne Siegbahn Institute, Freskativagen 24, 10405 Stockholm, SWEDEN

ABSTRACT. Connection between the structure of the electron spectrum of an isolated atom or a cluster and corresponding crystal is investigated in the case of exactly solvable models, constructed with the help of extensions theory for self-adjoint operators. Reflection and transmission coefficients for scattering on impurities are calculated.

1 Introduction.

The paper is devoted to the investigation of the spectral properties of clusters and crystals with the help of construction of exactly solvable models. Usually differente mathematical languages are used to investigate crystals and few atoms systems. Method of zero range potentials allows us to construct model operators which describe these two physical systems using the same mathematical objects. Moreover, constructed models are exactly solvable and all the characteristics can be calculated analytically without any approximations. All the considered models are collected in monographs (3,4). But in this paper we shall focus our attention on the case of low-dimensional structures in three-dimensional space, which are under interest during the last years.

The second section is devoted to the description of the simplest model of the system electron-atom. In the third section this model is generalized to the case of cluster. The band spectrum of the chain and connections with cluster spectrum are discussed in the fourth section. The last section is devoted to the scattering on impurities in a homogeneous chain.

2 Isolated Atom.

The model operator can be constructed with the help of restriction-extension procedure for self-adjoint operators. We shall briefly describe a formal construction here. A precise mathematical explanation can be find in (1). One can define Laplace operator on the set of smooth functions with a following singularity at one point x_0:

$$\Psi(x) \sim \frac{\Psi_-}{4\pi|x-x_0|} + \Psi_0 + o(1), x \to x_0, \Psi_-, \Psi_0 \in \mathbf{C}. \tag{1}$$

P. Jena et al. (eds.), Physics and Chemistry of Finite Systems: From Clusters to Crystals, Vol. I, 617–622.

Here the point x_0 markes the position of the atom. Self-adjoint perturbations of the background operator with an additional internal operator give us energy dependent boundary conditions :

$$\Psi_-/\Psi_0 = D(E), \tag{2}$$

with an analytical function $D(E)$, which has a positive imaginary part in the upper half-plane of energy E. An approximate graph of such function is presented on Fig. 1. This function accumulates all the spectral characteristics of the center. For example the continuous spectrum eigenfunctions can be presented as a sum of a plane wave (eigenfunction for unperturbed problem) and outgoing spherical wave:

$$\Psi(\vec{k}, x) = \frac{1}{\sqrt{2\pi}} \left(e^{i<\vec{k},\vec{x}>} + \rho(\vec{k}) \frac{e^{ik|x-x_0|}}{4\pi|x-x_0|} \right). \tag{3}$$

The amplitude $\rho(\vec{k})$ is calculated from the boundary conditions:

$$\rho(\vec{k}) = \frac{D(k^2)}{1 - \frac{ik}{4\pi} D(k^2)}. \tag{4}$$

Now the S-matrix can be calculated, it is non-trivial in the s-channel only:

$$S(E) = \frac{\frac{ik}{4\pi} + D^{-1}(E)}{\frac{ik}{4\pi} - D^{-1}(E)}, k = \sqrt{E}. \tag{5}$$

The dispersion equation on the binding energy is:

$$\left(\frac{ik}{4\pi}\right)^{-1} = D(E) \tag{6}$$

For simplicity, we shall consider the functions $D(E)$ with a finite nimber of singularities situated on the real axis only. Then each negative zero of D produces eigenvalue of the operatór, while the positive zeroes define resonances - singularities of S-matrix on non-physical sheet of energy (see Fig.1).

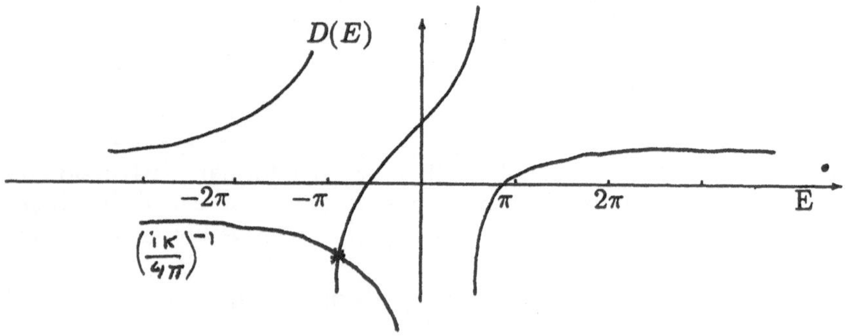

Fig.1 Graphic solution of the equation (6).

3 Cluster.

Model Hamiltonian of electron in a cluster of n atoms can be obtained by considering of energy dependent boundary conditions of type (2) at the points $x_1, x_2, ..., x_n$ marking the positions of atoms:

$$\Psi_-/\Psi_0 \mid_{x=x_j} = D_j(E), \tag{7}$$

Continuous spectrum eigenfunctions are:

$$\Psi(\vec{k}, x) = \frac{1}{\sqrt{2\pi}} \left(e^{\imath<\vec{k},\vec{x}>} + \sum_{i=1}^{n} \rho_i(\vec{k}) \frac{e^{\imath k|x-x_i|}}{4\pi|x-x_i|} \right). \tag{8}$$

We need to consider n-dimensional vectors $\vec{\Psi}_0$, $\vec{\Psi}_-$, $\vec{\rho}(\vec{k})$ and $\vec{\ell}$, where $\vec{\ell}_i = e^{\imath<\vec{k},x_i>}$, and $n \times n$ matrices $\mathbf{D} = diag\{D_i(E)\}$ and $\mathbf{B}$:

$$\mathbf{B}_{ij} = \begin{cases} \frac{\imath k}{4\pi}, i = j; \\ \frac{e^{\imath k|x_i-x_j|}}{4\pi|x_i-x_j|}, i \neq j \end{cases}. \tag{9}$$

The scattering amplitudes can be calculated:

$$\vec{\rho}(\vec{k}) = (1 - \mathbf{DB})^{-1} \mathbf{D}\vec{\ell} \tag{10}$$

The equation for binding energy is:

$$\det \left(\mathbf{D}^{-1} - \mathbf{B}\right) = 0. \tag{11}$$

One can prove that $\mathbf{B}$ is a matrix with a positive imaginary part in the upper half plane E. Then the solutions of (11) are situated on the negative half axis of energy and on non-physical sheet. The solutions of the equation coincide with the singularities of the analytical continuation of the scattering matrix.

4 One-dimensional Chain.

A model of a crystal can be obtained on the same way as the cluster model. The only difference is that one need to consider an infinite number of centers, situated in a regular way. We shall study the simplest case here - one-dimensional chain of atoms. we suppose atoms to be identical. Hence one has to consider identical boundary conditions (7) at the points : $x_j = j\vec{e}, j \in \mathbf{Z}$, where $\vec{e}$ is vector of the period of the chain. The spectrum is purely continuous, corresponding to scattered waves and waveguide functions. Eigenfunctions of the first type describe process of electron scattering on the chain:

$$\Psi(\vec{k}, x) = \frac{1}{\sqrt{2\pi}} \left(e^{\imath<\vec{k},x>} + \rho(\vec{k}) \sum_{j\in\mathbf{Z}} \frac{e^{\imath k|x-x_j|}}{4\pi|x-x_j|} e^{-\imath j<\vec{e},\vec{k}>} \right) \tag{12}$$

Waveguide functions describe electron transport trough the chain and define band spectrum of the operator. Its are some generalization of the bound state wavefunctions. Like all the solid state wavefunction these eigenfunctions can be written in a Bloch representation with a quasimomentum t :

$$\Phi(t,x) = \sum_{j\in \mathbf{Z}} \frac{e^{ik|x-x_j|}}{4\pi|x-x_j|} e^{ijt} \tag{13}$$

Substitution into the boundary conditions allows us to calculate the scattering amplitude ρ for the scattered waves:

$$\vec{\rho}(\vec{k}) = \frac{D}{1 - D\tilde{B}(k,k_{\|})}, \; k_{\|} = < \vec{k}, \vec{e} >, \tag{14}$$

and dispersion equation, connected energy and quasimomentum, for waveguide functions:

$$D^{-1}(E) = \tilde{B}(k,t) \tag{15}$$

Here $\tilde{B}(k,t)$ is a Fourier transformation of the infinite-dimensional matrix $\mathbf{B}(k)$:

$$\tilde{B}(k,t) = \frac{ik}{4\pi} + \sum_{j\neq 0} \frac{e^{ik|x_j|}}{4\pi|x_j|} e^{itj} = \frac{1}{4\pi} \ln \frac{1}{2(\cos k - \cos t)} \tag{16}$$

We shall concentrate our attention by the case of waveguide functions only. Solutions of the equation (15) are situated in a region $E < 0$ or $t > |k|$. Approximate lines of solutions are marked on the Fig. 2.

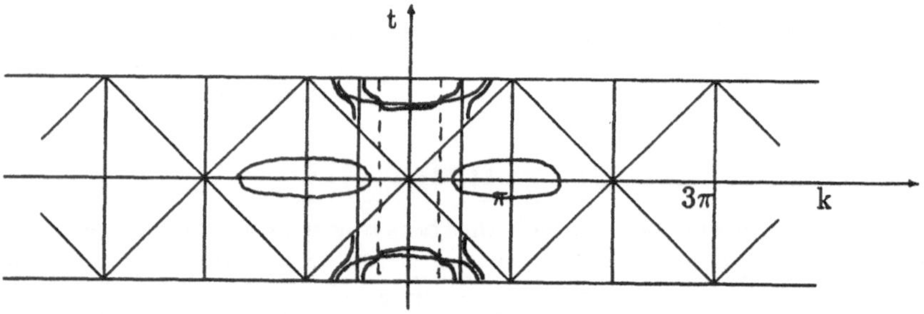

Fig.2 Dispersion curves for one-dimensional chain.

Inclined straight lines mark singularities of $\tilde{B}(k,t)$. Ovals correspond to zeroes of $\tilde{B}(k,t)$. Vertical straight lines are lines of zeroes of D, vertical dotted lines - singularities of D. The band structure of the spectrum can be obtained by projection of the solutions on the energy axis. Narrow gaps in the spectrum are obtained near the zeroes of D, which corresponds to resonances of isolated atom. Hence, the resonances of isolated atom determine narrow gaps in the band spectrum of the problem. Eigenstates of isolated atom produce small bands of continuous spectrum of the chain.

For completion of the investigation of the homogeneous chain we shall calculate the resolvent kernel for the operator. The Green function will be usefull in the next section during investigation of impurities in a homogeneous chain. The kernel of the resolvent can be presented in the form:

$$G(x,x') = \frac{e^{ik|x-x'|}}{4\pi|x-x'|} + \sum_n \frac{e^{ik|x-x_n|}}{4\pi|x-x_n|} a_n \tag{17}$$

Coefficients a_n form an ℓ_2 element: $\mathcal{A} = \{a_n\}_{n\in\mathbf{Z}}$. By considering vector $\mathcal{L} = \{\frac{e^{ik|x_n-x'|}}{4\pi|x_n-x'|}\}_{n\in\mathbf{Z}}$ one can find all the unknown coefficients:

$$\mathcal{A} = \left(D^{-1}(E) - \mathbf{B}(k)\right)^{-1} \mathcal{L} \tag{18}$$

Operator at the right side of equality (18) is an operator of multiplication by function $\left(D^{-1}(E) - \tilde{B}(k,\varphi)\right)^{-1}$ in Fourier representation.

5 Defect in the chain.

This section is devoted to the investigation of scattering on a defect in a one-dimensional chain of equivalent atoms. Defect can be produced by dislocation of one of the atoms and by substitution of one of atoms by an atom of another type. Both cases can be modeled with the help of described method. Continuous spectrum of the perturbed operator coincide with continuous spectrum of unperturbed one. We shall restrict our consideration by the band part of the spectrum.

a) **Scattering by heterogeneous atom.**

To construct the model one can substitute boundary conditions at point zero by boundary conditions with a differente dispersion function $D_0(E)$:

$$\Psi_-/\Psi_0 \mid_{x=0} = D_0(E). \tag{19}$$

Band spectrum eigenfunction for this problem can be presented in the form:

$$\Theta(t,x) = \Phi(t,x) + \rho G(x,0). \tag{20}$$

where Φ is a band spectrum eigenfunction for homogeneous chain. The representation of Green function can be simplified for $x' = 0$ and for the energies from the band spectrum. Really, Fourier transformation of vector $\mathcal{L}$ is:

$$\tilde{\mathcal{L}}(\varphi) = \frac{1}{\sqrt{2\pi}} \tilde{B}(k,\varphi).$$

We shall fix the dispersion dependence by condition $\pi > t(k) > 0$, then the coefficients a_n are:

$$n > 0 : a_n = \frac{\tilde{B}(k,t(k))}{\frac{\partial}{\partial k}\left(D^{-1}(E) - \tilde{B}(k,t(k))\right)} e^{it(k)n}$$

$$n<0: a_n = \frac{\tilde{B}(k,t(k))}{\frac{\partial}{\partial k}\left(D^{-1}(E)-\tilde{B}(k,t(k))\right)} e^{-it(k)n}$$

The amplitude ρ is calculated from the boundary conditions at zero point:

$$\rho = \frac{1-D_0(E)D^{-1}(E)}{(D_0(E)D^{-1}(E)-1)\,a_0-1}. \tag{21}$$

Asymptotics of wave functions contain incoming and outgoing waves for homogeneous chain. One can introduce transmission and reflection coefficients on the usuall way :

$$\mathbf{R}(t) = \frac{D_0(E)-D(E)}{D^2(E)\frac{\partial}{\partial k}\left(D^{-1}(E)-\tilde{B}(k,t(k))\right)-(D_0(E)-D(E))} \tag{22}$$

$$\mathbf{T}(t) = \frac{D^2(E)\frac{\partial}{\partial k}\left(D^{-1}(E)-\tilde{B}(k,t(k))\right)}{D^2(E)\frac{\partial}{\partial k}\left(D^{-1}(E)-\tilde{B}(k,t(k))\right)-(D_0(E)-D(E))} \tag{23}$$

Singularities of the dispersion function $D_0(E)$ define zero transmission coefficient. One obtain zero reflection coefficient when the dispersion function of impurity coincide with the dispersion function of atoms. Energies of bound states coincide with singularities of reflection and transmission coefficients.

b) **Scattering by dislocation.**

Dislocation can be modelled by changing of the position of one of the centers, for example $x_0 \neq 0$. Then an eigenfunction has the form:

$$\Theta(t,x) = \Phi(t,x) + \rho_0 G(x,0) + \rho_1 G(x,x_0). \tag{24}$$

Amplitudes ρ_0 and ρ_1 are calculated from the boundary conditions at points $x=0$ and $x=x_0$:

$$\Theta_- \mid_{x=0} = 0;$$

$$\Theta_-/\Theta_0 \mid_{x=x_0} = D(E).$$

For amplitudes ρ_0 and ρ_1 we have a system of linear equations:

$$\begin{cases} (1+a_0(0))\rho_0 + a_0(x_0)\rho_1 = -1 \\ G(x_0,0)\rho_0 + \frac{ik}{4\pi}\rho_1 = -\Phi(t,x_0) \end{cases}. \tag{25}$$

Asymptotic analysis of the solutions allows us to calculate the scattering matrix for the operator.

References:

1. Kurasov P.B., Pavlov B.S.(1988) 'Electron in a Homogeneous Crystal of Point Atoms with Internal Structure II', Teor.i Mat. Fiz., v74, 82-93.

2. Kurasov P.B. (1989) 'One-electron Model of a Long Molecule', Leningrad Univ. Vestnik, ser.4, v3(nr18), 8-15(in Russian).

3. Pavlov B.S. (1987) 'Electron in a Homogeneous Crystal of Point Atoms with Internal Structure I', Teor.i Mat. Fiz., v72, 403.

4. Demkov Yu.N., Ostrovskii V.N. (1988) 'Zero-range potentials and their applications in atomic physics', Plenum Press, NY.

5. Albeverio S., Gesztesy F., Hoegh-Krohn R., Holden H (1988) 'Solvable models in quantum mechanics', Springer-Verlag, Berlin.

CALCULATION OF THE ELECTRONIC STRUCTURE AND MAGNETIC PROPERTIES OF IRON MICROCLUSTERS

L. I. KURKINA, O. V. FARBEROVICH
Voronezh State University
University sq., 1, SU-394693, Voronezh, USSR

V. S. STEPANYUK, A. A. KATSNELSON
Moscow State University
Lenin Hills, SU-117334, Moscow, USSR

A. SZASZ
Eotvos University
Muzeum krt. 6-8, H-1088, Budapest, Hungary

ABSTRACT. Calculations of a spin-polarized electronic structure and magnetic moment of small iron clusters within the framework of the local spin-density approach and the model "atom in jellium" have been presented. It is shown that the occupation of localized d-states of atoms in iron clusters changes nonmonotonically depending on the atom position in relation to the cluster surface as well as on the cluster size. The nonmonotonic change of the electronic structure results in the nonmonotonic size dependence of iron cluster magnetic properties.

1. INTRODUCTION

Recently the properties of small transition clusters have been studied intensively. With development of methods to produce clusters of controlled size it has become possible to perform experiments on their electron and magnetic properties. For example, the magnetic moment [1], chemisorption reactivity [2] and ionization potential [2] of such clusters have been measured.

Theoretical investigations of cluster physical properties are carried out actively too. A variety of theoretical models has been proposed for calculations of a cluster electronic structure. However, most precise models (such as the

P. Jena et al. (eds.), Physics and Chemistry of Finite Systems: From Clusters to Crystals, Vol. I, 623–628.

self-consistent-field X_α-scattered wave method or some other methods) permit to obtain an electronic structure for clusters with low number of atoms. For large atomic aggregations these methods cannot be applied successfully, and simpler models are employed in this case. Thus, for study of electronic properties of sp-bonded metal clusters the jellium model is used, and it gives an excellent description of the various experimentally observed data.

In this work we have calculated an electronic structure and magnetic moment of spherical iron clusters containing up to 130 atoms within the framework of the model "atom in jellium". (Models of this type have been used in the investigation of, e.g., chemisorption on metallic surface, impurities in bulk metal and small clusters of simple metals, optical properties of pure metals.)

2. MODEL AND METHOD

In our calculations atoms of a cluster are mimiced by atoms embedded in the centre of spherical-finite jellium. The jellium sphere radius R is determined according to the position of the real atom in relation to the cluster surface: R defines the short cut between the atom and the cluster surface.

The electronic stricture of an atom embedded in spherical jellium is obtained using the self-consistent Kohn-Sham equations (in atomic Rydberg units):

$$[-\nabla^2 + V^s(r)]\,\psi_{nls}(\vec{r}) = E_{nls}\psi_{nls}(\vec{r}), \qquad (1)$$

$$V^s(r) = -\frac{2Z}{r} + 2\int \frac{n^-(\vec{r}\,') - n^+(\vec{r}\,')}{|\vec{r} - \vec{r}\,'|}\, d\vec{r}\,' + V^s_{xc}(r), \qquad (2)$$

$$n^-(\vec{r}) = \sum_{nls}^{occ} |\psi_{nls}(\vec{r})|^2 \qquad (3)$$

with the Vosko form of the local spin-polarized exchange-correlation potential $V^s_{xc}(r)$. Here the spin s is $\uparrow, \downarrow$; Z is the central nuclear charge. The rigid positive charge distribution of the jellium background is given by

$$n^+(r) = n_o^+\,\theta(R-r)\,\theta(r-r_o), \qquad (4)$$

where $\theta(x)$ is the unit step function, R and r_o being, respectively, the outer and inner radii of the positive background sphere; n_o^+ is taken to be the same as the mean valence electron density of the corresponding bulk metal:

$$n_o^+ = \frac{3\ N}{4\pi\ (R^3 - r_o^3)}\ , \tag{5}$$

$$R = (N/N_a + 1)^{1/3}\ r_o, \tag{6}$$

r_o is the Wigner–Seitz radius, N is the number of electrons in the jellium sphere, and N_a is the number of valence electrons in an atom of the cluster. We have employed for Fe $r_o = 2.67$ a.u., $N_a = 2$.

The electron configuration of ground state of "atom in jellim" was determined from following reasoning. Rules of energy level occupation separately in jellium (in a spherical potential well) and in a free atom are known. Clearly, insertion of an atom in the jellium sphere centre does not change the atom and jellium field symmetry, and, consequently, the symmetry of their electron states is not changed, too. It was supposed therefore that the number of electron states with the same symmetry in the jellium cluster with the central atom is equal to the sum of such symmetry states of the jellium cluster and atom. The order of the energy level is obtained by solving the spin-polarized self-consistent Kohn-Sham equation (1-3) with the angular l and principal n quantum numbers. The total energy of the Fe atom in the jellium cluster is computed for various occupation numbers of the upper levels. As the ground state we have chosen the electron configuration with the lowest total energy.

It has been obtained that the five lowest energy levels of "atom Fe in jellium" are identical with those of an iron core, the rest of them are similar to the energy spectrum of a spherical potential well. This order is broken by the localized d-state in the upper part of the occupied energy region.

3. RESULTS

The electronic structure of iron atom embedded in the jellium sphere centre has been calculated as a function of the jellium size. It was found that the electronic structure of "atom Fe in jellium" alters periodically with the increase of the jellium sphere radius (or of the number of electrons in the

TABLE 1. The electron configurations of an iron atom embedded in jellium spheres containing from 10 to 60 electrons

N	majority spin	minority spin
	$1s^1 2s^1 1p^3 3s^1 2p^3 4s^1 3p^3 1d^5$...	$1s^1 2s^1 1p^3 3s^1 2p^3 4s^1 3p^3 1d^5$...
10	...$2d^0$	...$2d^0$
12	...$2d^2$	...$2d^0$
16	...$2d^5 5s^1$	...$2d^0 5s^0$
32	...$2d^5 5s^1 1f^7$	...$5s^1 1f^7 2d^1$
50	...$5s^1 1f^7 4p^3 2d^5 1g^9$	...$5s^1 1f^7 4p^3 2d^4 1g^0$
54	...$5s^1 1f^7 4p^3 2d^5 1g^9$	...$5s^1 1f^7 4p^3 2d^5 1g^3$
	...$5s^1 1f^7 4p^3 2d^5 1g^9$...	...$5s^1 1f^7 4p^3 2d^5 1g^9$...
60	...$3d^0$	...$3d^0$

jellium N according to (6)). Each period starts when the new empty d-level appears and is completed by its full occupation (see Table 1). The boundaries of the period are the jellium spheres with N = 10, 60, 188, 430, etc.

It was obtained as the next d-level is filled in turn, the previous d-state is extended. So, the most localized of d-states is upper d-level which has been fully or partially occupied. However, this state is more slowly extended than

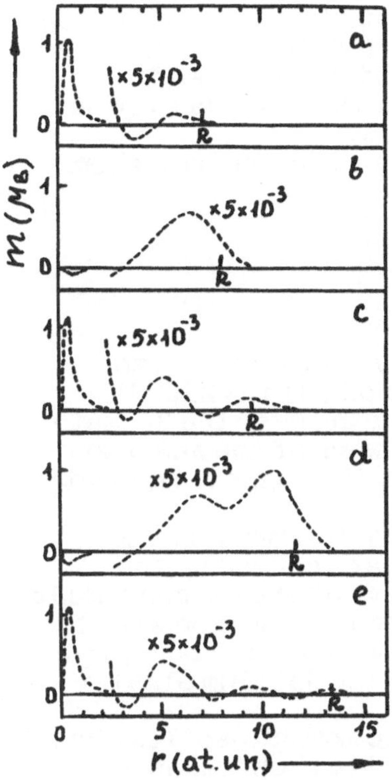

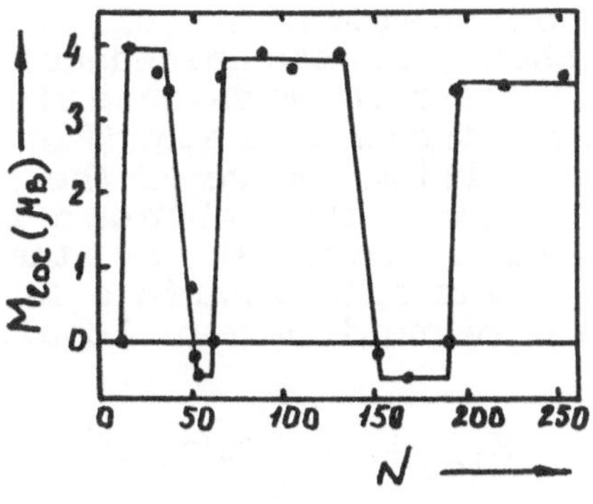

Figure 2. The magnetic moment localized near the central Fe atom in dependence on the number of electrons in the jellium sphere

Figure 1. Variation of the radial spin density distribution in the spherical-finite jellium containing the central Fe atom with the increase of the jellium sphere size: N = 32(a), 54(b), 88(c), 166(d), 252(e).

the d-shell of a free Fe atom, and the extent of the upper d-level increases with the rise of the jellium sphere radius.

Thus, the interaction between d-electrons of Fe atom and s-valence electrons of the Fe cluster has led to the periodical occupation of the localized d-state of the system "Fe atom in jellium" with the increase of the jellium sphere radius. Within the framework of our model this effect corresponds to the nonmonotonic change of the occupation of localized d-states with the moving away from the surface to the centre of the cluster.

The calculations of the radial spin-density distribution

$$m(r) = n_{\uparrow}^{-}(r) - n_{\downarrow}^{-}(r)$$

for the iron atom in the jellim spheres with various radii have shown that the spin-density oscillates due to the polarization of the valence electron gas by the magnetic moment of the central atom (Fig.1). The number of oscillations in the jellium sphere of the same period is equal and coincides with the nimber of d-levels which have been fully occupied before the start of the period. The spin density near the jellium sphere surface is positive.

The magnetic moment localized near the central atom

$$M_{loc} = 4\pi \int_0^{r_0} m(r)\, r^2\, dr$$

oscillates with the increase of the jellium cluster size (Fig.2). These results agree with those of [3] in which the mechanism of formation of magnetic ordering near the 3d-metal surface was considered within the framework of the Anderson periodic model, and it was obtained that the magnetic moment oscillates near the surface.

The negative local magnetic moments of some jellium spheres with a central atom (see Fig.2) are caused by the negative polarization of the spin density of the closed shells near the central atom due to the exchange interaction with the open shell. The d-states of these systems are fully occupied. The open shell (or shells) is (are) near the jellium cluster surface (see Fig.1, Table 1).

We suppose that these local magnetic moment oscillations contribute to the lack of the ferromagnetic properties of small iron clusters. (On the observation of the paramagnetic character of small Fe clusters see, e.g. [1,4].)

4. REFERENCES

[1] Cox D.M., Trevor D.J., Whetten R.L., Rohlfing E.A., and Kaldor A. (1985) 'Magnetic behavior of free iron and iron oxide clusters' Phys.Rev.B 32, 7290-7298.

[2] Whetten R.L., Cox D.M., Trevor D.J., Kaldor A. (1985) 'Correspondence between electron binding energy and chemisorption reactivity of iron clusters', Phys.Rev.Lett. 54, 1494-1497.

[3] Kazanskii A.K., Kondratev A.S., and Uzdin V.M. (1988) 'Effect of surface in magnetic properties of metals', Fiz.metallov i metallovedenie' 65, 1068-1072 (in Russian).

[4] Vonsovskii S.V. (1971) Magnetizm, Moscow, Nauka (in Russian).

IONIZATION POTENTIALS OF FE MICROCLUSTERS

L. I. KURKINA, O. V. FARBEROVICH
Voronezh State University
University sq., 1, SU-394693, Voronezh, USSR

V. S. STEPANYUK, A. A. KATSNELSON
Moscow State University
Lenin Hills, SU-117334, Moscow, USSR

A. SZASZ
Eotvos University
Muzeum krt. 6-8, H-1088, Budapest, Hungary

ABSTRACT. Ionization potentials of small iron clusters have been obtained within the local spin-density approximation and model "atom in jellium". The nonmonotonic character of the size-dependence of an ionization energy with a cluster size has been shown.

1. INTRODUCTION

In recent year size-selective chemical and physical properties of small metal clusters (SMC) have attracted great attention. This interest is bound up with the effective use of microclusters in technological processes including a geterogeneous catalysis. SMC are excellent catalysts and it is well known that the activity and selectivity of such catalysts depend on the cluster size. Recently the strong correlation between the chemisorption reactivity of free SMC and their ionization energy has been shown [1].

We have made an attempt to study the size dependence of an ionization potential (IP) of spherical Fe-clusters containing up to twenty atoms. The IP of these atomic systems has been experimentally obtained by authors of [1-3].

2. RESULTS AND DISCUSSION

In the calculations we simulated atoms of a Fe cluster by Fe

P. Jena et al. (eds.), Physics and Chemistry of Finite Systems: From Clusters to Crystals, Vol. I, 629–631.

atoms embedded in the centre of the spherical-finite jellium of a various size. The jellium sphere radius R is defined by the short cut between the atom and the cluster surface.

IP of "atom in jellium" was obtained using the ground-state theory (the local spin-density approximation) by the self-consistent calculation and subtraction of total energies of the neutral and ionic ground states.

It was found that ionization thresholds of atoms of the Fe cluster are different and depend on positions of the atoms in relation to the cluster surface. The IP of the cluster

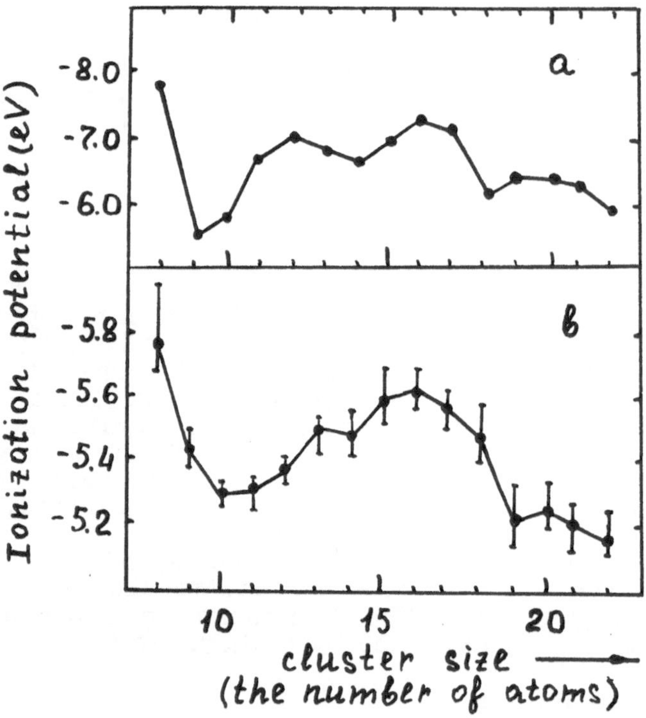

Figure 1. The size dependence of ionization potentials of iron clusters: theory (present work) (a) and experiment [1] (b).

atom having the least ionization energy of all was chosen as that of the Fe cluster.

IPs of Fe clusters with BCC and FCC geometry structures (cluster atoms are placed around the central atom like BCC or FCC coordination spheres of bulk iron) have been calculated. IP was obtained that IPs of iron clusters containing up to twenty atoms are determined by those of the central atoms (that is the IP of the cluster is coincided with that of Fe atom embedded in the centre of the jellium sphere with the radius of the cluster), and ones do not depend on the cluster geometry structure.

Our model calculations give the size dependence of IP which is in good qualitative agreement with the well-known experimental data (see Fig. 1). But our results predict the much larger amplitude of IP oscillations than the experiment. The oscillations on the calculated IP-curve are arisen from shell character of the model and the nonmonotonic occupation of the localized d-state of atoms of Fe clusters in dependence on the atom position in relation to the cluster surface as well as on the cluster size. (The occupation of localized d-states was chosen in accordance with the minimum of total energy of the system .)

Our calculations have shown that an IP of Fe microclusters consisting of up to 130 atoms does not reach the work function of bulk iron (4.5 - 4.8 eV). Apparently, the change from a Fe microclusters to solid state is occured for larger iron atomic aggregations.

3. REFERENCES

[1] Wetten R.L., Cox D.M., Trevor D.J., and Kaldor A. (1985) 'Correspondence between electron binding energy and chemisorption reactivity of iron clusters', Phys.Rev.Lett. 54, 1494-1497.

[2] Rohlfing E.A., Cox D.M., and Kaldor A. (1983) 'Photoionization measurement on isolated iron-atom clusters', Chem.Phys.Lett. 99, 161-166.

[3] Rohlfing E.A., Cox D.M., Kaldor A., and Johnson K.H. (1984) 'Photoionization spectra and electronic structure of small iron clusters', J.Chem.Phys. 81, 3846-3851.

MEASUREMENT OF IONIZATION POTENTIALS FOR Al_n AND Fe_n CLUSTERS

M. PELLARIN [1], J.L. VIALLE[1], J. LERME [1], F. VALADIER [1], B. BAGUENARD[1], J. BLANC[2] and M. BROYER [1]

[1] *Laboratoire de Spectrométrie Ionique et Moléculaire (URA CNRS 171) Université Lyon I, Bât. 205, 43 Bd du 11 Novembre 1918 69622 VILLEURBANNE Cedex, France*

[2] *Institut für Experimentalphysik, Frei Universität Berlin Arnimallee 14 - D 1000 BERLIN 33, Germany*

ABSTRACT. Using laser vaporization technique and standard Time-Of-Flight (TOF) spectrometric methods, Ionization Potentials (IP) of aluminum and iron clusters have been measured in extensive size ranges. These results are compared with recently published experimental works. For aluminium clusters, our results are extended to larger sizes and reveal new structures. For both types of clusters, the mass spectra recorded systematically for ionization energy just above the threshold reveal clear structures due to small effects in the ionization potential or in the ionization cross section. These effects are discussed in terms of the self consistent Spherical Jellium Background Model (SJBM).

1. Introduction

In simple metals like alkali metals, the electronic structure has been successfully interpreted in terms of the spherical jellium model. 'Magic numbers' observed in the size range from 10 to 3000 atoms are well explained in this frame [1-2-3] and very recently supershell structures have been obtained in sodium clusters [4-5]. For other metals like trivalent metals and transitions metals, the situation is less clear [6-7]. In this paper we present recent results on aluminium and iron clusters studied mainly through ionization potential measurements. Our results are in agreement with recent previous but new features are observed and are discussed.

2. Experimental (figure 1)

A frequency doubled Nd:YAG (BMI), focused on the rod of the material of interest, produces a superheated and dense plasma. The laser-shots take place during the peak carrier gas flow (helium) generated by a synchronized pulsed gas valve, leading to partial neutralization and cluster growth inside the high pressure preexpansion zone. Subsequent free jet expansion quenches the nucleation and cools the metal clusters. Then the cluster expansion is skimmed (a few cm dowstream of the nozzle) to form a supersonic molecu-

P. Jena et al. (eds.), Physics and Chemistry of Finite Systems: From Clusters to Crystals, Vol. I, 633–638.

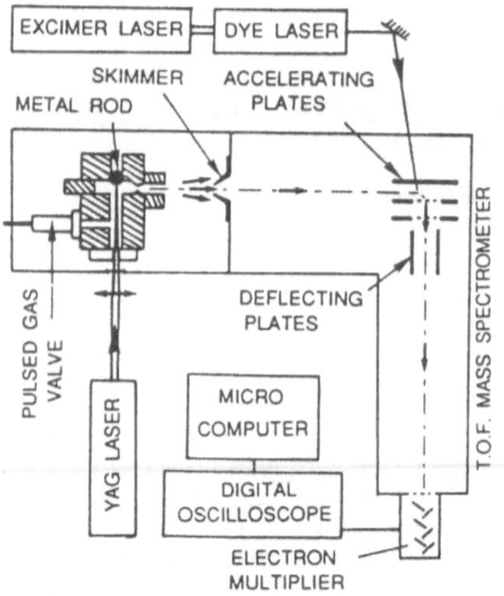

Figure 1. Schematic experimental set-up.

lar beam, and the neutral clusters are photoionized by a frequency doubled dye laser pumped by an XeCl excimer laser (Lambda Physik, crystal BBO) during passage through the accelerating region of a TOF mass spectrometer. The mass separated cluster photoions are detected by an electron-multiplier and the TOF spectra are recorded in a microcomputer via a fast digital oscilloscope (Lecroy 9400).

3. Results

In order to increase the signal to noise ratio a few hundred TOF spectra are averaged (typically 500-1000) for each fixed parameter set (TOF transmission parameters and ionizing laser frequency), although the mass spectrum from one laser-shot often exhibits the qualitative pattern of the integrated curve. Then this TOF spectrum is normalized to the average ionizing laser fluence. During the experiments the dye laser intensity was kept low enough to avoid multiphoton processes. From these TOF spectra the cluster Photoionization Efficiency (PIE) curves are generated and the IPs deduced from the photoionization thresholds obtained by linear extrapolation of the PIE curves. Note that this current procedure rests on implicit assumptions : low internal excitation, favorable Franck-Condon factors, no significant competitive processes as predissociation or dissociative ionization. The precision of our extrapolated IP values is about ± 0.03 eV.

ALUMINUM

Figure 2a shows a typical TOF spectrum. The pronounced large-scale fluctuations in the overall pattern and the striking high abundances for particular sizes is the manifestation — via the IP size-evolution — of the electronic shell structure [1]. The spectacular modification of the mass spectrum when the ionizing photon energy is decreased by a small amount (see figure 2b) rules out possible signature from the cluster abundance distribution in the incident supersonic beam (which is primarily governed by the nucleation-dissociation dynamics).

In the standard SJBM — worked out using the Kohn-Sham procedure in the Local Density Approximation formalism (LDA) — abrupt decreases in the IP curve occur after the filling — by the n_e valence electrons (n_e = N.v where v is the atomic valence and N

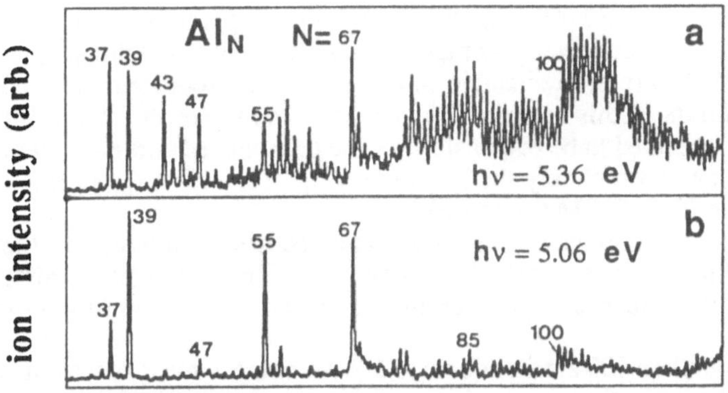

Figure 2. Aluminum cluster TOF spectra recorded with different ionizing photon energy.

the cluster size) — of the KS ($n\ell$) shells. Along this filling sequence [2] : $1s^2\ 1p^6 \mid 1d^{10}\ 2s^2 \mid 1f^{14} \mid 2p^6 \mid 1g^{18} \mid 2d^{10} \mid 1h^{22}\ 3s^2 \mid 2f^{14}\ 1i^{26}\ 3p^6 \mid 2g^{18}\ 1j^{30}\ 3d^{10}\ 4s^2 \mid$ ... particularly strong discontinuities, indicated by vertical bars, are expected in the predicted IP curve. The corresponding N-sizes (the so-called 'magic numbers') reflect the filling of the highest levels of the KS level bunches [2]. Some of these magic numbers (n_e = 8, 20, 34, 40, 58, 68, 92, 138, 198, 254, 338 ...) account very well for the strong dips observed at N = 47 (n_e = 138 + 3), 67 (198 + 3) and 85 (254 + 1) (see figure 3a). In addition mass spectra recorded with higher ionizing laser fluence (enhancing multiphoton excitations) exhibit two high peaks N = 7 (n_e = 20 + 1) and N = 14 (40 + 2), and a noti-

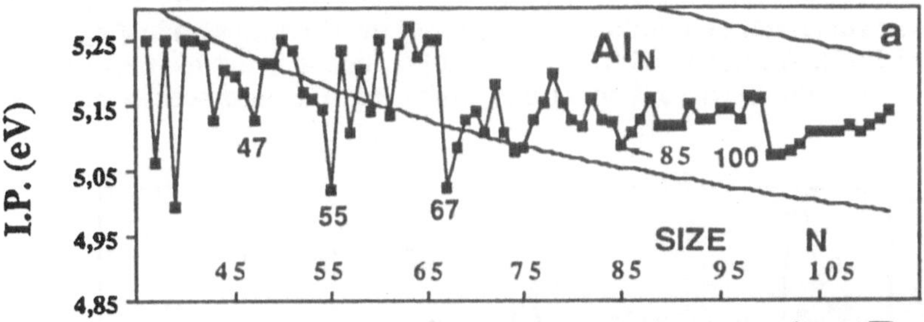

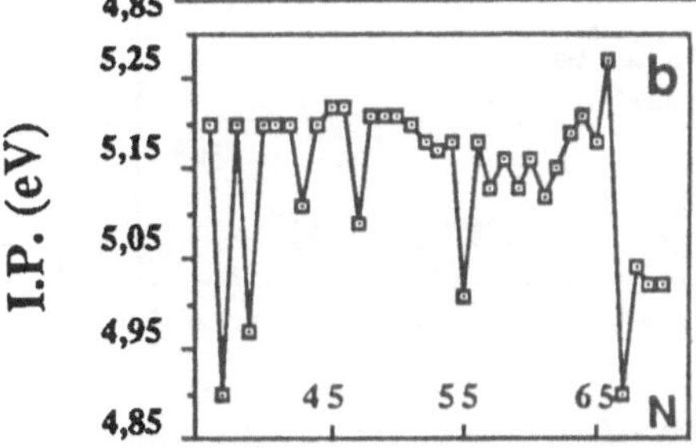

Figure 3. Aluminum cluster IP values.
a) Present work. All points at, or above 5.25 eV, are lower bounds of the corresponding cluster IP values. The two smooth curves correspond to the results of the classical Conducting Spherical Droplet model : IP (eV) = WF (eV) + α 14.4/R (Å). Aluminum bulk Work Function = 4.28 eV. α = 3/8 (lower curve ; ref. ..) or α = 1/2 (upper curve ; ref. 9), depending on the classical theoretical approach.
b) Results of Whetten et al. (ref. 6).

ceable discontinuity at N = 23 (68 + 1), three indications supporting the partial reliability of the SJBM to explain some features of the IP curve. However additional apparent shell openings at N = 55, 100 (non-predictable whatever the local reordering of the KS levels is) point out the limits of this oversimplified model. Moreover the shape-deformation or/and spin effects assumed to be responsible for the pronounced odd-even alternations in the N = 55-62 range and probably the large lowering of the N = 37, 39 IPs, cannot be accounted for in the LDA-SJBM framework.

Qualitatively our measurements are in excellent agreement with those of Whetten et al. [6] (figure 3b). Minor quantitative deviations are probably attributable to differences in the procedure for the linear extrapolation of the PIE curves, or/and in internal cluster temperature.

For clusters larger than 100 atoms and up to 1000 atoms the mass spectrum is strongly structured when the ionization energy is just above the threshold. This effect is due to small difference in the ionization potential and may be interpreted as electronic shell effect like in Na_n clusters. However the KS level bunches are different in aluminium clusters. The number of shell k may be plotted as a function of $n_e^{1/3}$ where n_e is the number of electrons. We obtain a straight line $k = \alpha\, n_e^{1/3}$. The slope α is about two times the slope obtained for sodium clusters. No supershell structure [4-5] is observed. The interpretation of these effects is in progress in our laboratory.

IRON

For iron clusters, the IP values are reported in figure 4. They are consistent with the results obtained by Yang et al [7]. The systematic shift of our photoionization thresholds must probably be attributed to the low temperature of the liquid nitrogen-cooled source employed in ref. 7. Except for n = 23 and n = 26 the IP curve seems rather smooth. Moreover the IP values does not follow the 1/R law of the classical Conducting Spherical Droplet model [8-9]. As discussed by Yang et al [7], this is usual for transition metal clusters. However if the IP curve is not clearly structured, the mass spectra recorded for

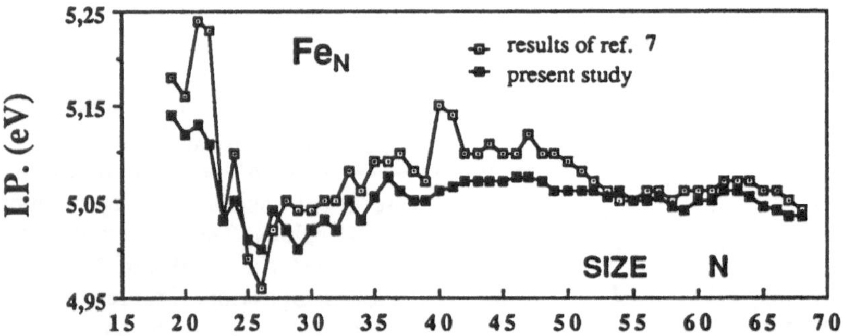

Figure 4. Iron cluster IP values. The results of the classical Conducting Spherical Droplet model are far above the experimental curves in this size range. WF = 4.50 eV (bulk polycrystalline Work Function).

photon energy just above the ionization threshold are well structured (figures 5 and 6). The observed numbers (n = 23, 40, 55, 65, 78, 85, 96, 106, 113, etc...) are very close to those obtained in aluminium clusters (n = 23, 37-39, 47, 55, 67, 78, 85, 100, 113). While in Al_n clusters clear steps are observed at a given size, the structure seems more blurred in iron clusters and leads to very small effects on the IP values. The blurred character of this structure may be explained by the existence of isomers having different IP values [7]. However the structure observed is reproducible and could support an electronic shell structure (with a valence number v = 3) or specific geometric structures in iron clusters, complicated by the existence of isomers. Further investigations are necessary in order to obtain convincing interpretation of the experimental mass spectra.

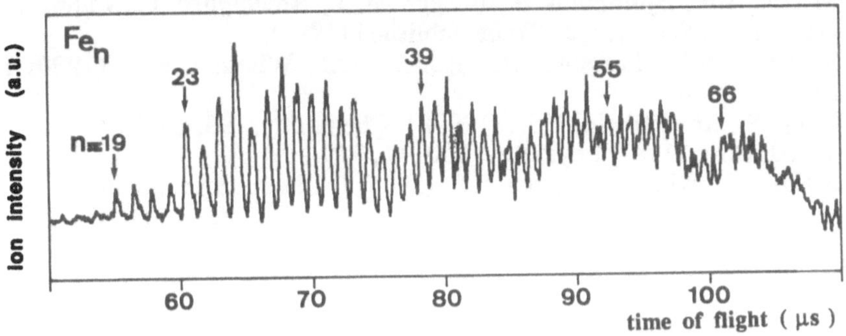

Figure 5. Mass spectrum of iron clusters for an ionization energy of 5.20 eV.

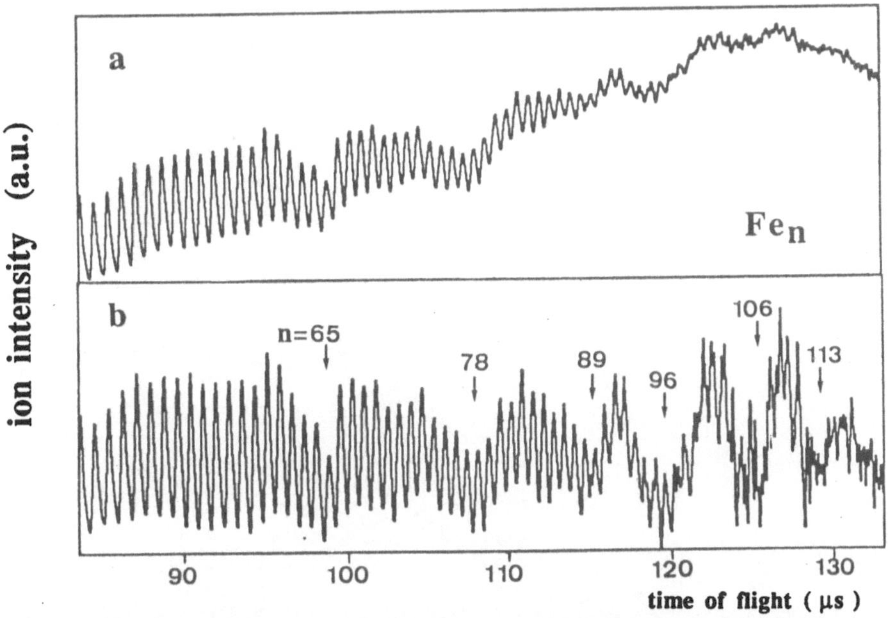

Figure 6. Mass spectrum of iron clusters for an ionization energy of 5.17 eV.
a) obtained mass spectrum. b) same mass spectrum with suitable mathematic treatment in order to make more apparent the structures at large size.

References

[1] Knight, W.D., Clemenger, K., de Heer, W.A., Saunders, W.A., Chou, M.Y., and Cohen, M.L. (1984) Phys. rev. Lett. 52, 2141.

[2] Bjornholm, S., Borggreen, J., Echt, O., Hansen, K., Pedersen, J., and Rasmussen, H.D. (1990) Phys. Rev. Lett. 65, 1627.

[3] Martin, T.P., Bergmann, T., Göhlich H., and Lange T. (1990) Chem. Phys. Lett. 172, 209.

[4] Pedersen, J., Bjornholm, S., Borggreen, J., Hansen, K., Martin, T.P., and Rasmussen, H.D., To be published in Nature (1991)

[5] Martin, T.P., Bjornholm, S., Borggreen, J., Brechignac, C., Cahuzac, Ph., Hansen, K., Pedersen, J., To be published (1991).

[6] Schriver, K.E., Persson, J.L., Honea, E.C., Whetten, R.L. (1990) Phys. Rev. 64, 2539.

[7] Yang, S., Knickelbein, M.B. (1990) J. Chem. Phys. 93, 1533.

[8] Wood, D.M. (1981) Phys. Rev. Lett. 46, 749.

[9] Makov, G., Nitzan, A., Brus, L.E. (1985) J. Chem. Phys. 88, 5076.

ON THE STRUCTURES AND STABILITIES OF SMALL MIXED CLUSTERS

KRISHNAN RAGHAVACHARI
MOHAMMAD A. AL-LAHAM
AT&T Bell Laboratories
Murray Hill, NJ 07974

ABSTRACT. Electronic structures and stabilities of stoichiometric Al_nP_n, Ga_nAs_n and Mg_nS_n clusters (n=1–3) have been explored by means of accurate quantum chemical calculations. MgS clusters are dominated by ionic considerations while AlP and GaAs clusters exhibit both covalent and ionic bonding. Two different planar rhombic structures represent the low energy forms of the A_2B_2 clusters. The A_3B_3 clusters, on the other hand, show a variety of local minima with a planar structure, a bicapped tetrahedron, and a distorted trigonal-prism being the principal low-energy isomers. Our results show that the electronegativity difference between the two constituents of the mixed cluster is the major factor in determining the nature of the ground state isomer.

1. Introduction

In recent years, there have been numerous experimental and theoretical investigations of small clusters, especially those formed from a single element. Carbon, silicon, and alkali metal clusters have received the lion's share of attention in such studies while transition metal clusters and other nonmetallic clusters have been investigated to a lesser extent. All these studies have shed some light on the structures, stabilities, reactivities, and the chemical bonding in such clusters, and have led to a better understanding of the relationship between these small clusters and the bulk.

In contrast to elemental clusters, much less attention has been paid to mixed clusters. Computationally, this is due to the complications resulting from the large number of structural and permutational (resulting from the presence of more than one element) isomers that need to be considered in the search for the ground state structures of these clusters. In addition, the nature of the bonding can also show a large variation ranging from ionic to covalent to metallic character. Recently, however, a few theoretical studies have been reported on small clusters of GaAs, AlP, MgS and NaCl [1-5]. In this manuscript we discuss the structures and stabilities of small clusters of AlP, GaAs, and MgS (as prototypes for III-V and II-VI group mixed clusters). We also relate our results to those of isoelectronic Si clusters [6]. This comparison helps us to understand the influence of the electronic configurations, atomic sizes, and the electronegativity differences of the constituents on the structures and stabilities of these clusters.

2. Theoretical Methods

Ab initio molecular orbital techniques have been used throughout. All electron calculations [7] were performed for AlP and MgS clusters, while effective core potentials (ECP) [8] were used to calculate the energies of GaAs clusters. Structures representing the local minima were obtained by optimizing the geometry for several different starting arrangements of atoms. The reported

P. Jena et al. (eds.), Physics and Chemistry of Finite Systems: From Clusters to Crystals, Vol. I, 639–644.

geometries in Table I were obtained at the Hartree-Fock (HF) level of theory using a polarized double-zeta basis set for AlP and MgS clusters [3], and a polarized triple-zeta basis set for GaAs clusters [4]. These geometries were then used to perform electron correlation calculations using quadratic configuration interaction as well as complete fourth order Møller-Plesset perturbation theory [7]. Atomic charges were calculated using the natural population analysis [9]. A more detailed description of the theoretical methods employed can be found elsewhere [3,4].

Molecular structures representing the calculated low-energy local minima are shown in Figure 1. In this figure, the electropositive atoms (Al, Ga, or Mg) are represented by dark circles and the electronegative atoms (P, As, or S) by open circles. Solid lines have been drawn to represent internuclear distances that are within about 20% of the bulk bond lengths.

3. Results

A summary of our calculated results is given in Table I. The geometries and atomic charges in the table correspond to those structures shown in Figure 1. In this section we consider some of our important conclusions for each of the three cluster systems.

3.1. AlP CLUSTERS

Diatomic AlP molecule (structure **1** in Figure 1) has a $^3\Sigma^-$ ground state. It has two π-electrons, similar to the ground state of the isoelectronic Si_2 molecule. As shown in Table I, AlP has a bond length of 2.43 Å, somewhat larger than the bulk distance of 2.36 Å. There is also a low-lying $^3\Pi$ state, containing three π-electrons, which has a shorter bond length of 2.27 Å. As in the case of Si_2, the $^3\Pi$ state of AlP lies only 1 kcal/mol above the ground state.

Our search for the low energy isomers of Al_2P_2 led to two different rhombic structures (**2** and **3**). As shown in Figure 1, two P atoms are located along the short diagonal in **2** while two Al atoms occupy those positions in **3**. Both structures are planar with D_{2h} symmetry. The bonding orbitals in **2** are analogous to those in isoelectronic Si_4 [6] and the atomic charges (±0.61e) are rather small. The rhombic structure **3**, on the other hand, has atomic charges that are almost twice as large (±1.10e), and contains bonding orbitals similar to those in ionic Na_2Cl_2 [5]. **2** is more stable than **3** by 39 kcal/mol.

We have considered many different structural candidates for the ground state of Al_3P_3. We have identified a rich and varied set of local minima among which the low-energy forms are displayed in Figure 1. The structure given as **4** is planar with D_{3h} symmetry. As seen from the atomic charges in Table I, it is highly ionic. We have also identified a trigonal-prism-like structure **5** to be a local minimum. This structure (C_s symmetry) can be described as a triangle of Al atoms which are capped on one side by a P atom and on the other side by two bonded P atoms.

The bicapped tetrahedral structures shown as **6** and **7** can be called Si_6-like due to their resemblance to the ground state geometry of Si_6 [6]. The difference between **6** and **7** arises from interchanging the positions of two of the atoms (labeled 2 and 6) and allowing the resulting structures to relax. As shown from the atomic charges in Table I, both structures can be considered to be principally covalently bonded. Our calculations indicate that **6** is more stable than **7** by 13 kcal/mol. This energy difference seems to be mainly due to the replacement of one Al–P "bond" in the former with one Al–Al "bond" in the latter.

TABLE 1. Optimized molecular geometries (Å and degrees) and Atomic charges for GaAs, AlP, and MgS clusters.

Structure	Point Group	Geometry				Natural Charge			
		Parameter	GaAs	AlP	MgS	Atom	GaAs	AlP	MgS
1	$C_{\infty v}$	r(1−2)	2.61	2.43	2.15	1	+0.52	+0.53	+1.23
						2	−0.52	−0.53	−1.23
2	D_{2h}[a]	r(1−2)	4.92	4.64	4.24	1	+0.58	+0.61	+0.87
		r(3−4)	2.29	2.04	2.27	3	−0.58	−0.61	−0.87
		r(1−3)	2.71	2.53	2.44				
3	D_{2h}	r(1−2)	2.47	2.37	2.65	1	+0.84	+1.10	+1.54
		r(3−4)	4.16	3.87	3.80	3	−0.84	−1.10	−1.54
		r(1−3)	2.42	2.27	2.32				
4	D_{3h}	r(1−2)	2.96	2.85	3.30	1	+0.93	+1.14	+1.58
		r(1−4)	2.35	2.21	2.29	4	−0.93	−1.14	−1.58
		θ(2−6−3)	78°	80°	92°				
5	C_s[b]	r(1−2)	3.00	2.82	3.22	1	+0.49	+0.81	+1.58
		r(2−3)	2.81	2.55	2.65	2	+0.59	+0.75	+0.87
		r(2−6)	2.56	2.40	2.57	4	−0.44	−0.64	−0.83
		r(1−6)	2.41	2.29	2.29	6	−0.77	−1.02	−1.65
		r(4−5)	2.54	2.36	2.17				
		r(1−4)	2.50	2.32	2.40				
		r(2−4)	2.51	2.35	2.64				
6	C_s	r(1−5)	2.78	2.59		1	+0.62	+0.65	
		r(1−4)	3.11	2.91		2	+0.50	+0.57	
		r(5−6)	2.52	2.24		3	+0.53	+0.50	
		r(4−5)	2.47	2.21		4	−0.40	−0.42	
		r(2−3)	3.04	2.85		5	−0.62	−0.65	
		r(2−4)	2.53	2.35					
		r(2−6)	2.95	2.70					
		r(3−6)	2.87	2.71					
7	C_1	r(1−2)	2.98	2.93		1	+0.56	+0.57	
		r(2−3)	2.85	2.93		2	+0.16	+0.22	
		r(1−5)	2.75	2.55		3	+0.48	+0.57	
		r(2−4)	2.46	2.31		4	−0.21	−0.24	
		r(2−5)	2.97	2.60		5	−0.58	−0.56	
		r(3−5)	3.16	3.95		6	−0.40	−0.56	
		r(4−5)	2.48	2.21					
		r(2−6)	2.72	2.60					
		r(3−6)	2.73	2.55					
		r(4−6)	2.49	2.21					
		r(5−6)	2.47	2.17					

[a] Mg_2S_2 structure deviates slightly from D_{2h} symmetry (C_{2v}).
[b] Mg_3S_3 structure deviates slightly from C_s symmetry (C_1).

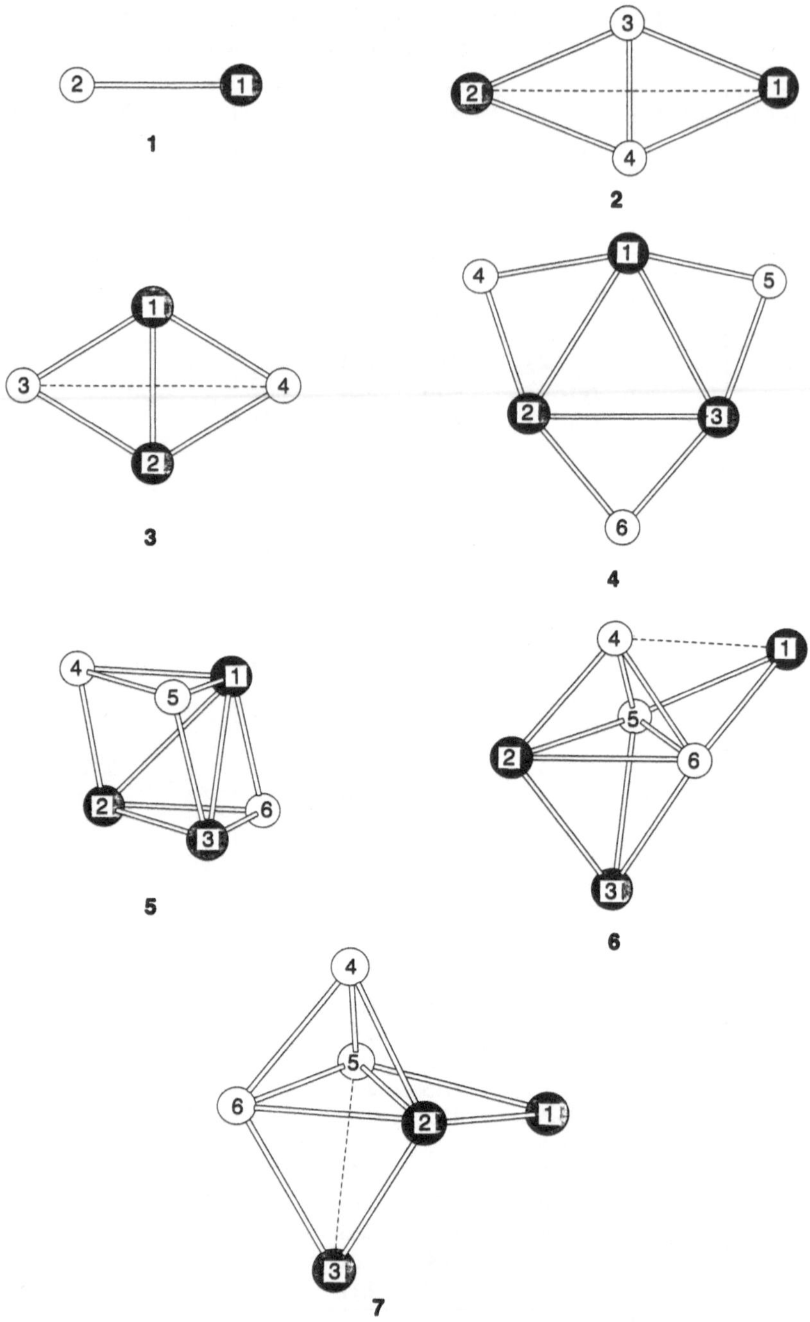

FIGURE 1. Low energy structures for Al_nP_n, Ga_nAs_n and Mg_nS_n clusters (n=1–3). Dark circles represent Al, Ga or Mg while open circles represent P, As or S.

Our calculations indicate that **4**, **5** and **6** are close in energy. Using our highest level of theory, we find that the Si_6-like structure (**6**) is 9 kcal/mol more stable than the planar structure (**4**) and 13 kcal/mol more stable than the trigonal-prism-like structure (**5**).

3.2. GaAs CLUSTERS

As in the case of AlP, the ground state of GaAs is $^3\Sigma^-$. The low-lying $^3\Pi$ state lies about 5 kcal/mol higher in energy. This difference is somewhat larger than the corresponding difference in AlP or Si_2 ($\approx$ 1 kcal/mol).

The ground state of Ga_2As_2 is the rhombic structure **2**. This structure, with the two As atoms along the short diagonal, is similar to that of the ground state of Al_2P_2 and has a ^1A_g electronic state. A triplet state (^3A_u) of the rhombus lies 18 kcal/mol higher in energy. In close agreement with the energy difference between the corresponding Al_2P_2 isomers, structure **3** for Ga_2As_2 is 40 kcal/mol less stable than **2**. These results for GaAs and Ga_2As_2 are in good agreement with other available theoretical and experimental results [1].

Based on the similarities in the valence electronic configurations of GaAs and AlP, we expect Ga_3As_3 to have structures similar to those of Al_3P_3. Indeed, our calculations indicate the presence of four analogous low energy structures for Ga_3As_3 also. Again, the lowest energy structure is the Si_6-like structure **6**. Using a local density functional approach, Lou *et al.* have recently identified a structure similar to **7** to be the lowest energy form of Ga_3As_3 among the structures they considered [2]. While we found **7** to be a local minimum our results clearly show that **7** is 13 kcal/mol higher in energy than **6**. The stability of **6** over **7** can be attributed to the formation of stronger Ga–As "bonds" in **6**. The planar structure **4** was also found to be a local minimum, lying 22 kcal/mol higher in energy than the ground state. Finally, the trigonal-prism-like structure **5** also lies 22 kcal/mol higher in energy than **6**.

Atomic charges in Ga_3As_3 isomers show the importance of ionic bonding in **4** and the covalent bonding tendency of the other three 3-dimensional structures. Generally, Ga_3As_3 isomers appear to be slightly more covalent than the corresponding Al_3P_3 isomers. Overall, the geometries and atomic charges in Table I show that the optimized structures for AlP and GaAs are similar except for the expected differences resulting from the larger atomic radii in GaAs.

3.3. MgS CLUSTERS

MgS clusters behave considerably different from AlP or GaAs clusters. The ground state of MgS diatomic molecule is $^1\Sigma^+$. It has a bond length of 2.15Å, in very good agreement with the experimental value of 2.14Å [3]. The molecule is considerably ionic, as can be seen from the atomic charges of ±1.23e. The low-lying $^3\Pi$ state is 8 kcal/mol higher in energy.

As in the case of Al_2P_2 and Ga_2As_2 clusters, Mg_2S_2 also has two rhombic structures. However, the energy ordering of the two isomers is now reversed, clearly demonstrating the dominance of ionic terms in Mg_2S_2. **2** represents the more covalently bonded rhombus where the two S atoms are along the short diagonal. For Mg_2S_2, the optimized structure for **2** deviates slightly from D_{2h} symmetry. As seen earlier, the other structure **3** (with the Mg atoms along the short diagonal) leads to a more ionic bonding arrangement. In contrast to Al_2P_2 and Ga_2As_2 clusters, Mg_2S_2 cluster favors the ionic isomer (**3**) over the more covalent isomer (**2**) by 47 kcal/mol.

We also investigated various structural arrangements for Mg_3S_3. The planar 2-dimensional structure **4** was found to be the most stable isomer by a large margin. In **4**, the Mg and S atoms have charges of ±1.58e. The 3-dimensional trigonal-prism-like structure **5** (deviating slightly from C_s symmetry) was also found to be a local minimum, but lies 68 kcal/mol less stable than **4**. We also attempted to optimize a structure analogous to **6**. The resulting optimized structure, which was 125 kcal/mol less stable than **4**, did not show much resemblance to **6**.

4. Discussion

The calculated structures and atomic charges for the different isomers show that ionic forms favor 2-dimensional geometries due to the effectiveness of charge alternation in such arrangements. On the other hand, covalent isomers favor 3-dimensional structures in order to provide maximum overlap of the bonding orbitals. Ionic factors are clearly dominant in MgS clusters (one stable isomer). AlP clusters are intermediate between the ionic MgS clusters and the covalent Si clusters with both ionic and covalent isomers being local minima (four low lying isomers). GaAs clusters are slightly more covalent than AlP clusters. The degree of ionicity (or covalency) can be estimated based on the energy difference between the 2-dimensional and 3-dimensional iosmers. These results can be easily correlated with the electronegativity differences between the constituent elements of the mixed clusters.

In an attempt to relate these small clusters to the bulk, we tried to account for the energy difference between structures **6** and **7** using bond energies obtained from the bulk cohesive energies [10]. In this analysis, we use values of 0.57, 2.29, 2.15, 0.47, 1.97, and 1.63 eV for the energies of Al–Al, P–P, Al–P, Ga–Ga, As–As, and Ga–As bonds, respectively. As shown in Figure 1, structure **6** has one more Al–P (or Ga–As) "bond" and one less Al–Al (or Ga–Ga) "bond" than **7**. Using the bond energies given above we predict **6** to be more stable than **7** by 36 kcal/mol for Al_3P_3 and by 26 kcal/mol for Ga_3As_3. These values are twice as large as the values calculated using our accurate *ab initio* techniques. This overestimation can probably be attributed to the formation of much stronger Al–Al and Ga–Ga bonds in these clusters than those in the bulk where the coordination number is much larger.

5. References

1. Meier, U., Peyerimhoff, S. D., and Grein, F. (1991), Chem. Phys. 150, 331; Balasubramanian, K. (1990), Chem. Phys. Lett. 171, 58.
2. Lou, L., Wang, L., Chibante, L. P. F., Laaksonen, R. T., Nordlander, P., and Smalley, R. E. (1991), J. Chem. Phys. 94, 8015.
3. Al-Laham, M. A., Trucks, G. W., and Raghavachari, K. (1991), J. Chem. Phys. (in press).
4. Al-Laham, M. A., and Raghavachari, K. (1991), Chem. Phys. Lett. (in press).
5. Phillips, N. G., Conover, C. W. S., and Bloomfield, L. A. (1990), J. Chem. Phys. 94, 4980.
6. Raghavachari, K. and Rohlfing, C. M. (1988), J. Chem. Phys. 89, 2219.
7. Hehre, W. J., Radom, L., Schleyer, P. v. R., and Pople, J. A. (1986), Ab Initio Molecular Orbital Theory, John Wiley, New York.
8. Wadt, W. R. and Hay, P. J. (1985), J. Chem. Phys. 82, 284.
9. Reed, A. E., Weinstock, R. B., and Weinhold, F. (1986), J. Chem. Phys. 84, 2428.
10. Cohesive energies are obtained from the heat of formation and the atomization energies. See Weast, R. C., Ed., (1985), The CRC Handbook of Chemistry and Physics, 65th ed., CRC Press, Florida.

A HARTREE-FOCK STUDY OF SMALL CARBON CLUSTER ANIONS

A. K. RAY
Department of Physics, University of Texas, Arlington, Texas 76019

B. K. RAO
Department of Physics, Virginia Commonwealth University, Richmond, Virginia 23284

ABSTRACT. Results from Hartree-Fock calculations on optimized C_n^- clusters are presented for $n \leq 7$. An extended basis set has been used for these calculations to obtain correct electron affinities. Equilibrium geometries are all linear. The geometries, binding energies, and electron affinities are discussed in the light of available theoretical and experimental data.

Introduction

We have witnessed in recent years, a widespread interest in the theoretical and experimental investigations of atomic and molecular clusters [1-5]. Of these systems, carbon clusters are of particular interest because of their importance in diverse processes such as catalysis, interstellar gases, and superconductivity. A summary of the available theoretical and experimental results on neutral and cationic carbon clusters has been presented earlier by Ray [6] and Islam and Ray [7]. A particular form of carbon, namely C_{60}, has recently received widespread attention from the scientific community [8-11] . However, theoretical studies of evolution of electronic structure and associated properties of neutral and ionized carbon clusters with growing size are indeed rare. In this paper we present our work on the study of small carbon cluster anions.

We first present some of the latest available results on carbon cluster anions. Yang et al [12] have reported ultraviolet photoelectron spectra of negative carbon clusters in the size range from 2 to 29 atoms. Their data indicate that the clusters in the size range of two to nine atoms take the form of linear chains. In particular, the even numbered chains having open shell electronic structures show high electron affinity. Size selected photofragmentation of negative carbon ions C_n^- ($4 \leq n \leq 20$) has been studied by Deluca and Johnson [13]. For most clusters smaller than n = 14, fragmentation was found to occur via loss of neutral C_3 clusters. This is similar to the behavior found in carbon cations. Raghavachari [14] has determined the geometries and vibrational frequencies of C_3, C_3^- and C_3^+ with the quadratic configuration interaction technique using the polarized double-zeta (6-31G*) basis set. In this calculation C_3^- was found to have linear geometry, in agreement with the

P. Jena et al. (eds.), Physics and Chemistry of Finite Systems: From Clusters to Crystals, Vol. I, 645–650.

results of Yang et al [12]. Watts et al [15] have used many-body perturbation theory and coupled cluster theory to compute the vertical ionization potentials of linear and rhombic C_4^-. They have concluded that the ionization potentials of linear C_4^- give a good account of the observed photoelectron spectrum . In a recent paper, Schauer et al [16] have reported the production of carbon cluster ions with a double negative charge by sputtering graphite where ions as small as C_7^{--} were detected.

Calculations

The present work deals with the calculations on C_2^- - C_7^- clusters. As implied before, our calculations are based on accurate all electron *ab initio* self-consistent field (SCF) molecular orbital techniques within the Hartree Fock approximation. One of the primary considerations involved in these calculations is the determination of the type of basis set to be used. Anions, in particular, present a computational challenge because of the necessities of polarization and diffuse functions to adequately represent the electronic properties. For carbon, we have used a (13s9p1d/5s4p1d) basis set. This basis set was chosen because it reproduces the electron affinity (EA) of the carbon atom accurately. In particular, we find that the EA at the Hartree-Fock (HF) level of calculation is 0.47 eV. Inclusion of correlation effects at the fourth-order Möller-Plesset many body perturbation theory (MP4) yields an EA of 1.09 eV. This compares very well with the experimental EA [17] of 1.12 eV. All the calculations were done on a Cray Y-MP computer using the program GAUSSIAN 88 [18] and geometries were optimized by the use of analytic gradient techniques within the framework of the HF theory.

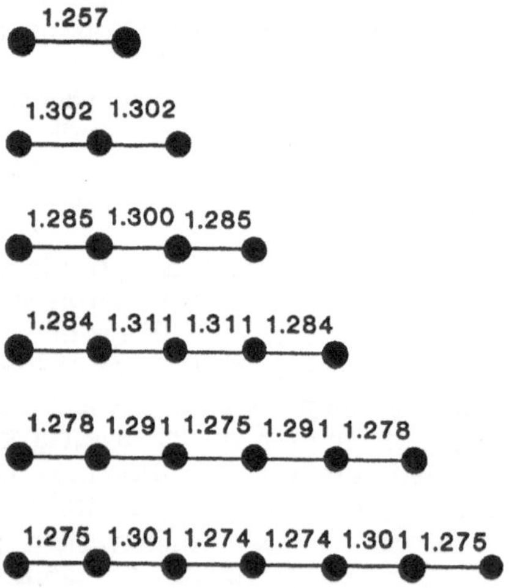

Fig. 1. Optimized geometries of carbon cluster anions. The bond lengths are given in Å.

Results and Discussions

In the study of C_2^- we note first that it is one of the few negative ions with bound valence states. There has been a series of calculations on the low-lying states of C_2 and C_2^- with theoretical EA values ranging from 2.84 to 3.75 eV [19] while the experimental values of Nichols and Simon [20] range from 3.374 to 3.408 eV. The theoretical values represent calculations using different basis sets and different approximations for the inclusion of correlation effects. Our optimized bond length of C_2^- is 1.257 Å with a total energy at the HF level of -75.559115 a. u. This gives us an electron affinity of 2.55 eV, in fair agreement with the value of obtained by Dupuis and Liu [19]. The ground state was found to be a doublet, in agreement with published results in the literature.

For C_3^-, apart from the work of Raghavachari [14], photoelectron experiments [21] indicate an electron affinity of 1.98 ± 0.02 eV for C_3. Sunil et al [22] have concluded that C_3^- has a linear ground state geometry. Our results (Fig. 1) indicate that C_3^- is linear, with a $^2\Pi_g$ ground state. We also note that the C-C bond stretches by 0.045 Å from C_2^- to C_3^-. The HF electron affinity of C_3^- is found to be 1.85 eV. The optimized C-C bond length is 1.302 Å. These values are in good agreement with the bond length of 1.322 Å and EA of 2.0 eV reported by Raghavachari [14].

TABLE 1. Total energies and electron affinities of C_n^- ($1 \leq n \leq 7$) clusters.

n	Total Energy (hartree)	Electron Affinity (eV)
1	- 37.708420	0.47
2	- 75.559115	2.55
3	-113.444750	1.85
4	-151.304364	5.52
5	-189.074380	2.97
6	-227.001668	2.63
7	-264.864382	3.65

For C_4^-, as mentioned before, the photoelectron spectra of Yang et al [12] suggest a linear structure while the Coulomb explosion experiments of Algranati et al [23] seem to indicate a rhombic C_4^- structure. Our calculations indicate that C_4^- has a linear structure with bond lengths of 1.285 and 1.300 Å respectively (Fig 1). The optimized linear geometry of Watts et al [15] using a 5s4p1d basis and many-body perturbation theory upto fourth order predicts bond lengths of 1.314 Å and 1.283 Å respectively. We note that the large basis set used in our work is expected to give more accurate bond lengths in our calculations and in our case the outer bond length is shorter than the central bond. The electron affinity of Watts et al [15] for the linear C_4^- ranges from 3.3 to 3.6 eV depending upon the method used for incorporating the correlation effects. Our HF value of 5.52 eV is somewhat higher. However, the better representation of the diffuse and polarization functions in our case is expected to give more electron affinity.

For the C_5^-, C_6^- and C_7^- clusters, we have not found any theoretical study in the literature on the electronic and geometric structure of these species, presumably because of computational difficulties. Our calculations indicate that linear structures with doublet ground states are preferred for all these systems. We also note that for C_5^-, the outer bonds decrease in length by about seven per cent while for C_7^-, the bond lengths alternate. The electron affinities of C_5^- - C_7^- clusters are found to lie in the range of 2.6 - 3.7 eV (see table 1).

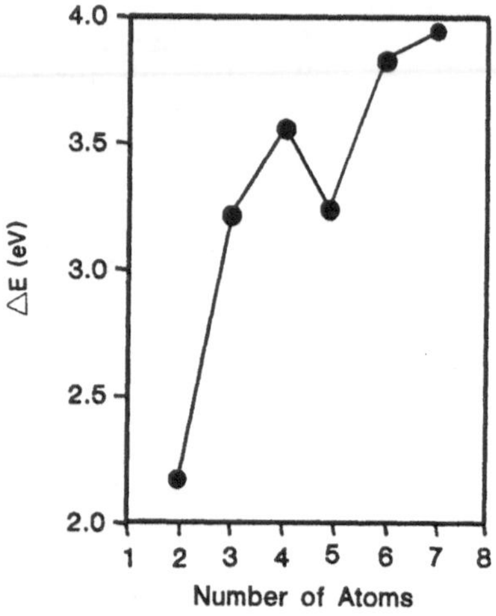

Fig. 2. Binding energy per atom for carbon cluster anions in eV.

We have also studied the stabilities of these clusters in the separated atom limit. We define the binding energy per atom for a cluster of c_n^- as

$$\Delta E = [\ E(C^-) + (n-1)E(C) - E(C_n^-)\ ]\ /\ n$$

The plot of ΔE against the number of carbon atoms in the cluster has been presented in Figure 2. It is easy to see that all the clusters are found to be stable, with a general monotonic increase in the binding energies as the cluster size grows.

Mulliken population analysis has been performed for all the systems studied. Obviously the results for $n \leq 2$ are not interesting and will not be presented here. We notice for C_3^- that the extra electron gets distributed unevenly among the three carbon atoms. The central carbon atom retains 14% of the extra charge while the end atoms get 42% each. One can interpret this as the mutual Coulomb repulsion which send most parts of the charge to the end atoms. In the case of the C_5^- the end atoms get 20% each while the central atom gets 26%. The other two atoms get 17% each. In this case, obviously the charge distribution is rather more uniform. This should be expected because with the increasing length of the carbon chain the repulsion decreases due to the shielding of the intermediate atoms.

It is clear that while it is possible to obtain reasonable information on the geometries, electronic structure, and electron affinities of Cn^- clusters using the present basis set one must include correlation corrections to better utilize the capability of this basis set and to get accurate results. Preliminary calculations show that better electron affinity is obtained when one includes correlation corrections at the MP4 level. These calculations are in progress at present.

Acknowledgements

AKR was supported by the Welch Foundation, Houston, Texas (Grant No. Y-1092) and BKR was supported in part by the U. S. Department of Energy (DE-FG05-87ER45316). Computational support from the University of Texas System Center for High Performance Computing is also gratefully acknowledged.

References

1. P. Jena, B. K. Rao, and S. N. Khanna (1987) Proc. Int. Conf. On the Physics and Chemistry of Small Clusters (Plenum, New York).
2. S. Sugano, Y. Nishina, and S. Ohnishi (1987) Proc. First NEC Symposium on Fundamental Approaches to New Material Phases (Springer-Verlag, Berlin).
3. G. Benedek, T. P. Martin, and G. Pacchioni (1988) Proc. 13th International School, Erice, Italy (Berlin, Springer-Verlag).
4. C. Chapon, M. F. Gillet, and C. R. Henry (1989) Proc. 4th Int. Meeting on Small Particles and Inorganic Clusters (Springer-Verlag, Berlin).
5. R. S. Averback, D. L. Nelson, and J. Bernholc (1991) Clusters and Cluster Assembled Materials, MRS Proceedings.
6. A. K. Ray (1987) J. Phys. B 20, 5233.
7. M. S. Islam and A. K. Ray (1989) J. Phys. B 22, 2071.
8. H. W. Kroto, J. R. Heath, S. C. O'Brien, R. F. Curl, and R. E. Smalley (1985) Nature 318, 162.
9. R. F. Curl and R. E. Smalley (1988) Science 242, 1017.
10. H. W. Kroto (1988) Science 242, 1139.
11. S. Saito and A. Oshiyama (1991) Phys. Rev. Lett. 66, 2637.
12. S. Yang, K. J. Taylor, M. J. Craycraft, J. Conceicao, C. L. Pettiette, O. Cheshnovsky, and R. E. Smalley (1988) Chem. Phys. Lett. 144, 431.
13. M. J. Deluca and M. A. Johnson (1988) Chem. Phys. Lett. 152, 67.
14. K. Raghavachari (1990) Chem. Phys. Lett. 171, 249.
15. J. D. Watts, I. Cernusak, and R. J. Bartlett (1991) Chem. Phys. Lett. 178, 259.
16. S. N. Schauer, P. Williams, and R. N. Compton (1990) Phys. Rev. Lett. 65, 625.
17. CRC Handbook of Chemistry and Physics (1981-82), 2nd Edition, pp. E64 (CRC Press, Boca Raton).

18. M. J. Frisch, M. Head-Gordon, H. B. Schlegel, K Raghavachari, J. S. Binkley, C. Gonzalez, D. J. DeFrees, D. J. Fox, R. A. Whiteside, R. Seeger, C. F. Melius, J. Baker, R. L. Martin, L. R. Kahn, J. P. Stewart, E. M. Fluder, S. Topiol, and J. A. Pople (1988) 'GAUSSIAN 88' (Gaussian Inc., Pittsburgh).
19. M. Dupuis and B. Liu (1980) J. Chem. Phys. 73, 337.
20. J. A. Nichols and J.J. Simons (1987) J. Chem. Phys. 86, 6972.
21. J. M. Oakes and G. B. Ellison (1986) Tetrahedron 42, 6263.
22. K. K. Sunil, A. Orendt, K. D. Jordan, and D. J. DeFrees (1984) Chem. Phys. 89, 245.
23. M. Algranati, H. Feldman, D. Kella, E. Malkin, E. Miklazky, R. Naaman, Z. Vager and J. Zajfman (1989) J. Chem. Phys. 90, 4617.

ELECTRONIC STRUCTURE AND ADDUCT GEOMETRIES IN THE CHEMISORPTION OF NITRIC OXIDE IN SMALL GAS-PHASE COBALT CLUSTERS

J. ROBLES[1], A. MARTINEZ[1], A. VELA[1] and M. P. IÑIGUEZ[2].

1 Department of Chemistry. Universidad Autonoma Metropolitana-Iztapalapa. Apdo.Postal 55-534, C.P. 09340, D.F. MEXICO.

2 Department of Theoretical Physics. Universidad de Valladolid, E-47011, Valladolid, ESPAÑA.

ABSTRACT. A Theoretical study of the structures of Co_xNO^+ and of the mechanism for the reaction Co_x^+ + NO with x=2,3,4, is presented. Combining a Simplified Molecular Mechanics, SMM, (to produce input geometries) with Extended Huckel Molecular Orbital, EHMO, electronic structure calculations, we obtain the more stable geometries and bonding schemes in each adduct. It is shown that the ionization potential of the bare cluster is an excellent parameter to predict the stabler adduct showing that charge transfer plays an important role in these systems. We present results for the adduct enthalpy of formation and global hardness for some reaction paths. We find that the maximum hardness principle is useful to establish the more probable reaction path, and also produces more realistic geometries than those obtained from direct minimization of the EHMO energy.

I. INTRODUCTION

The understanding of the processes that occur during the chemisorption of small molecules in metal clusters and surfaces is of interest not only in basic research but also of paramount importance in technological aspects such as pollution control. Recent experimental advances have made important contributions towards the comprehension of the changes that take place during the chemisorption of molecules in small metal clusters.

In particular, Klaassen and Jacobson [1] (KJ) have studied the chemisorption of NO on small cobalt clusters (Co_x, x=2,3,4). From the displacement reactions of Co_xNO^+ with $^{18}O_2$ they were able to distinguish between molecular and dissociative chemisorption. They conclude that for Co_2^+ molecular chemisorption is preferential, occurring in the 55 % of their reaction yield, while the remaining 45 % consist of unreactive structures which may be dissociated. On the other hand, in Co_3^+. and Co_4^+, NO is mainly attached to the cobalt clusters through dissociative chemisorption.

The objective of the present work is to study, from a theoretical point of view, the more probable structures of Co_xNO^+ (x=2,3,4) and the possible reaction mechanism of the chemisorption process.

P. Jena et al. (eds.), Physics and Chemistry of Finite Systems: From Clusters to Crystals, Vol. I, 651–656.

II. METHODS

The more feasible geometries of the adducts are obtained through a systematic procedure that consists in generating several adduct Co_xNO^+ (x=2,3,4) geometries with a Simplified Molecular Mechanics [2] (SMM), which minimizes total energies using as atomic parameters the van der Waals and covalent radii, and the atomic valence states. During the minimization process, the molecular connectivity is kept constant. These final geometries are used as input for the electronic structure calculation that, in our case is done with the Extended Huckel Molecular Orbital [3] (EHMO) method. The most stable structures are obtained by comparison of the total EHMO electronic energies [4] and the bonding scheme is established from the analysis of the Mulliken overlap populations.

On the other hand, the study of the possible reaction mechanism is also done with the EHMO method. We select and fix the Co-Co bond length to 2.32 Å, and for the NO molecule two distances (1.45 Å and 1.7 Å) are considered. The first two distances are obtained from the SMM procedure while the third one empirically turned out to be the best one for our EHMO study.

To scan the rather complex potential energy surface we choose six possible reaction paths for each adduct, guided by symmetry and chemical intuition [5] Here we present results for one typical path, studying the dependence on the NO bond length.

III. RESULTS AND DISCUSSION

With the procedure described in the previous section, we find some stable structures for Co_2NO^+, Co_3NO^+ and Co_4NO^+. From the net atomic charges one can see that in all adducts, Co clusters undergo electron transfer towards the NO. An example will be given below, in Fig. 3. This is confirmed in the analysis of all molecular orbitals diagrams (not shown here). From the enthalpy of formation (ΔH_f) values, one can see that the larger the transfer, the better chemisorption seems to be achieved. Since the initial electron transfer is extensively determined by the pure cluster ionization potential (I), one might expect that a lower cluster I will determine stronger chemisorption of NO. I is computed from

$$I = E(Co_xNO^+) - E(Co_xNO) \quad (1)$$

while

$$\Delta H_f = E(Co_xNO^+) - E(Co_x^+) - E(NO) \quad (2)$$

where x=2-4

In Table 1, we show the values of I and ΔH_f obtained for each cluster size.

Table 1. Number of cobalt atoms, pure clusters I and adduct ΔH_f for the more stable adducts.

n	I (eV)	$-\Delta H_f$ (eV)
2	11.026	14.175
3	11.337	10.403
4	10.833	14.464

In figure 1 we plot the pure cluster I (upper section) and the ΔH_f (lower section) versus n (number of atoms in pure cluster). It is evident that minima and maxima are parallel in both curves, confirming the previous idea.

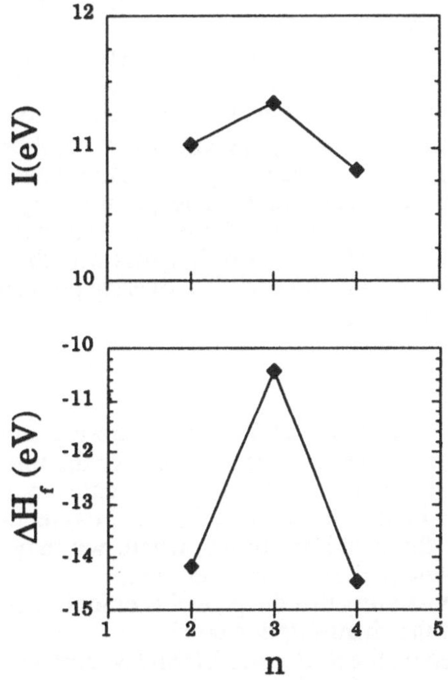

Figure 1. Upper section: the pure clusters ionization potentials(I) versus n(number of atoms in pure clusters). Lower section: adduct formation enthalpy (ΔH_f) versus n. (Dots are joined with straight segments only to emphasize behavior).

We have chosen one possible reaction path to illustrate how some of the adduct structures might be formed . Within the EHMO method, we compute both ΔH_f at each point of the reaction coordinate, and the global hardness (η), calculated within the finite-differences approximation and using Koopmans theorem,

$$\eta = \varepsilon\,(\text{LUMO}) - \varepsilon\,(\text{HOMO}) \tag{3}$$

The latter is computed since the recently proved maximum hardness principle states that "molecules arrange themselves in such a way as to be as hard as possible", under the constraint of constant chemical potential and at finite temperature [6]. We indeed find that the principle is useful to predict the probable structures along a reaction path for chemisorption of NO in small Co clusters.

The chosen path, which we term the *parallel attack* is shown in Figure 2. ΔH_f and η are plotted versus the reaction coordinate (ξ), which is chosen as the internal coordinate given by the y-axis, because in this path the Co clusters and the NO are in the xy plane. Co(1) and Co(2) are in the x-axis, and the N-O bond is always parallel to it. The NO is moved towards the Co cluster along the y-axis.

In Fig. 2a results are shown for the case when the NO bond length is fixed at 1.7 Å while in Fig.2b it is fixed at 1.45 Å.

One can see that the ΔH_f curves predict that the stable structures occur when ξ is about 0.5Å. This is unrealistic since it is a very short distance. This is a feature that normally occurs within the EHMO method due to its failure to take account of many inter-electronic repulsions. However, from the η curves, the NO····Co_x^+ equilibrium separations happen at 0.7, 1.3 and 1.7 Å for x=2,3 and 4 respectively. These are indeed more realistic results. On the other hand, the dependence on the NO fixed length is strong. One can see that the one to be preferred is 1.7 Å since for this NO length the stable structures obtained corresponds to the NO dissociatively chemisorbed adduct which agrees with the experimental information.

This dissociation is evident from the Mulliken overlap populations for these equilibrium structures, which are shown in Fig. 3.

IV CONCLUSIONS.

In this paper, we presented a theoretical study of the electronic structure, the adducts geometries and some plausible reaction mechanisms for the formation of clusters Co_xNO^+ (x=2,3,4). The theoretical method employed indicates a general trend for NO to dissociatively chemisorbe in all cluster sizes. This is in agreement with the experimental results of KJ, except for the smallest cluster, where we may be predicting dissociated structures that are to be found in the 45% yield portion.

We have discussed the possibility that the pure cluster ionization potential might be a good indicator of the strength of the chemisorption bond.

Our suggestions are based on the SMM plus EHMO electronic calculations which allow us to compute the ΔH_f and η at each point of the reaction path. At least within the EHMO method, we present some numerical evidence that the maximum hardness principle might be more useful to study a cluster reaction mechanism than using total energies. This of course does not conflict with the variational principle, since the problem in EHMO is that total energies are poorly computed. We performed calculations of the chemical potential for our chosen path and found that indeed there were no significant deviations from being a constant along the reaction path. This is a formal condition for validity of the maximum hardness principle.

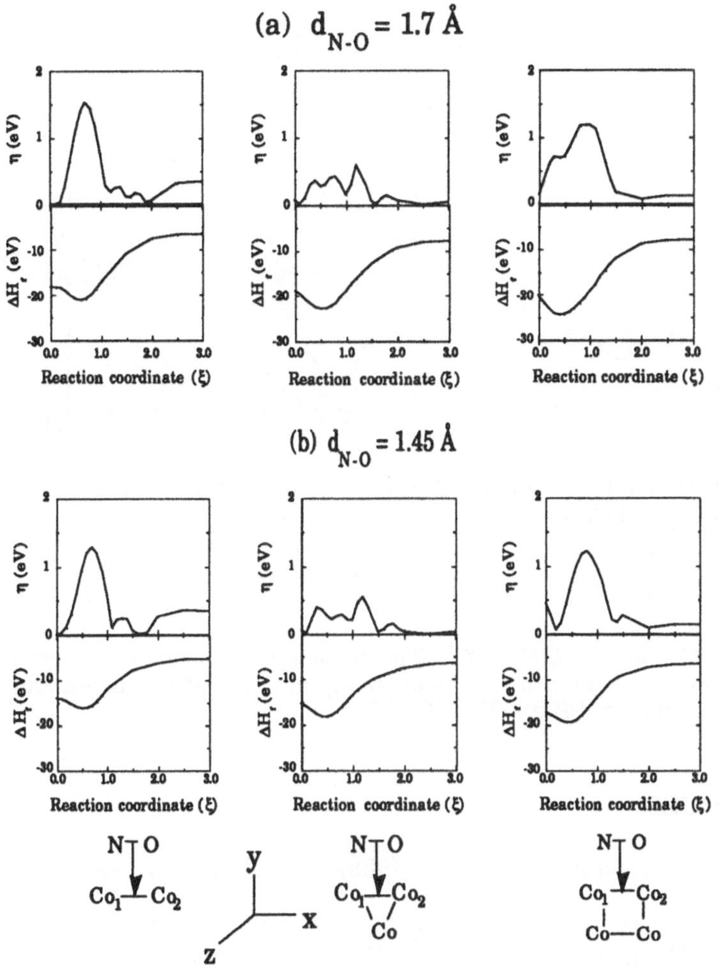

Figure 2. (a) ΔH_f and η versus the reaction coordinate ξ (defined in text) at fixed NO bond length of 1.7 Å. (b) ΔH_f and η versus the reaction coordinate ξ at fixed NO bond length of 1.45 Å.

$$d_{N\text{-}O} = 1.7\ \text{Å}$$

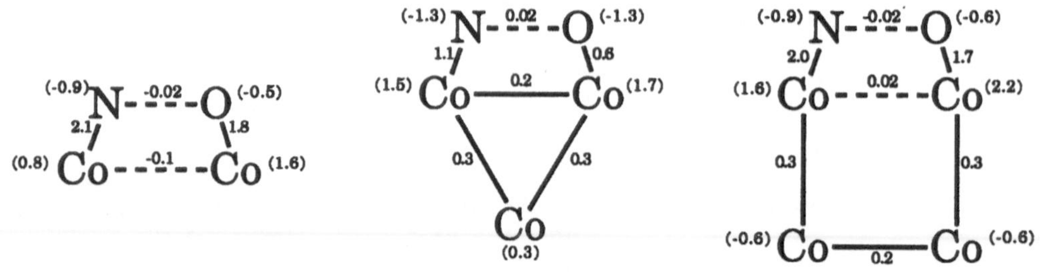

Figure 3. Mulliken overlap populations for the equilibrium structures obtained at the η curves maxima in Fig.2.

ACKNOWLEDGEMENTS

J.R. and A.M. thank CONACYT. The former for financial support to assist to this Symposium and the latter for the scholarship granted. M.P.I. is grateful for a CAICYT grant (project number PB89-0352-C02-01).

REFERENCES

1.- Klaaseen, J. J. and Jacobson, D.B..(1988) Journal of the American Chemical Society, 110, 974-975.
2.- Rubenstein, M. and Rubenstein, S. (1988) Program Chem 3D: "The molecular modeling system", v.2.0. Cambridge Scientific Computing, USA.
3.- a) Hoffmann, R. and Lipscomb, W.M. (1962) Journal of Chemical Physics, 36, 2179.
b) Hoffmann, R. (1963) Journal of Chemical Physics, 39, 1397.
c) Hoffmann, R. et al. (1977) QCPE, 11, 344.
d) All parameters taken from Vela, A. and Gazquez, J.L. (1988), Journal of Physical Chemistry, 92, 5688. Also see discussion on the parameters issue in Ref. 4.
4.- Robles, J. : Martinez, A.; Vela, A. and Iñiguez, M.P. (1991) Journal of Physical Chemistry, submitted.
5.- Martinez, A.; Vela, A. and Robles, J. (1991) Journal of Physical Chemistry, submitted.
6.- a) Parr, R. and Yang, W. (1989) "Density Functional Theory of Atoms and Molecules". Oxford University Press, USA.
b) Parr, R. and Chattaraj, P. (1991) Journal of the American Chemical Society, 113, 1854.
c) Harbola, M. (1991) Proc. Natl. Acad. Sci. USA. submitted.

Stability, Ionization Potential and Electronic Affinity of Small Silver Clusters calculated in the LSD approximation.

V. RUSSIER, C. MIJOULE and J. LANGLET.
Laboratoire de Dynamique des Interactions Moléculaires,
Université P. et M. Curie, 4 Place Jussieu 75252 Paris Cedex 05, France.

ABSTRACT. We present a theoretical study of small Ag_n clusters (n = 1-15). The electronic calculations are done by using the LSD approximation. The cohesion energy, as well as the stability $D_n^{\pm}$ of the Ag_n^+ and Ag_n^- ionic clusters ($D_n^{\pm} = (n-1)E(Ag) + E(Ag^{\pm}) - E(Ag_n^{\pm})$), are calculated. The characteristic odd -even alternation is obtained for the increase $\Delta_n^{\pm} = D_n^{\pm} - D_{n-1}^{\pm}$ in the stability when going from the ($n-1$)- to the n- atoms cluster. The ionization potential IP(n) and the electronic affinity EA(n) present also an odd-even alternation behavior. Finally we present a first estimate of the electrostatic part of the solvation free energy of the ionized clusters, in view of an understanding of the behavior of the electrochemical potential of solvated Ag_n^+ ions.

1. Details of the calculations

The calculations have been performed by using the spin - polarized version of the local density approximation (LSD) of the density functional theory of Hohenberg and Kohn. The molecular orbitals are expanded by a linear combination of Gaussian type orbitals (LCGTO) [1] and the core electrons of Ag are represented by a model potential of Huzinaga type [2]; the number of electrons explicitely treated is Z = 17. The corresponding basis sets are given in [3]. The exchange and correlation energy functional is that of Vosko *et al.* [4].

2. Results

For the determination of the stable structure, for each value of n, we have considered a few reasonable shapes, for both the planar and a compact (3 -dimensional) geometry . For the Ag - Ag distance we find $d_2 = 4.80$ au for $n = 2$ and $d_n \approx 5.00$ au for $n = 4$ to 10. This agrees with the results of Ref. [5] while in Refs. [6,7] a slightly larger distance is found, $d_n \approx 5.40$. For $n = 3$ to 5, $n = 7$ and 10 the same equilibrium structure as in Ref [8], where the Huckel approximation is used to represent metal clusters, is found. For $n = 6$ we find the planar goemetry (equilateral triangle) to be the stable structure, in opposite to the result of Ref. [8]. However the difference between the two corresponding cohesive energies is less than 10% of E_{coh}. Because of this good agreement, the equilibrium structure that we have taken for Ag_9 and Ag_{13} is that of Ref. [8].Thus the equilibrium structure corresponds to a planar geometry for $n < 8$ and to a 3 - dimensional one for $n \geq 8$. Our results for the

P. Jena et al. (eds.), Physics and Chemistry of Finite Systems: From Clusters to Crystals, Vol. I, 657–662.

structure are in general agreement with, on the one hand those concerning Ag_n for $n \leq 5$ [5,7], Na_n [9] and Li_n [10] clusters from *ab -initio* methods and on the other hand with those concerning Na_n clusters from the LSD approximation [11]. In figure 1 we compare the dimensionless quantity $E_{coh}(n)/E_{coh}(2)$ as calculated in the present work to the results corresponding to (a):Na_n clusters [9], calculated from an *ab -initio* method, (b): Na_n clusters [11,12] from the LSD approximation and (c): the Huckel approximation [8]. The results of Ref. [6] for $n = 4, 5$ are undistinguishable from ours on the scale of Fig.1. The different sets of results agree well for $n \leq 10$. A similar agreement is obtained with the results relative to Li_n [10,13]. From the comparison of figure 1 we can deduce that, at least for small values of n, Ag_n clusters behave as "one electron per atom" kind of clusters, and moreover that the Huckel approximation is valid for the description of such clusters when they are neutral.

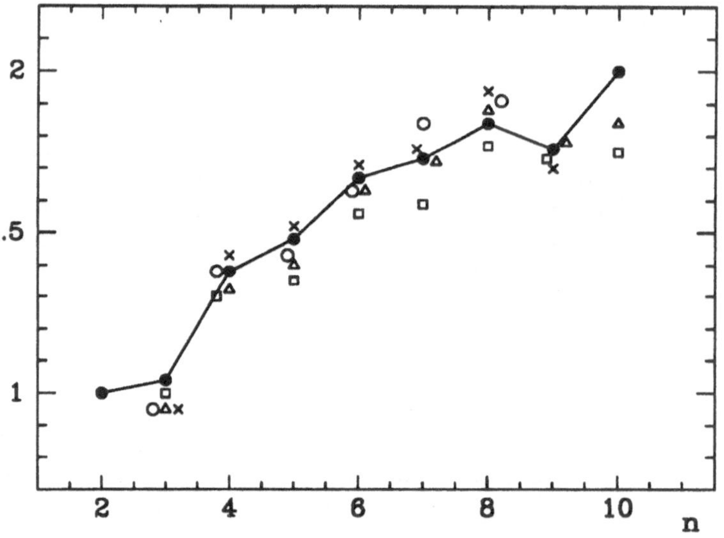

Figure 1: $E_{coh}(n)/E_{coh}(2)$ in terms of n. Full circles: Ag_n clusters from this work. Open triangles [11]; Open circles [12]; Crosses [9]; Open squares: Huckel approximation [8]. The symbols are shifted from integer values of n for clarity.

Furthermore we examine the stability of two Ag_n clusters compared to that of the combination of the Ag_{n-1} and Ag_{n+1} clusters. This is done through the second difference function: $\Delta_2(n) = [2nE_{coh}(n) - (n-1)E_{coh}(n-1) - (n+1)E_{coh}(n+1)]$. The important result is the odd - even alternation behavior, $\Delta_2(n)/ < 0$ for n odd and $\Delta_2(n) > 0$ for n even, showing that $[Ag_{n-1} + Ag_{n+1}]$ is more stable than $[Ag_n + Ag_n]$ when n is odd and vice-versa when n is even. The dimensionless function $\Delta_2(n)/E_{coh}(2)$ is in qualitative agreement with the results of *ab -initio* calculations for Na_n [9] and Li_n [10,13] up to $n = 9$, with the results of the LSD approximation for Na_n of Ref. [11] up to $n = 9$ and of Ref. [12] up to $n = 5$.

We find here that the stable structure of the monovalent ionic clusters $Ag_n^{\pm}$ is the same as that of the neutral clusters, excepted for $n = 6$ where we find the pentagonal pyramid for the stable structure of the positive ion while for that of the neutral cluster we find the equilateral triangle. In any case the contribution to the ionization potential due to this structure change is negligible (0.05 eV). The stability of the ionic clusters is characterized by the function $D_n^{\pm}$. This is a monotonic increasing function of n which indicates that the $Ag_{n-1}^{\pm}$ cluster in presence of a Ag atom will form the $Ag_n^{\pm}$ ion. The comparison of the stability of the combination $[Ag_n^{\pm} + Ag_n^{\pm}]$ to that of the combination $[Ag_{n+1}^{\pm} + Ag_{n-1}^{\pm}]$ is done by the means of the function $\Delta^{\pm}{}_2(n) = [2\, D_n^{\pm} - (n-1)\, D_{n-1}^{\pm} - (n+1)\, D_{n+1}^{\pm}]$, the analogue of the function $\Delta_2(n)$ for the neutral clusters. $\Delta^{+}{}_2(n)$ and $\Delta^{-}{}_2(n)$ both exhibit an odd - even behavior as is the case for the neutral clusters; however the combination $[Ag_n^{\pm} + Ag_n^{\pm}]$ is favoured with respect to $[Ag_{n+1}^{\pm} + Ag_{n-1}^{\pm}]$ for the odd values of n in opposite to the case of the neutral clusters, in agreement with Ref. [10]. This alternation behavior is thus relative to the number of electrons. In a same way we find an odd - even altrernation behavior for the increase in $D_n^{\pm}$; the function $\Delta_n^{\pm} = [D_n^{\pm} - D_{n-1}^{\pm}]$ takes high values for n odd and small values for n even. This may explain the alternation found in the experimental mass spectra of ionized Ag_n (and Cu_n) clusters [14].

Concerning the ionization potential IP(n) and the electronic affinity EA(n) we also find an odd-even alternation behavior: IP(n) is high and AE(n) is small for n even and *vice versa* for n odd. The overall variation of IP(n) is not a monotonic decreasing function for n in between 3 and 8, and presents important discontinuities between $n = 2$ and 3 and between $n = 8$ and 9. Our results parallel roughly those of Ref. [6] for $2 \leq n \leq 5$. This behavior can be understood in the framework a shell filling model, as is found in the spherical jellium model [15] and does not correspond to the behavior of the classical metal droplet where IP(n) decreases monotonically from IP(1) to the metal work function Φ[16]. This can be seen by comparing IP(n) for our Ag_n clusters with IP(n) of the spherical jellium model of Ref [15]. We have not calculated IP(n) for the Jellium model and in Ref. [15] the only value of r_s considered is $r_s = 4.00$ corresponding to the bulk density of Na. Having in mind mainly the comparison between qualitative trends we have adapted the results of Ref. [15] as follows. By assuming that the variation of IP(n) with n is roughly scalled on the value of $IP(n = \infty) = \Phi$, we deduce from Ref. [15] a jellium like approximation adapted to our problem

$$IP^J(n) = \Phi\,(1 + (IP(n,r_s{=}4) - \Phi(\,r_s{=}4))/\Phi(\,r_s{=}4)) \qquad (1)$$

where $(IP(n,r_s{=}4) - \Phi(\,r_s{=}4))$ is taken from Ref [15] and we use Φ = 4.40 eV for silver. In figure 2 we compare our results for the ionization potential of Ag_n to both the classical metal droplet IP and to $IP^J(n)$. As we can see the overall behavior of IP(n) for Ag for $3 \leq n \leq 9$ is closer to that of $IP^J(n)$ than that of the classical metal droplet. This shows that quantum mechanical effects are important and that the odd - even alternation behavior occurs in addition to the shell filling type of behavior which is characteristic of small jellium spheres (where no spin polarization is taken into account). This jellium - like behavior of the Ag_n clusters ionization potential is to put in parallel with the alkaline metal atoms clusters like behavior that we find for the stability of the neutral Ag_n.

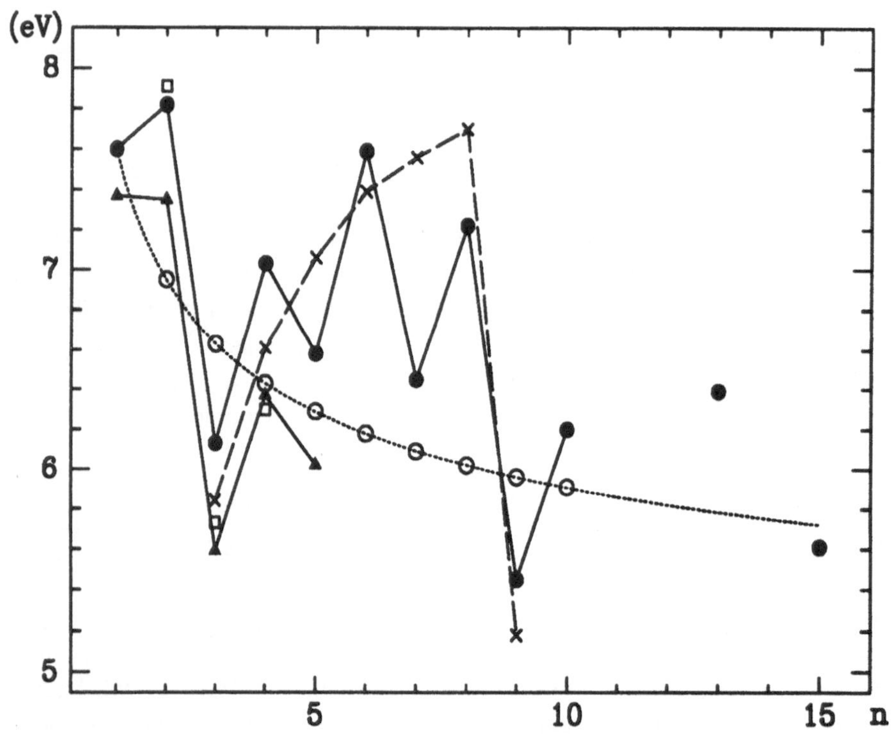

Figure 2: Full circles: Ionization potential of Ag_n clusters. Open circles: classical metal droplet [15]. Crosses: $IP^J(n)$ from Eq. (1). Full triangles: [6]; Open squares: [5].

The determination of the solvation free energies of both the neutral Ag_n and the ionic Ag_n^+ clusters is of interest since recent experiments show that a critical size is required for positive ionic clusters to accept an electron from another solvated species S^- [17]. Such an electron transfer is possible only if the corresponding electrochemical potentials satisfy: $E^0[Ag_n^+/ Ag_n] > E^0[S/S^-]$. One can show that the size dependence of $E^0[Ag_n^+/ Ag_n]$ is governed by

$$IP_s(Ag_n) = IP(Ag_n) + F_{sol}(Ag_n) - F_{sol}(Ag_n^+) \qquad (2)$$

$IP(Ag_n)$ is the ionization potential in vacuum of the Ag_n cluster, presenting the same nuclear structure as in solution, and $F_{sol}(M_n)$ is the solvation free energy of M_n. In this preliminary study, we consider the simplest determination of $F_{sol}(Ag_n) - F_{sol}(Ag_n^+)$. The structure of the Ag_n cluster in solution is assumed to be compact (3 -dimensional for $n > 3$) and we use the primitive model (dielectric continuum) to represent the solvent in the same way as in Ref. [18]. The solvated cluster is represented by a cavity, built from a collection of, possibly overlapping, hard spheres located at the nuclei positions; the radius of the spheres is the Van der Waals radius of Ag (R_w = 2.38 au). The difference $F_{sol}(Ag_n) - F_{sol}(Ag_n^+)$ includes only the electrostatic contribution, $- F^{es}_{sol}$, of $- F_{sol}(Ag_n^+)$, which coincides with the Born solvation free energy in the special case of a spherical cavity. F^{es}_{sol} is a function of the radius R_w and by replacing R_w by λR_w, one can adjust the location of the dielectric cut relative to the external surface of the actual cavity. The deviation of λ from the value $\lambda = 1.00$ allows to take into account steric effects of the solvent: in a hard spheres like picture the solvent - solute minimum distance of approach is represented by $\lambda R_w = R_w + R_s$ where R_s is the radius of the solvent molecules. In addition one expects an overpolarization of the solvent in the vicinity of the ion, which can be taken into account by reducing the value of λ. Here we use the parameter λ in order to fit the experimental value of the solvation free energy of the Ag^+ ion (4.9 eV) and this leads to $\lambda = 1.15$. This very crude model reproduce roughly the experimental result: $IP_s(Ag_n) < \Delta E_c$ for $n \leq n_c$ and $IP_s(Ag_n) > \Delta E_c$ for $n > n_c$, excepted for $n = 2$ and $n = 9$, because of the strong discontinuities of $IP(Ag_n)$. We find $n_c = 5$ and $\Delta E_c = 3.9$ eV while the experimental values are $n_c = 4$ and $\Delta E_c = 4.0$ eV.

3. Conclusion

The cohesion of Ag_n clusters is similar to that of alkaline metal atoms clusters and this can be well reproduced by the Huckel approximation. The ionization potential depends on n according to the shell filling behavior of jellium spheres in addition to an odd -even alternation and thus it is not a simple monotonic decreasing function. Finally by taking into account the solvent polarization effect in the simplest way (primitive model) and by fitting only one parameter (λ) on the experimental solvation free energy of Ag^+ one can reproduce qualitativelly one feature of the experimental behavior of the difference between the solvation free energies of the ionic Ag_n^+ and of the neutral Ag_n clusters. However the precise understanding of these experiments still necessitates an extensive work.

References

1. B.I. Dunlap, I.W. Connolly and J.R. Sabin (1979), J. Chem. Phys. **71**, 3386
2. S. Huzinaga, M. Klobukowski and Y. Sakai (1984), J. Chem. Phys. **88**, 488
3. J. Andzelm, E. Wimmer and D.R. Salahub (1985), J. Chem. Phys. **83**, 4573
4. S.H. Vosko, L. Wilk and N. Nusair (1980), Can. J. Phys. **58**, 1200
5. J. Flag, G. Igel-Mann, H. Preuss and H. Stoll (1984), Chem. Phys. **90**, 257
6. C. W. Bauschilcher, S.R. Langhoff and H. Partridge (1990), J. Chem. Phys. **93**, 8133
7. K. Balasubramanian and P.Y. Feng (1990), J. Phys. Chem. **94**, 1536
8. D.M. Lindsay, Y. Wang and T.F. George (1987), J. Chem. Phys. **86**, 3500
9. V. Bonacic-Koutecky, P. Fantucci and J. Koutecky (1988), Phys. Rev. B **37**, 4369
10. I. Boustani, W. Pewestorf, P. Fantucci, V. Bonacic-Koutecky and J. Koutecky (1987), Phys. Rev. B **35**, 9437
11. U. Rothlisberger and W. Andreoni (1991), J. Chem. Phys. **94**, 8129
12. J.L. Martins, J. Buttet and R. Car (1985), Phys. Rev. B **31**, 1804
13. A.S. Hira and A.K. Ray (1991), Surf. Sci. **249**, 199
14. P. Joyes and P. Sudraud (1984), Surf. Sci. **156**, 451
15. W. Ekardt (1984), Phys. Rev. B **29**, 1558
16. D.M. Wood (1981), Phys. Rev. Letters **46**, 749
17. M. Mostafavi, J.L. Marignier, J.Amblard and J. Belloni (1989), Radiat. Phys. Chem. **34**, 605
18. J. Langlet, P. Claverie, J. Caillet and P. Pullman (1988), J. Phys. Chem. **92**, 1617

STABILITY OF HYDROGEN CLUSTERS NEAR A TRANSITION METAL SURFACE

L. Stauffer [1] , M. Habar [1] and H. Dreyssé [2]

[1] Laboratoire de Physique et Spectroscopie Electronique
4 rue des Frères Lumière, 68093 Mulhouse, France.
[2] Laboratoire de Physique des Solides,
BP 237, 54506 Vandoeuvre-les-Nancy, France.

ABSTRACT: We propose an unified framework to obtain the electronic structure of impurities clusters in a metallic host. The electronic structure and the Cluster Interactions are determined in a self-consistent way within the tight-binding model using generalized phaseshifts. This real-space calculation is well suitable to describe any atomic configuration.

This approach is applied to H/Pd system. Pairs, trimers, tetramers Hydrogen arrangements, near the (001) surface, are investigated. The electronic structure and the Cluster Interactions with different orientations relative to the surface are exactely computed. Good agreement is found with experimental results. Moreover, this approach allows to describe the initial growth stage of Hydrogen near the surface.

1. Introduction

Small interstitial impurities clusters, immersed in a metallic host, can be well described within a lattice gaz model /1/. To study the equilibrium positions at T = 0°K (ground state) , two ways are possible : the first one would be to obtain the total energy of a cluster configuration as the sum of the energies of individual atoms. This simple conceptually approach suffers of numerical difficulties, since the relevant energies are difference of large numbers which have been computed separately. Thus another way consists to determine the Cluster Interactions. The starting point of this approach is that the total energy of any system can be written in the basis of the correlation functions; the Cluster Interactions are then proportional to the coefficients in this expansion /2/. The most important Cluster Interactions are the Pair but higher order clusters are often necessary /3/.

The determination of the stability of small interstitial clusters is a delicate problem: on one hand the discrete nature of the impurities needs careful study , on the other hand the extended size of the host must be taken into account. For this reason few attempts have been perfomed to obtain Cluster Interactions beyond Pair Interactions. Let us mention the pioneering work

P. Jena et al. (eds.), Physics and Chemistry of Finite Systems: From Clusters to Crystals, Vol. I, 663–668.

of Einstein /4/ which determined, in a simple framework, trio interactions for the O/W(110) system. More recently, Einstein et al /5/ computed pair and trio interactions using the Embedded Atom Method. But these authors did not obtain agreement with experimental results for H/Ni(111) as for H/Pd(111). Muscat has studied extensively Hydrogen Pair Interactions near a transition metal surface. His first results for trio , indicated that the trio interactions can be decomposed in the sum of the constituting pairs. Good agreement with experimental results have been reached for H/Ni(111) but not for H/Pd(111) /6 /.

2. Theoretical framework

Bulk alloy phase stability is an intense research field.We shall not discuss here the various approaches to compute the Cluster Interactions in a concentrated alloy /3/. Here we shall use an approach dedicated to dilute alloy exclusively. In this framework /7/ the Cluster Intercations (C.I.) are computed as **global** quantities. The C.I. are defined as usually by $W^N(i) = E_{tot}(i) - E_{Tot}(\infty)$

where E_{Tot} refers to the total energy of

- the host and N interactiong impurities in a given cluster configuration at sites k (k=1,2, ...,N).
- the host and one impurity at site k, all the other impurities being in the reference level; this last term $E_{Tot}(\infty)$ is then the sum of N terms. The sign of these C.I. will indicate if the cluster or isolated impurities in the host are most stable. Using a statistical model, these C.I. give the phase diagram /2/.

The tight binding formalism provides an efficient framework to study such problem. The finite size of the perturbation, induced around the cluster atoms, allow to use generalized phase-shifts /7/. Details of this aproach can be found in ref. 7.

3. The H/Pd(001) system

This approach will be applied to the well experimentally studied H/Pd(001) system. In our previous study, a reduced set of pair and trio interactions have been considered. However good agreement with the experimental results have been obtained /7/ :

- for the isolated impurities, the fourfold "hollow" position is slightly more attractive than the twofold "brige" position.
- the inclusion of pair interactions, and later trio, stabilizes the "hollow" position .
- no "subsurface" chimisorption occurs.
- the c(2x2) structure is obtained.

- the asymmetry of the phase diagram has been explained by the values of the trio interactions.

In this work, we improve the formalism. For the first time, as far we know, interactions with atoms in different planes parallel to the surface are computed. In order to study the stability of small Hydrogen clusters near the surface, all possiblities must be explored: atoms and/or in tetrahedral positions (as other positions as saddle-points). In a recent study, we found that , for H/Pd(001) system, the octahedral sites are the most stable positions /8/. Pair Interactions with two Hydrogen, located one on octahedral site, the other one on the nearest tetrahedral site, are strongly repulsive. Thus in this paper, we shall concentrate on the Cluster Interactions for Hydrogen atoms on octahedral positions, corresponding to a fcc lattice.

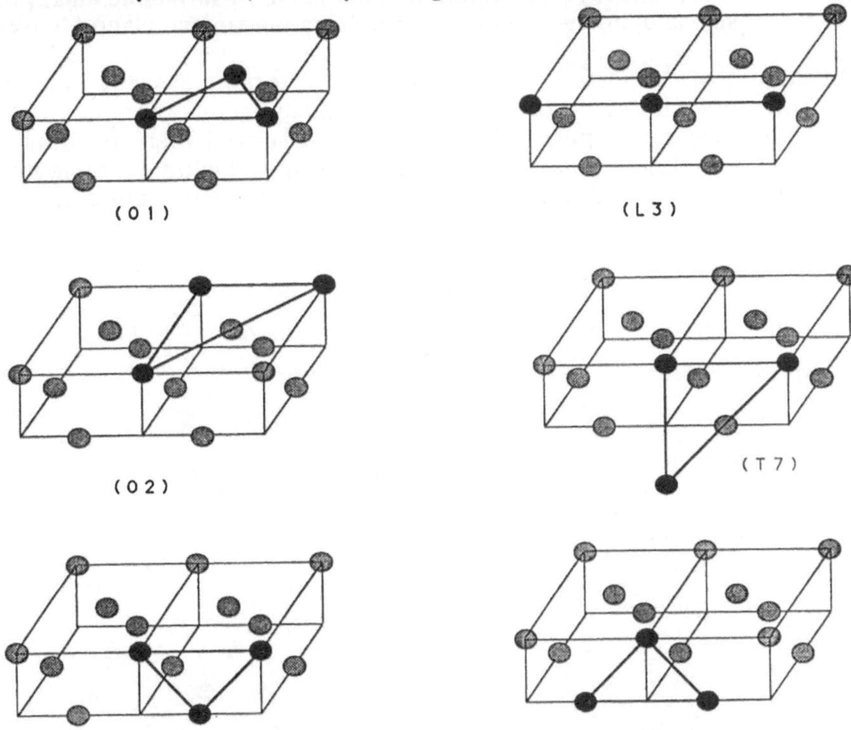

Figure 1: Trimer interactions, the surface plane is the upper plane.

Let us recall the adsorption energies of isolated Hydrogen in octahedral positions near the Pd (001) surface: - 2.32 eV for the on-top position (n = -1),

-2.69 eV for the "hollow" position (n = 0), - 2.44 eV for the subsurface (n = 1) and -2.29 on the following plane ; the bulk adsorption value is - 2.36 eV. The difference of the isolated impurity adsorption energy between the surface plane (n = 0) and the two adjacent planes(n = - 1 , out of the host and n = 1 , in the host), is large. This will greatly contribute to stabilize the adsorbates in the surface plane, if the temperature is not too high and no significative diffusion into the bulk occurs .

V1(0,0)	V2(0,0)	V2(0,2)
-0.016	-0.665	-0.604

Table I: Pair Interactions: V1 (V2) refers to n.n. (n.n.n.) pair interactions, (n_1, n_2) indicate the positions of the two atoms , n = 0 being the surface plane (in eV).

On table 1 are reported Pair Interactions. The biggest values are found when all atoms are in the surface plane. A relatively small anisotropy is found for V2. Since the n.n.n. pair is largely more important than the other pair interactions, we shall present trimer and tetramer including al least one n.n.n. bound. We have found that the C.I. without n.n.n. bound are much smaller /8/.

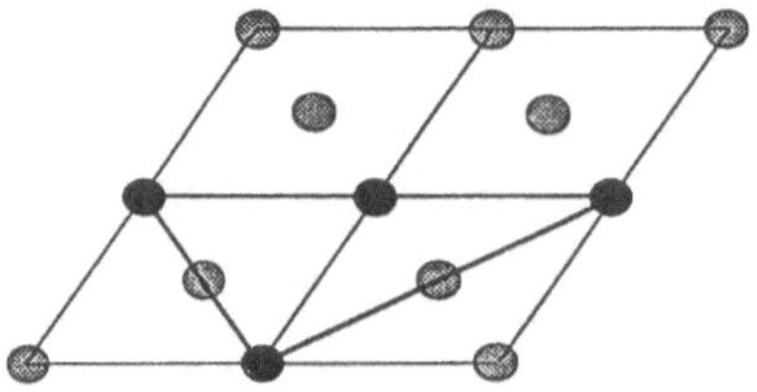

(T E 11)

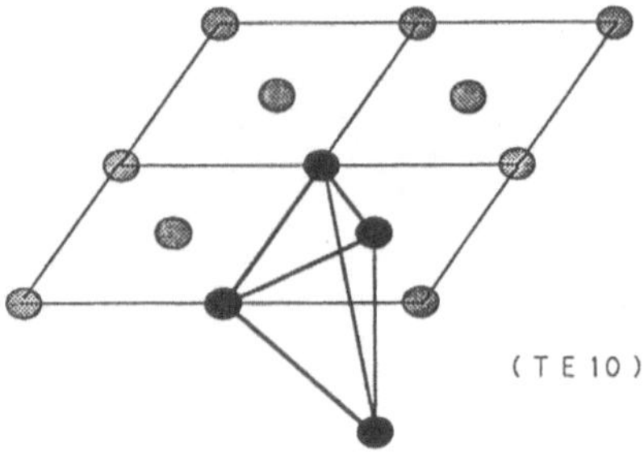

(T E 10)

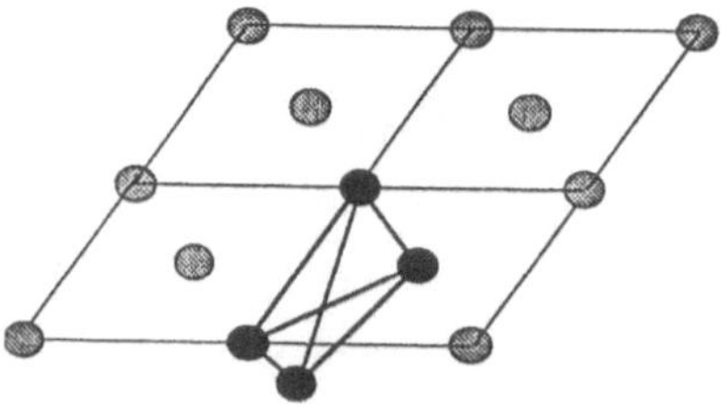

(T E 8)

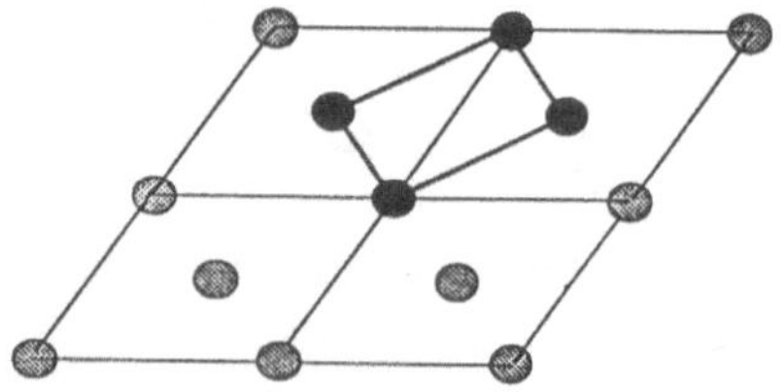

(S S)

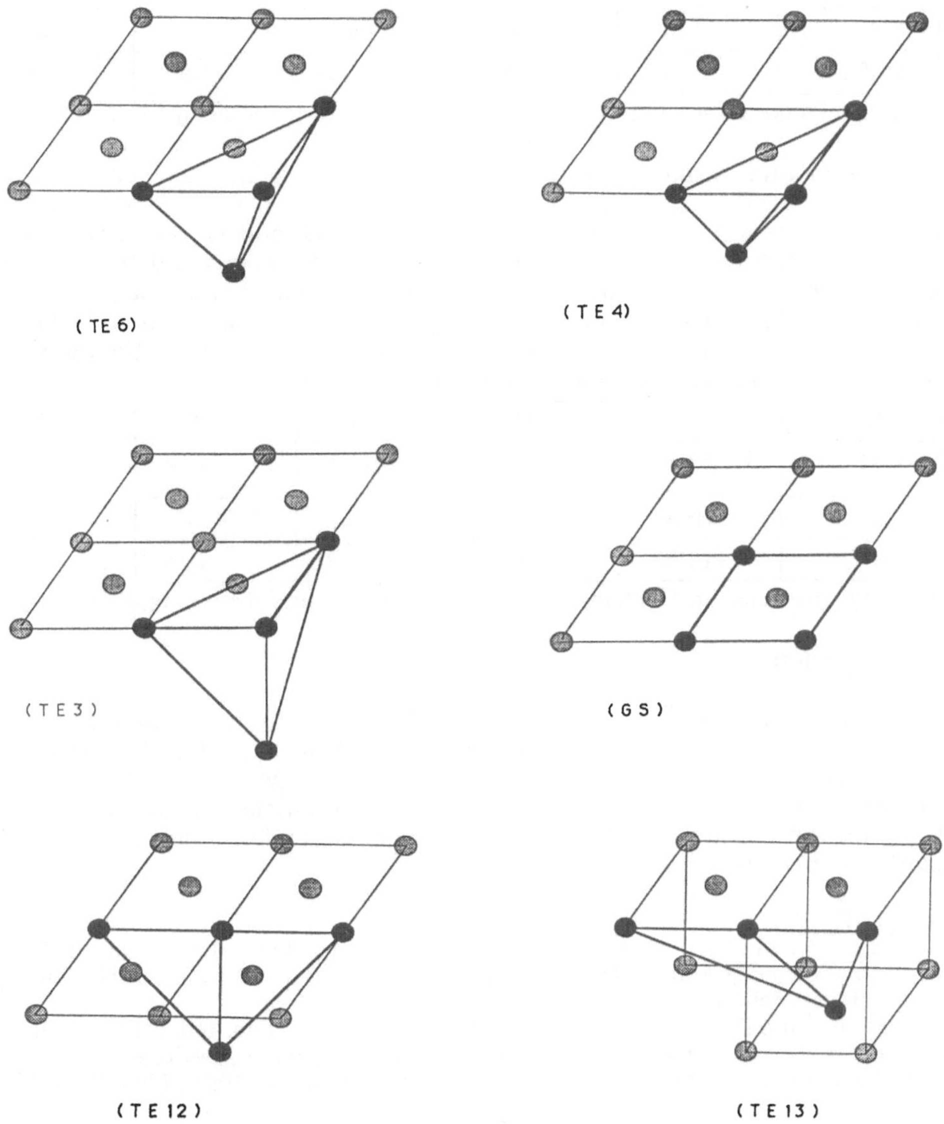

Figure 2: Tetramer interactions, the surface plane is the upper plane.

O1	L3	O2	T7	T5	T2
-0.684	-1.030	-1.101	-0.995	-0.650	-0.404

Table II: Trimer Interactions The various positions are given in fig.1 (in eV)

Trimer and tetramer C.I. are reported in tables 2 and 3. For Hydrogen atoms in the surface plane, the O_2 configuration with two n.n.n. bounds is dominant. The anisotropy of the C.I. is weak : for O_2 and T7, as for O_1 and T5. The value obtained for T3 (- 0.404 eV) must be compared to the value obtained for O_1 configuration with all atoms in plane below the surface (- 0.456 eV). Anisotropy is more present for tetramers (for exemple TE11 and TE12) . The important difference between SS and TE8 is related the diference of one n.n.n. bound between these two configurations.

SS	GS	TE3	TE4	TE6
-1.317	-1.684	-1.292	-1.063	-1.078
TE8	TE10	TE11	TE12	TE13
-0.644	-1.252	-1.370	-1.250	-1.016

Table III: Tetramer Interactions The various positions are given in fig.2 (in eV)

4. Conclusion

This approach allows to describe any cluster configuration , immersed in a host and has been recently used to study the H/Ni(001) system /8/, with good agreement with experimental results. It open the way to study the initial growth mode of Hydrogen adsorbed on a transition metal. Barrier heigths for the diffusion can be also determined. Other interstitial atoms (as O, N ..) can be considered.

REFERENCES

/1/ Van Hove M.A. , Tong S.Y. and Weinberg W.H. (1986) in "L.E.E.D.: Experiment,Theory And Surface Structure Determination" (Springer Series in Surface Science,Vol 6)

/2/ de Fontaine D. (1987), in Electronic Band Structure and its Applications, vol 283 of Lectures Notes in Physics, edited by M. Yussouff (Springer, Berlin)

/3/ Wolverton C. , Asta M. , Dreyssé H. and de Fontaine D. (1991), Phys. Rev. B 44, 4914.

/4/ Einstein T.L. (1979), Surf. Sci. 84, L497.

/5/ Einstein T.L. , Daw M.S. , and Foiles S.M. (1990) , Surf. Sci. 227 114.

/6/ Muscat J.P. (1986), Phys. Rev. B33, 8136.

/7/ Stauffer L. , Riedinger R. and Dreyssé H. (1990), Surf. Sci. 238 83.
Stauffer L. and Dreysse H. (1991), Surf. Sci. 251-252, 1148.

/8/ Stauffer L. and Dreysse H. , to be published .

THEORETICAL INVESTIGATION OF LOCATIONS OF $^{19}F^*$ AND NUCLEAR QUADRUPOLE INTERACTIONS IN SILICON AND DIAMOND

S. SWINGLE NUNES, S.B. SULAIMAN, N. SAHOO and T.P. DAS
Department of Physics
State University of New York at Albany
Albany, New York 12222

ABSTRACT. Available nuclear quadrupole interaction data for the excited nuclear state $^{19}F^*$ for fluorine in silicon and diamond are used together with results of Unrestricted Hartree-Fock cluster investigations to identify possible sites for fluorine impurities in these materials. Fluorine atoms are embedded in carefully chosen clusters of host atoms to simulate the impurity system. Allowing for lattice relaxation, optimal locations of the fluorine atom in the intrabond, antibonding (interstitial), and substitutional sites are obtained through total energy minimization. Wave functions calculated for the optimized configurations are used to evaluate the quadrupole parameters, e^2qQ and η. The results are compared with experimental parameters to ascertain the sites responsible. In silicon, the two observed sets of experimental data are identified with intrabond and antibonding sites. The same appears to be true for diamond where our results are of a preliminary nature.

Introduction.

The Time Differential Perturbed Angular Distribution (TDPAD) technique[1] has been used to study the nuclear quadrupole interaction of fluorine impurities in silicon[2] and, more recently, in diamond[3]. In both materials, two distinct sets of quadrupole parameters were observed, pointing to the existence of two stable sites for the implanted fluorine impurities. As the values of e^2qQ and depend on the electronic distribution about the nuclei[4], comparison between the values observed experimentally and those obtained from theoretical electronic wave functions can provide insights into the structure of these systems.

P. Jena et al. (eds.), Physics and Chemistry of Finite Systems: From Clusters to Crystals, Vol. I, 669–676.

The experimental values obtained for e^2qQ and η at sites 1 and 2 in both silicon and diamond are presented in Table I. For silicon[2], it is important to note that the zero value of η observed at both sites indicates that the electric field gradient (efg) tensor is axially symmetric at both sites. In diamond[3], this is the case for only site 2.

TABLE I. Summary of Experimental Results for $^{19}F^*$ Implanted in Crystalline Silicon and Diamond

Host Material	Sites	e^2qQ (MHz)	Asymmetry Parameter	Orientation of efg
Silicon[2]	1	34.9	0	<111>
	2	23.1	0	random
Diamond[3]	1	59	0.17	<111>
	2	25	0	<111>

We have attempted to identify the two sites 1 and 2 in each material which give rise to the observed quadrupole interaction parameters. Three distinct locations in the lattice are considered. These locations, which are illustrated in figure 1, are the intrabond site (IB) between two "bonded" host atoms, the interstitial or antibonding (AB) site which is located along a line passing through the hexagonal and tetrahedral interstitial positions, and the substitutional (S) site in which fluorine completely replaces a host atom.

For each site, minimum energy configurations are obtained using clusters of various sizes. Relaxation of the lattice in response to the presence of the impurity atom is incorporated during this process by allowing the positions of various groups of equivalent host atoms surrounding the impurity of vary along with the position of the impurity itself. The Unrestricted Hartree-Fock (UHF) procedure[5] is optimized clusters. These are used to obtain the efg tensors at the sites of the $^{19}F^*$ nuclei, which in turn yield[4] the quadrupole coupling parameters. These theoretical results are compared with the experimentally observed parameters. Stable sites are associated with the configurations for which the two correspond.

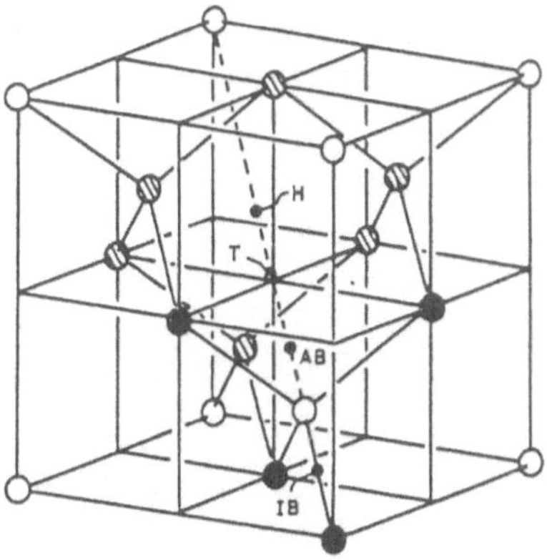

Figure 1. Illustration of the three lattice sites (labeled IB, AB, and S) studied as possible locations of $^{19}F^*$ in both diamond and silicon.

Procedure.

For each of the locations described above, a finite cluster of atoms was chosen to represent the surrounding lattice. Where possible, more than one cluster was used to test the convergence of the result with respect to cluster size. Interaction of the cluster with the rest of the infinite solid is accounted for by saturating the dangling bonds with hydrogen atoms. The host atom-hydrogen bond lengths were taken to be the covalent bond distances, 1.48A in silicon and 1.092A in diamond. Lattice relaxation effects were incorporated by using the Fletcher-Powell procedure[6] to minimize the total energy of the cluster with respect to variation in the positions of the fluorine atom and various groups of equivalent neighboring host atoms. The total energy used in the minimization process was calculated using the AM1 procedure[7].

The IB site corresponds to the location of a fluorine along a Si-Si or C-C bond. The site was modeled using two clusters, Si_2H_6F (C_2H_6F) and $Si_8H_{18}F$ ($C_8H_{18}F$), as illustrated in Figure 2. In both systems, when the fluorine atom was constrained to move along the <111> axis between its two nearest host atom neighbors and the rest of the host atoms in the cluster (excluding the saturator hydrogens) were allowed to relax, the optimized position of fluorine was found to be closer to one of its neighbors, the corresponding Si-F and C-F distances being

1.68A and 1.38A. These distances are close to the covalent bond distances of 1.6A and 1.34A.

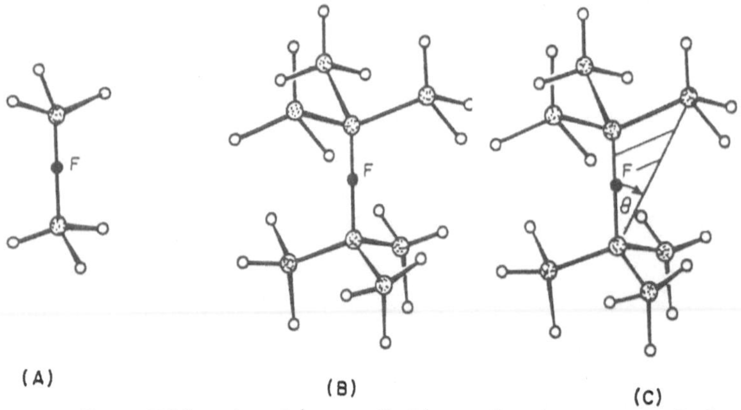

Figure 2. Illustration of the clusters used to study the IB site. (a) Si_2H_6F or C_2H_6F (b) $Si_8H_{18}F$ or $C_8H_{18}F$ (c) Illustration of the θ variation included to test the possibility of an off-axis position for $^{19}F^*$ in diamond.

In view of the rather short C-F distance obtained for the on-axis position in diamond and the fact that site 1 exhibits a non-zero η , we decided to permit the F atom to move off the <111> axis during the optimization process. Using the optimized lattice positions obtained with full lattice relaxation of the 27 atom cluster, the fluorine atom was allowed to vary off-axis in the manner illustrated in Figure 3c. A minimum energy configuration was obtained for $\theta = 8.0°$, corresponding to a nearest neighbor distance of 1.323A and a C_2-C_1-F angle of 8.0°.

The 27 atom cluster illustrated in figure 3 was used to model the AB site. It is comprised of the ten host atoms surrounding an impurity when it is located in the interstitial region. The position which is equidistant from the four atoms labeled A,J,L, and N is referred to as the tetrahedral interstitial position (T) while the exact center of the six atoms labeled I,J,K,L,M and N defines the hexagonal site.

In all optimizations, the positions of fluorine, and host atom A were varied along the <111> axis only. Relaxation of atoms J,L, and N was accomplished by allowing them to vary radially with respect to T. When this is done, the fluorine atom is found to move slightly away from T toward host atom A, its distance from A being 1.713A in silicon and 1.52A in diamond.

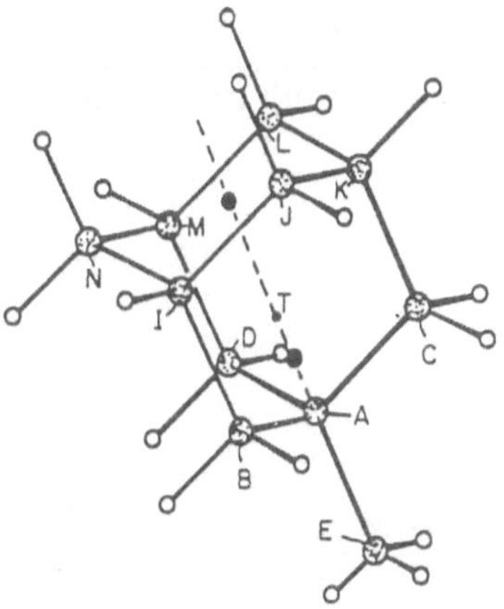

Figure 3. The basic 30 atom cluster used to study the interstitial sites. Atoms A,I,K, and M are exactly one lattice length from the tetrahedral interstitial position.

In order to test convergence with respect to cluster size in diamond, a double structure was designed by adding a four carbon tetrahedron directly above atoms J,L, and N to establish a second tetrahedral interstitial site.

The total energies of the original and expanded clusters are minimum when the fluorine atom is located 1.47A and 1.46A, respectively, from the carbon at position A in figure 3. As a further comparison the configuration of the expanded cluster was optimized with respect to the positions of the fluorine atom, and the four carbon atoms which define the tetrahedral site as had already been done in the smaller cluster. The final C_A-F distance is 1.51A as compared to 1.52A in the smaller cluster. These results demonstrate that the smaller cluster represents the infinite solid as well as the larger one.

When fluorine is at a substitutional (S) site, the T_d point group symmetry will lead to a vanishing efg. For a non-vanishing efg, the fluorine atom must move from the substitutional site. Should it move towards one of the four nearest neighbor host atoms along a <111> axis, the T_d symmetry will reduce to C_{3v}, leading to an axially symmetric efg ($\eta = 0$). This condition is necessary in the silicon system as suggested by the experimental data in Table I.

In order to study the efg at the substitutional site in the silicon system, a 26 atom cluster, ($Si_7H_{18}F$), illustrated in figure 4 was used. An analysis of the electronic structur of the neutral cluster revealed that the highest occupied state would be doubly degenerate with

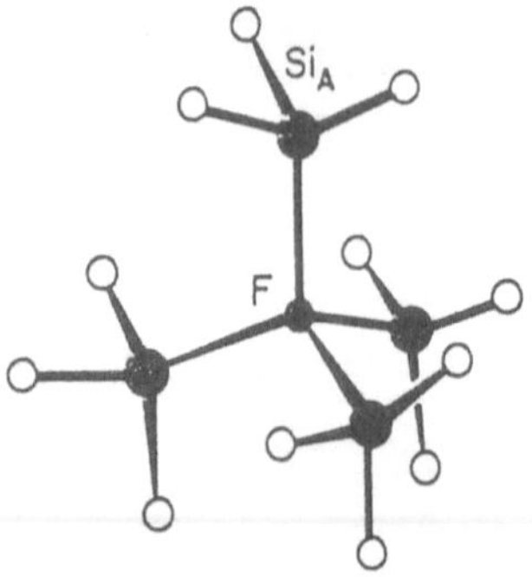

Figure 4. The 17 atom cluster used to study the substitutional site ($Si_7H_{18}F$).

only one electron. In this situation, the system would be unstable and undergo spontaneous Jahn-Teller distortion, reducing the symmetry to C_{2v} and thus removing the spatial degeneracy. This would lead to a non-zero η, in contradiction with the experimental result in Table I. However, for a cluster with a charge of +1 or +2, the degeneracy associated with the C_{3v} symmetry can be removed. The optimized Si_A-F distance in figure 4 was found to be 1.78A and 1.86A for the +1 and +2 charged clusters, respectively. Relaxation of other silicon atoms was not considered at this time. The substitutional site in diamond is still under investigation.

UHF calculations were carried out to obtain the electronic wave functions for all three sites in the silicon system using the optimized geometries discussed above. The components of the efg tensor at the $^{19}F^*$ site were calculated using these wave functions. Three different atomic basis sets were used for the first set of calculations on the silicon IB cluster. These basis sets[8] are the STO-3G and 3-21G Gaussian basis sets and the larger, more flexible D-95 one. Polarization functions were added to all atoms in the IB model. The value of $Q(^{19}F^*)$ was taken to be 0.72 barns[9].

In diamond, field gradient calculations have been undertaken only for the on-axis optimized position of fluorine in the IB cluster. The same three basis sets were used but polarization functions were not included.

Results and Discussion

The values of e^2qQ in MHz for the IB model in silicon for the cluster $Si_8H_{18}F$ and Si_2H_6F using the STO-3G, 3-21G, and D-95 basis sets were found to be (49.64, 49.68), (35.68, 35.50), and (33.77, 32.71) respectively. The values of e^2qQ for the large and small clusters are very close indicating rapid convergence with cluster size. The results obtained using the split valence and more extensive D-95 basis sets differ by only six and eight percent for the large and small clusters, respectively. Hence we can safely quote within 10% confidence limit that e^2qQ = 35.7 MHz for the IB model. This is quite close to the experimental value of 34.9 MHz associated with site 1 in Table I. Based upon the closeness of the D-95 and 3-21G results, only the D-95 basis set will be utilized for the AB and S sites.

The values of e^2qQ in MHz obtained for the optimized position of fluorine along the <111> axis in the IB model in diamond, using the C_8H_{18} cluster with STO-3G, 3-21G and D-95 basis sets are (66.1, 55.2 and 59.0) respectively. These values compare favorably with the experimental result of 59.0 MHz observed for site 1. This is on the order of the value expected for a fluorine atom covalently bonded to carbon. As described earlier, our preliminary work indicates that the IB fluorine atom in diamond may actually move off the <111> axis leading to a non-vanishing η . A radical change in e^2qQ is not expected since the fluorine atom is likely to move closer to the "normal" C-F bond distance if it moves off-axis to relieve crowding.

The theoretical value of e^2qQ for the AB model in silicon is 25.25 MHz which is close to the experimental value of 23.1 MHz obtained for site 2. For the substitutional model, the values of e^2qQ are 17.4 and 14.23 MHz for the +1 and +2 charged clusters, respectively, which are rather small compared to either experimental value.

Thus, our analysis of the quadrupole interaction of $^{19}F^*$ in silicon at three different locations shows that the IB and AB sites could explain very well the two experimentally obtained values for the coupling constant. Both symmetry considerations and the small values of e^2qQ compared to experiment rule out the substitutional site as a possible location for the fluorine impurity.

A detailed study of fluorine in diamond similar to that in silicon has not yet been completed. However,our work so far reveals several similarities to the silicon results. Preliminary calculations at the IB_2 site indicate that the higher experimental value of e^2qQ in Table I may well be associated with this site. Work is in progress on the AB and S sites. For the latter, Jahn-Teller distortions for the neutral cluster would lead to non-zero η. From Table I, one cannot therefore rule out the S-site from symmetry consideration as in the case of silicon and quantitative results for e^2qQ to compare with experiment will be necessary to make any conclusions in this respect.

REFERENCES:

/1/ H. Haas, Phys. Scr., 11, 221 (1975).

/2/ K. Bonde-Nielsen, H.K. Schou, T. Lauritsen, G. Weyer, I. Stensgaard, J.W. Petersen, and S. Damgaard, J. Phys. C: Solid State Phys., 17, 3519 (1984).

/3/ S. Connell, K. Bharuth-Ram, J.P.F. Sellschop, M. Stemmet, H. Appel, and J.E. Lowther, Nuclear Instr. Meth., B35, 423 (1988).

/4/ T.P. Das and E.L. Hahn, "Nuclear Quadrupole Resonance Spectroscopy", Academic Press, New York, 1958.

/5/ J.A. Pople and R.K. Nesbet, J. Chem. Phys., 22, 571 (1954).

/6/ R. Fletcher and M.J.D. Powell, Comput. J., 6, 163 (1963).

/7/ H. Bernhard Schlegel, J. Comput. Chem., 3, 214 (1982).

/8/ STO-3G: W.J. Hehre et. al, J. Chem. Phys. 52, 2769 (1970); 3-21G: M.S. Gordon et. al, J. Am. Chem. Soc. 104, 2797 (1982); D-95: T.H. Dunning and P.J. Hay in Modern Theoretical Chemistry 3: Methods of Electronic Structure Theory, ed. by H.F. Schaefer III (Plenum Press, New York, 1977).

/9/ K.C. Mishra, K.J. Duff and T.P. Das, Phys. Rev. B25, 3389 (1982).

SIMULATION OF IONIC CRYSTALS FOR ELECTRONIC PROPERTIES USING THE HARTREE-FOCK CLUSTER PROCEDURE

DALE W. MITCHELL,[a] S. B. SULAIMAN,[a] N. SAHOO,[b] P. C. KELIRES,[c] and T. P. DAS[a]
(a) Dept. of Physics, State University of New York at Albany, Albany, NY 12222
(b) Dept. of Medical Physics, Memorial Sloan-Kettering Cancer Center, New York, NY 10021
(c) Dept. of Physics, University of Crete, 71409 Iraklion, Crete, GREECE

ABSTRACT. The Unrestricted Hartree Fock cluster procedure is being used by our group for studying nuclear quadrupole and magnetic hyperfine properties and other related properties of ionic crystals. The method being employed involves using a finite cluster of ions centered about the ion, the hyperfine properties of whose nucleus are being investigated. The influence of the ions in the rest of the lattice outside of the chosen cluster is incorporated through the inclusion of their Coulomb potential acting on the electrons in the cluster. Results are given for the calculated nuclear quadrupole coupling constants for Al_2O_3, Fe_2O_3, Cu_2O, ZnF_2, ZnO and $ZnAl_2O_4$ and compared to experiment.

I. Introduction.

Intuitively, one expects that for localized properties of solid state systems, instead of using band structure approaches appropriate for the infinite solid, it would be quite reasonable to carry out a direct space study of a finite number of atoms with the effects of the rest of the infinite system taken into account in a self-consistent manner. Such a cluster approach to the simulation for electronic properties in ionic solids has the advantage that one can utilize the vast array of methods available from molecular orbital theory for the study of solids. Since one deals with direct space wave functions in the cluster molecular orbital approach a further advantage over band theoretical methods is that it is possible to study defects in solid state systems where the translational symmetry of the perfect solid is no longer valid. This has been indeed found to be the case[1] from several applications to ionic, semi-conductor and metallic systems. Here we present results from our group on the calculation of nuclear quadrupole coupling constants in the perfect lattice of several relatively simple ionic systems using the Unrestricted Hartree-Fock (UHF) cluster procedure[2] with the effects of the Madelung potential due to the rest of the infinite solid taken into account self-consistently.

Hyperfine interactions[3,4] are very important experimentally for the study of solids in that they provide precise information on microscopic properties arising from the splitting of nuclear energy levels due to the interaction of a nuclear moment and the electric field-gradients and magnetic fields due to the surrounding electrons. These hyperfine interactions, being localized in nature[3,4] are ideal properties for study by the UHF cluster procedure. In particular the results presented here on the experimentally measured nuclear quadrupole coupling constants[4] e^2qQ are the product of a nuclear quadrupole moment eQ and an electric field gradient eq due to the surrounding charge distribution. The eQ are known from independent measurements[5]. The UHF cluster procedure is used to calculate the electric field gradient tensor at the nuclear site of interest. A comparison between theoretically

P. Jena et al. (eds.), Physics and Chemistry of Finite Systems: From Clusters to Crystals, Vol. I, 677–682.

calculated and experimentally measured values of e^2qQ can provide a valuable test of the calculated wavefunctions especially the anisotropy of the charge distributions they lead to. In this paper, we describe the procedure for calculation of the electric field gradient tensor used in obtaining e^2qQ, our results[6-12] for several ionic systems and finally a discussion of these results.

II. Procedure.

Ionic solids can be described as wide gap systems typically possessing narrow bands which indicates that the ground state wave-functions can be considered to be built out of tight-binding functions[13] involving an array of functions in real space localized over a limited number of atomic centers. The UHF cluster procedure deals with these localized wave-functions and considers a limited number of atoms, chosen as the cluster to be solved using the all-electron self-consistent UHF method. The chosen cluster involves the central atom containing the nuclear site of interest and as many atoms as is computationally practicable, including at least the first nearest neighbors. The Madelung potential has been incorporated[14] in the all-electron UHF-cluster procedure we have used by replacing the atoms comprising the surroundings of the cluster with ionic point charges at lattice sites out to about 10 to 12 angstroms from the central atom. The charges on the ions in the outermost shells are adjusted to exactly reproduce the Madelung potential of the true solid at all cluster atoms as well as requiring that charge neutrality is maintained for the entire system of cluster plus point charges. In this way the influence of the ions in the rest of the lattice outside of the chosen cluster is incorporated through inclusion of their Coulomb potential acting on the electrons in the cluster. This allows one to carry out electronic structure investigations in effect over the infinite crystal making it of comparable viability as band structure calculations for localized phenomena.

We next briefly describe the calculation of the EFG tensor V_{ij} (i,j = x,y,z) by the UHF cluster procedure[2]. Once the molecular orbital wavefunctions ψ_μ for the cluster are determined self-consistently the components of the EFG tensor at the nucleus of interest can be calculated[3,6,10,11] using the expression

$$V_{ij} = V_{ij}^{N} + V_{ij}^{e} + V_{ij}^{ext} \tag{1}$$

where V_{ij}^N and V_{ij}^e correspond to the contributions to the ij component of the EFG tensor from the nuclear charges ξ_N and the electrons in the cluster while V_{ij}^{ext} represents the contribution due to the point charges ξ_I for the ions external to the cluster. These contributions are given in general by

$$V_{ij}^{N} = \sum_N \xi_N\, e(3X_{iN}X_{jN} - r_N^2\delta_{ij})/r_N^5 \tag{2}$$

$$V_{ij}^{e} = -\sum_\mu \int e\psi_\mu^*\psi_\mu \, d\tau \,(3X_iX_j - r^2\delta_{ij})/r^5 \tag{3}$$

$$V_{ij}^{ext} = \sum_I \xi_I\, e(3X_{iI}X_{jI} - r_I^2\delta_{ij})/r_I^5 \tag{4}$$

The X_i in Eq. (3) represent the ith component of the radius vector **r** of an electron with respect to the nuclear site at which the EFG is calculated. Due to the r^{-3} dependence of V_{ij}, the main electronic contributions arise from the molecular orbital on the central atom and its nearest neighbors. By the inclusion of the nearest neighbors in the cluster, covalency and charge transfer effects on V_{ij} are included in a first-principles manner. Further, since this is an all-electron calculation, the core and valence electrons in the ions are all perturbed by the sources of EFG internal and external to the cluster and therefore Sternheimer antishielding effects are directly included without the need to introduce antishielding factors.

Since V_{ij} is a symmetric tensor, V_{ij} is diagonalizable with the usual convention[3] $V_{zz} \geq V_{yy} \geq V_{xx}$. In addition, since Laplace's equation holds here, only two components are independent[3], related to the experimental parameters e^2qQ and η respectively, eq being the largest EFG component V_{zz} and η the asymmetry parameter. The latter is defined as[3] $\eta = (V_{xx} - V_{yy})/V_{zz}$, and is a measure of the deviation of V_{ij} from cylindrical symmetry, taking the values $1 \geq \eta \geq 0$ with $\eta = 0$ corresponding to a cylindrically symmetric EFG tensor.

III. Results and Discussion.

Here we tabulate results for e^2qQ and η for several perfect ionic solids calculated[6-12] by our group utilizing the UHF cluster procedure. The systems investigated are the following: Al_2O_3 and α - Fe_2O_3 with the rhombohedral corundum structure[6,7]; hexagonal[8] Li_3N; cubic[9] Cu_2O; tetragonal[10] ZnF_2; wurtzite[11] ZnO; and $ZnAl_2O_4$ with the spinel structure[12]. Only lattice sites with point symmetry lower than cubic can have a nonzero EFG and only those sites with available[16-22] experimental data are considered. In Table I we give the systems, clusters, nuclear site of interest, and the calculated and experimental e^2qQ and η for those sites. All of the clusters contain at least the nuclear site of interest and the first coordination shell.

Table I: Comparison of Theoretical and Experimental Nuclear Quadrupole Coupling Constants in Ionic Crystals

			e^2qQ(MHz)		η	
System	Nucleus	Cluster	Theory	Exp.	Theory	Exp.
Al_2O_3	^{27}Al	$(AlO_6)^{9-}$	1.76	$\pm$2.393	0	0
Al_2O_3	^{17}O	$(Al_4O_5)^{2+}$	2.40	2.167	0.56	0.517
Fe_2O_3	^{57m}Fe	$(FeO_6)^{9-}$	2.71	10.2(2)	0	0
Li_3N	$^7Li(1)$	$(Li_{15}N_2)^{9+}$	0.90	$\pm$0.582(2)	0	0
Li_3N	$^7Li(2)$	$(Li_{10}N_3)^{1+}$	-0.47	$\pm$0.285(2)	0	0
Li_3N	^{14}N	$(Li_8N)^{5+}$	0.52	$\pm$0.505(2)	0	0
Cu_2O	^{63}Cu	$(Cu_7O_2)^{3+}$	61.7	$\pm$54.04	0	0
ZnF_2	^{67}Zn	$(ZnF_6)^{4-}$	8.42	7.57	0.17	0.24
ZnF_2	$^{19}F^*$	$(Zn_3F)^{5+}$	-14.2	$\pm$17.41	0.59	0.43
ZnO	^{67}Zn	$(ZnO_4)^{6-}$	2.50	2.401(4)	0	0
$ZnAl_2O_4$	^{27}Al	$(AlO_6)^{9-}$	-3.08	$\pm$3.68	0	0

In Table I, the $\pm$ refers to experimentally undetermined sign. The values of Q used in the calculation of e^2qQ are ^{27}Al(0.165), ^{67}Zn(0.150), $^{19}F^*$ (0.072), ^{63}Cu(-0.18), ^{57m}Fe(0.082), ^{17}O(-0.0257), ^7Li(-0.041) and ^{14}N(0.016) where the respective Q are given in parenthesis in units of barns. The references to the theoretical results are Refs. 6-12 and to the experimental results are Refs. 16-22.

The agreement between measured and calculated values for e^2qQ in Table I is generally satisfactory. The correct sign, where experimentally determined[16,17,18,20,21], is obtained in all systems. Excellent agreement is found for the ^{17}O site in Al_2O_3 and the ^{67}Zn site in ZnO between theory and experiment. The case of ^{17}O in Al_2O_3 is one of just a few systems[16,23] where data including the sign is available for the oxygen site. The only case where poor quantitative agreement is found[7] is the ^{57m}Fe site in Fe_2O_3. In that study, magnetic hyperfine properties and isomer shift were found to be in excellent agreement with experiment[17] as opposed to the poor agreement for e^2qQ. A minimal basis set optimized for charge -2 in a Watson sphere[24,25] was used for the oxygen nearest neighbors. As discussed below, it is possible that better agreement may be obtained through two sources of improvement, with an extensive rather than minimal basis set for the oxygen atoms and a cluster involving a larger number of ions. For the two inequivalent Li sites in Li_3N, while the absolute signs for e^2qQ are not known experimentally, it has been determined[18] that the relative signs are opposite in the two sites. This is in agreement with the results of our theoretical UHF cluster calculations[8].

In analysing the nature of agreement between theory and experiment, several points should be considered in a UHF cluster investigation for e^2qQ in ionic systems. These include the choice of cluster, the incorporation of the influence of the lattice and the choice of basis sets for the cluster atoms. First of all, the cluster should include at a minimum the central atom of interest and its first nearest neighbors. This is because it would allow for covalency effects and charge overlap effects within the cluster, and polarization between the central atom and its neighbors. With the addition of the Madelung potential due to the ions at the lattice sites outside the cluster approximated as point charges, not only is one able to simulate the infinite lattice, one is also able to, as mentioned earlier, include the Sternheimer antishielding effect in a first-principles manner. There is thus no need to include[26] Sternheimer antishielding factors[15,25] in the calculations.

Gaussian-type basis sets[2] have been used in all of the calculations. It has been found important to use extensive high-quality basis sets for the central cluster atom where e^2qQ is calculated. We have also found it important to use extensive basis sets on the nearest neighbors which gives a better description of covalency effects in these systems. Furthermore, it has been found important to test the convergence of the results as the flexibility of the basis set increases, for example, by uncontracting the most diffuse exponents or by adding diffuse and polarization functions. The anions are rather diffuse, especially N^{3-} and O^{2-} which are unstable in free space. To improve convergence we have optimized the neutral ligand basis sets by use of the Watson sphere model[6,7,24,25]. In this approach one re-optimizes the basis set in a sphere of radius equal to the ionic radius and charge opposite to the formal ionic charge. However, we have found recently with our studies[11,27] in ZnO and other systems that neutral basis sets can give results nearly identical to ionic basis sets as long as extensive basis sets are used in the calculations. This is because the Madelung potential due to the point charges stabilizes the anions just like the Watson sphere and the use of extensive basis sets gives sufficient flexibility to converge to the same e^2qQ as one gets from the use of ionic basis sets. While an ionic basis set more closely approximates the atom in the ionic solid, a neutral basis set which is extensive (for example of double ζ quality[28] with possibly diffuse and polarization functions added) is sufficiently flexible to allow convergence to the Haretree-Fock limit in self-consistent UHF calculation of cluster plus external point charges. However, if convergence is not achieved or a less extensive basis set has to be used due to computational constraints then ligand basis sets optimized in a Watson sphere for both cations and anions are suggested. This we feel could be a possible avenue of improvement for the case of ^{57m}Fe nuclear quadrupole interaction in Fe_2O_3 where the theoretical value[7] appears to be about a quarter of experiment[17].

Another important question that should be considered for assessing the accuracy of the theoretical results for the nuclear quadrupole coupling constants is that of the convergence with repsect to cluster sizes. For the monovalent cation systems, for instance, in the case of ^{63}Cu in Cu_2O, the calculation has been carried out[9] using both $(CuO_2)^{3-}$ cluster involving only the nearest O^{2-} neighbors of the central Cu^{+} ion and the larger cluster shown in Table I involving the other Cu^{+} neighbors of the O^{2-} ions. The convergence was found to be very good, better than 5 per cent. A similar convergence has been found for Cu^{2+} ion[29] in the high-T_c system $YBa_2Cu_3O_7$. On the other hand, for the trivalent ion systems Al^{3+} in Al_2O_3 the convergence is not as good[6]. The likely reason for this is that the trivalent ion system has a tendency towards more covalency and hence a larger migration of charge from the negative ion ligands, a feature that is expected to be more prominent when the ligand is a diffuse negative ion like the doubly charged oxygen negative ion. In such a situation, there is a need for using larger clusters, including other positive ion neighbors for the negative ions besides the central positive ion. The situation is expected to be further aggravated when the trivalent ion involves an incomplete d-shell with more diffuse valence ions than the p-shell in Al^{3+}. Thus, the need for including more neighbors, especially other Fe^{3+} neighbors for the O^{2-} ions is expected to be more acute for Fe_2O_3, where there is very significant difference between theory[7] and experiment[17], and should be implemented in future to study if the agreement with experiment improves for the ^{57m}Fe quadrupole coupling constant without affecting the good agreement already found[7] for primarily isotropic properties such as the magnetic hyperfine field and isomer shift.

IV. Conclusion.

In summary, the Unrestricted Hartree-Fock Cluster procedure has been shown to be able to successfully explain the nuclear quadrupole interaction properties of a number of ionic crystals and thus demonstrated the capability of this procedure to explain the anisotropy of the electron distribution around the nucleus. A notable exception[7] is the case of Fe_2O_3, where more investigations are needed to examine the convergence with respect to basis set and size of cluster used, especially the latter. Additionally, this procedure has also been shown to be able to successfully explain[7] Mössbauer isomer shift data which involves the isotropic part of the charge density at the nucleus and the contact and dipolar hyperfine fields which involve the isotropic and anisotropic components of the unpaired spin density at the nucleus. These features suggest that the cluster procedure is capable of quantitatively explaining the major features of the electron distribution in ionic crystals, especially near the nucleus.

References.

1 N.Sahoo, S.K.Mishra, K.C.Mishra, A.Coker, C.K.Mitra, L.C.Snyder, A.Glodeanu and T.P.Das, Phys.Rev.Lett.50, 913 (1983); N.Sahoo, S.B.Sulaiman, K.C.Mishra and T.P.Das, Phys.Rev.B39, 13389 (1989); J.M.Vail, J.Phys.Chem.Solids, 51, 589 (1990); K.Jackson, M.R.Pederson and B.M.Klein, Phys.Rev.B43, 2364 (1990); J.Sauer, Chem.Rev.89, 199 (1989);A.B.Kunz and B.M.Klein, Phys.Rev.B17, 4614 (1978); N.W.Winter, R.M.Pitzer and D.K.Temple, J.Chem.Phys.87, 2945 (1987); G.L.Bendazzoli and O.Donzelli, J.Phys.Condens.Matter 1, 8227 (1989).

2 M. Seel and P.S. Bagus, Phys. Rev. B28, 2023 (1983); I.P. Batra, P.S. Bagus and K. Hermann, Phys. Rev. Lett. 52, 384 (1984); J.C. Malvido and J.L. Whitten, Phys. Rev. B26, 4458 (1982); S.M. Mohapatra, B.N. Dev, L. Luo, T. Thundat, W.M. Gibson, K.C. Mishra, N. Sahoo and T.P. Das, Rev. Solid State Sci. 4, 873 (1990).

3 T.P.Das and E.L.Hahn, Nuclear Quadrupole Resonance Spectroscopy, Academic Press, New York, (1957).

4 T.P. Das, Relativistic Quantum Theory of Electrons, Harper and Row, New York (1973).

5 N.F. Ramsey, Molecular Beams, Oxford University Press, New York, (1956).

6 P.C. Kelires, Ph.D. Thesis, State University of New York at Albany (1987) (Unpublished); P.C.Kelires, S.B.Sulaiman, N.Sahoo and T.P.Das, Abstract at Amer.Phys.Soc. March Meeting, Indianapolis (1992).

7 P.C. Kelires and T.P. Das, Hyp. Int. 34, 285 (1987).

8 P.C. Kelires, K.C. Mishra and T.P. Das, Hyp. Int. 34, 289 (1987).

9 J.Stein, N.Sahoo, S.B.Sulaiman and T.P.Das, Hyp.Int.60, 849(1990).

10 H.H. Klauss, N. Sahoo, P.C. Kelires, T.P. Das, W. Potzel, G.M. Kalvius, M. Frank and W. Kreische, Hyp. Int. 60, 853 (1990).

11 D.W. Mitchell, S.B. Sulaiman, N. Sahoo, T.P. Das, W. Potzel and G.M. Kalvius, Phys. Rev. B44, 6728 (1991).

12 D.W.Mitchell, S.B.Sulaiman, N.Sahoo, T.P.Das, W.Potzel, G.M.Kalvius, W. Schiessl, H.Karzel and M.Steiner, Abstract at Amer. Phys. Soc. March Meeting, Indianapolis (1992).

13 J.Reitz in Solid State Physics vol.1, F.Seitz and D.Turnbull eds., Academic Press, New York (1955).

14 J.Q. Broughton and P.S. Bagus, Phys. Rev. B30, 4761 (1984).

15 R.M. Sternheimer, Z. Naturforsch, 41a, 24 (1986).

16 R.V. Pound, Phys. Rev. 79, 685 (1980).

17 J.O. Artman, A.H. Muir and H. Wiedersich, Phys.Rev. 173, 337 (1968).

18 K. Differt and R. Messer, J. Phys. C13, 717 (1980).

19 H.L. Cox and D. Williams, J. Chem. Phys. 32, 633 (1960).

20 W. Potzel and G.M. Kalvius, Phys. Lett. 110A, 165 (1985); H. Barfuss et. al., J. Mol. Struct. 58, 503 (1980).

21 C. Schäfer, W. Potzel, W. Adlassnig, P. Pöttig, E. Ikonen and G.M. Kalvius, Phys. Rev. B37, 7247 (1988).

22 P. Ruegsegger, F. Waldner and E. Brun, Helv.Phys.Acta 43, 799 (1970).

23 Ch. Gabathuler, E.E. Hundt and E. Brun, XVII Congress Ampere, 1973, ed.V.Nove, North Holland Pub.Co., Amsterdam, Netherlands, page 499.

24 R.E. Watson, Phys. Rev. 111, 1108 (1958).

25 E. Paschalis and Al. Weiss, Theor.Chim.Acta 13, 189 (1967); P.C. Schmidt, K.D. Sen, T.P. Das, and Al. Weiss, Phys.Rev.B22, 4167 (1980).

26 B.G. Dick and T.P. Das, Phys. Rev. 127, 1053 (1962); R.R. Sharma and T.P. Das, J.Chem.Phys. 41, 3581 (1964); R.J. Quigley and T.P. Das, Phys. Rev. 164, 1185 (1967).

27 T.P. Das, D.W. Mitchell,S.B. Sulaiman, N. Sahoo, W. Potzel, G.M. Kalvius, H. Karzel, W. Schiessl and M. Steiner, Abstract at Amer. Phys. Soc. March Meeting ,Indianapolis (1992).

28 E.R. Davidson and D. Feller, Chem. Rev. 86, 681 (1986).

29 N. Sahoo, S. Markert, K. Nagamine and T.P.Das, Phys Rev.B41, 220 (1990).

THE ELECTRONIC STRUCTURE OF ICOSAHEDRAL BORON CLUSTERS

A. C. SWITENDICK
Solid State Theory, 1151
Sandia National Laboratories
Albuquerque, New Mexico 87185
U. S. A.

ABSTRACT. The one-electron energy levels of icosahedral boron clusters have been calculated as a function of intericosahedral spacing maintaining the intraicosahedral spacing of α-boron. For crystalline lattice constants greater than 1.25 times the equilibrium one, band overlap occurs with concomitant metallic behavior. At smaller lattice constants, orbitals(bands) associated with bonds to adjacent icosahedra are lowered and orbitals(bands) associated with "antibonds" are raised. Four bands which were three quarters full become empty, while three bands which were empty become filled. This leads to an energy gap between the filled states and the empty states which accounts for the experimentally observed insulating behavior of this elemental material with three valence electrons per atom.

1. Introduction

Many molecules and crystals contain clusters of boron atoms [1]. Molecular examples are hexaborane B_6H_{10}, and decaborane, $B_{10}H_{14}$. In solid state chemistry the metal hexaborides, MB_6, cubic dodecaborides, MB_{12}, and rhombohedral $B_{12}P_2$ are well known examples containing octahedra, cubo-octahedra and icosahedra, respectively. Carbon substitutions for boron often occur as in the carboranes, $C_2B_{10}H_{12}$ and boron carbide, $B_{12}C_3$. Icosahedral structures have been studied by many authors [2-13]. Isolated icosahedra [2-6], icosahedra in combination with other elements: either as real materials [10], or as simulations of solid state enviroments [6], as well as calculations of crystalline α- rhombohedral boron [8-13] have all been published.

In this paper I study the transition between the energy levels of "isolated" boron icosahedra and their crystalline counterparts in α-boron, B_{12}. I do this using the same procedure for both the separated icosahedra and the solid. I have used the linearized augmented plane wave (LAPW) method [14] with the local density approximation of Hedin and Lundqvist [15] for the exchange- correlation potential. The calculations were iterated to self-consistency. Although the muffin tin approximation[16] was used for the crystalline potential, charge density

P. Jena et al. (eds.), Physics and Chemistry of Finite Systems: From Clusters to Crystals, Vol. I, 683–690.

plots show that the wave function and charge density solutions are not restricted to this form. Calculations of the valence charge density for α-boron showed very good agreement with experimental measurements [17]. Full potential (FLAPW) [18] calculations are probably needed to get accurate total energies; although an evaluation using final total charge density and related potential may be adequate [19]. The concern of this paper is the nature of the interaction between the atoms of the icosahedron and between the icosahedra.

The crystal structure parameters of Morosin et al. [20] were used. An icosahedral cluster is shown in Fig. 1. Table I gives the relevant bond distances for the icosahedra and between the icosahedra. To simulate isolated icosahedra, the crystalline lattice parameters were expanded by factors of 1.1, 1.25 and 1.4, keeping the intraicosahedral distances fixed. The unit cell contains six polar borons, BP, forming

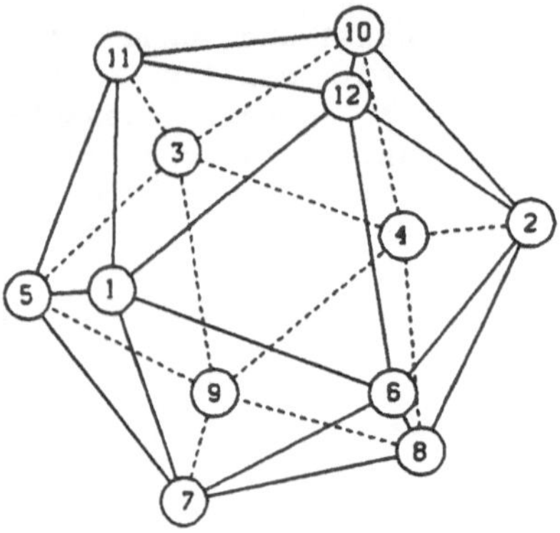

Figure 1. Boron cluster

TABLE 1. Bonds and their length

bond	figure	name	dist (Å)
intraicosahedral			
BP-BP	10-11,12	polar	1.75
BP-BE	10-4	short	1.80
BP-BE	10-2,3	long	1.81
BE-BE	1-5,6	equat.	1.78
intericosahedral(α-boron)			
BP-BP'	10-7'	sigma	1.67
BE-BE'	1-2',3'	delta	2.02

key: P-polar, E-equatorial, '-other icosahedra

triangular planes at the top and bottom of the icosahedra(atoms 7-12). In α-boron each of these polar atoms is bonded to two other polar atoms in its plane and to a polar atom in the icosahedron above or below it, e.g. 10 to 11 and 12 and to 7, respectively. Each polar boron is also bonded to three of the six equatorial borons, BE, atoms(1-6); which form a puckered hexagon(three in a plane above the equator and three in a plane below) around the middle of the icosahedra(10 to 2, 3 and 4). Each of these equatorial atoms is bonded to three polar atoms(1 to 7, 11 and 12) and two other equatorial atoms (1 to 5 and 6) and two somewhat more distant borons in neighboring icosahedra(1 to 2 and 3) at the same z-level. Thus each polar(equatorial) icosahedral atom forms five bonds with other atoms of the icosahedra and one(two) external bond(s). In Table 1 the exterior sigma polar bond is the shortest while the exterior equatorial three centered(1,2,3) or delta bond is significantly longer than the others.

2. Symmetry analysis

Because of translational symmetry, the icosahedron has lost its fivefold rotational axes and the icosahedra is no longer regular; the polar atoms and the equatorial atoms are distinct. The space group is $R\bar{3}m(D^5_{3d})$. The point group contains twelve operations with six irreducible representations. Since symmetry plays an important role in my analysis and results, I shall spend some time explaining the allowed symmetries and the implications on the interactions between icosahedra. Due to the lowered symmetry, the icosahedron of Figure 1 can be decomposed into two figures consisting of staggered triangular prisms with a common apex: one prism consisting of the equatorial atoms as shown in Figure 2 and one prism consisting of the polar atoms shown in Figure 3.

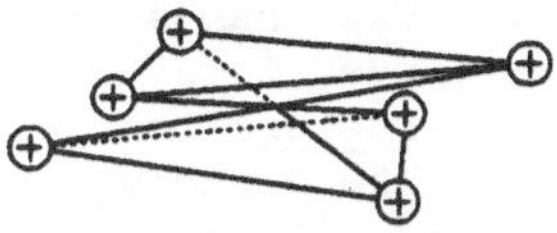

Figure 2. Equatorial prisms, Γ1 symmetry.

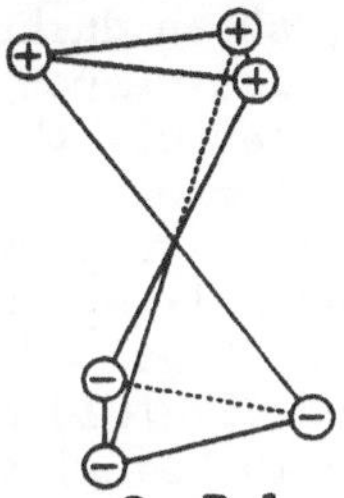

Figure 3. Polar prisms, Γ2' symmetry.

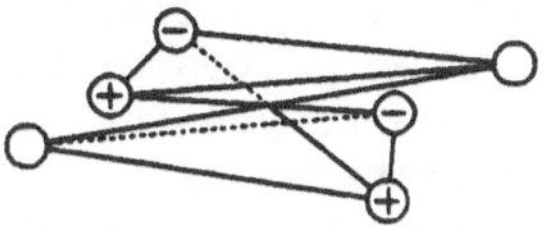

Figure 4. Γ3' symmetry.

Although the equatorial prism in Figure 2 has a short altitude and large base, and the polar prism in Figure 3 has a long altitude and smaller base, both figures display the full point group symmetry. These relative distances make interactions between the bases more important in the equatorial case than in the polar case. Fiqure 2 also shows a totally symmetric state of A_{1g}(Γ1) symmetry, where the roman designation is that of the chemists [21] and the greek is that of the physicists [22]. The overall symmetry is S-like. In Figure 3 the atoms in the lower prism have the opposite sign and display a state of A_{2u}(Γ2') symmetry: Z-like with one nodal plane. Figure 4 shows one of the doubly degenerate partners of E_u(Γ3') symmetry: here X-like. Figure 5 a partner of E_g(Γ3) symmetry: this has XZ-symmetry with two nodal planes. For the six s, p_z and radial orbitals on either the polar or equatorial atoms these constitute the only symmetries which can be formed giving one basis function of each type(partner). States of A_{1u}(Γ1') and A_{2u}(Γ2) require p-orbitals in the basal planes-tangential to the triangles. These are shown in Figures 6 and 7.

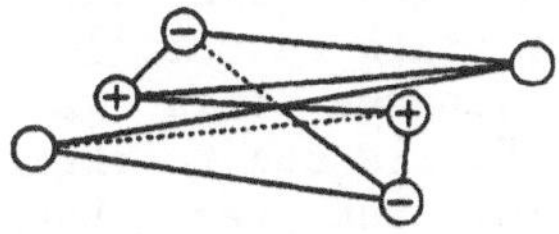

Figure 5. Γ3 symmetry.

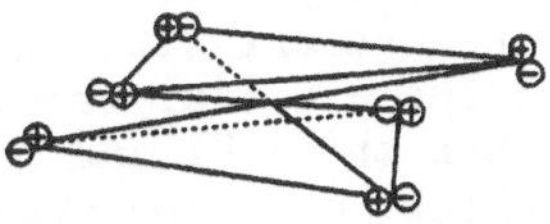

Figure 6. Γ1' symmetry.

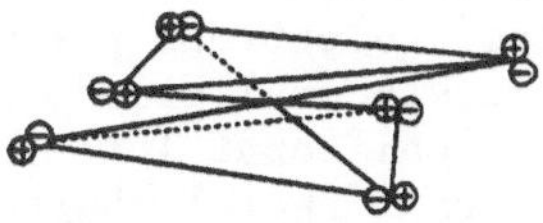

Figure 7. Γ2 symmetry.

3. Results

Figure 8 shows the results of the calculations for expansion factors of 1.4(left panel) and 1.0, α-boron(right panel). The energy bands are plotted using a symbol which shows the orbital character of the largest component in the wavefunction. For a circumscribed sphere to the atoms, we have radial pointing in, radial pointing out, tangential to a longitude, (tang. Z), and tangential to a latitude(tang. XY) for both the polar and equatorial atoms. This may be the dominant character or it may be only the largest marginally. Often this character changes with the k-vector(abscissa). The symmetry of the state at gamma, the charge character and plots of the orbital charge density in various planes allows one to make the connections between the panels. The zero of energy is the Fermi energy, E_F, shown dashed in both panels.

On the left the bands are narrow reflecting little interaction between the different icosahedra of the "solid". The bands form separated groups of bands. These groups contain one, three, five, seven bands corresponding to the overall symmetry and nodal structure of the wavefunction of S, P, D and F-like, respectively. The ordering and grouping of the levels follows that of many authors [2-13]. The splitting of the threefold and fivefold degenerate states reflects the lowering from icosahedral to rhombohedral symmetry; it is small. Only 29 levels are labeled. The lowest 25 represent the 13 internal and 12 external orbitals of Longuet-Higgins [2]. The icosahedral symmetry correspondences (S_g, T_{1u}, V_g, U_u, T_{2u}, U_g) are shown to the left of the actual rhombohedral ones in the left panel. Above the 25th band there is a gap, four bands, then a continuous manifold of bands. Because of band overlap, there is no gap at the Fermi energy and one would expect metallic behavior. Even at infinite separation, in both the icosahedral and rhombohedral case, the states at the Fermi level are degenerate, incompletely filled and one does not have a closed shell case with a gap to the first empty state. Considering only gamma and 36 electrons in the unit cell, the Fermi level falls in the middle of the fourth Γ3' which is the 18th and 19th state. The highest occupied and the lowest unoccupied state is the same. One might expect these orbitals to be involved in the bonding between the icosahedra. These states are largely of polar radial-out character and thus point in the right direction to neighboring icosahedra.

On the right the energy bands are considerably broadened reflecting the interactions between the icosahedra. In the lower part of the figure, one can still maintain correspondences between the left and the right panel. The major difference is the formation of a gap at E_F. There are 18 bands below this gap which accounts for the experimentally observed insulating behavior of this elemental material with three valence electrons per atom in contrast with aluminum which is metallic. The icosahedral structure and the external bonding imposes this gap. The significant lowering of the third Γ1 and second Γ2' is due to their equatorial radial character which overlaps that from the two other equatorial atoms in different icosahedra at the same z-level as shown in Figure 9. The three above(solid) and the three(dashed) below each have the same phase: all plus for Γ1 and plus above and minus below for Γ2'.

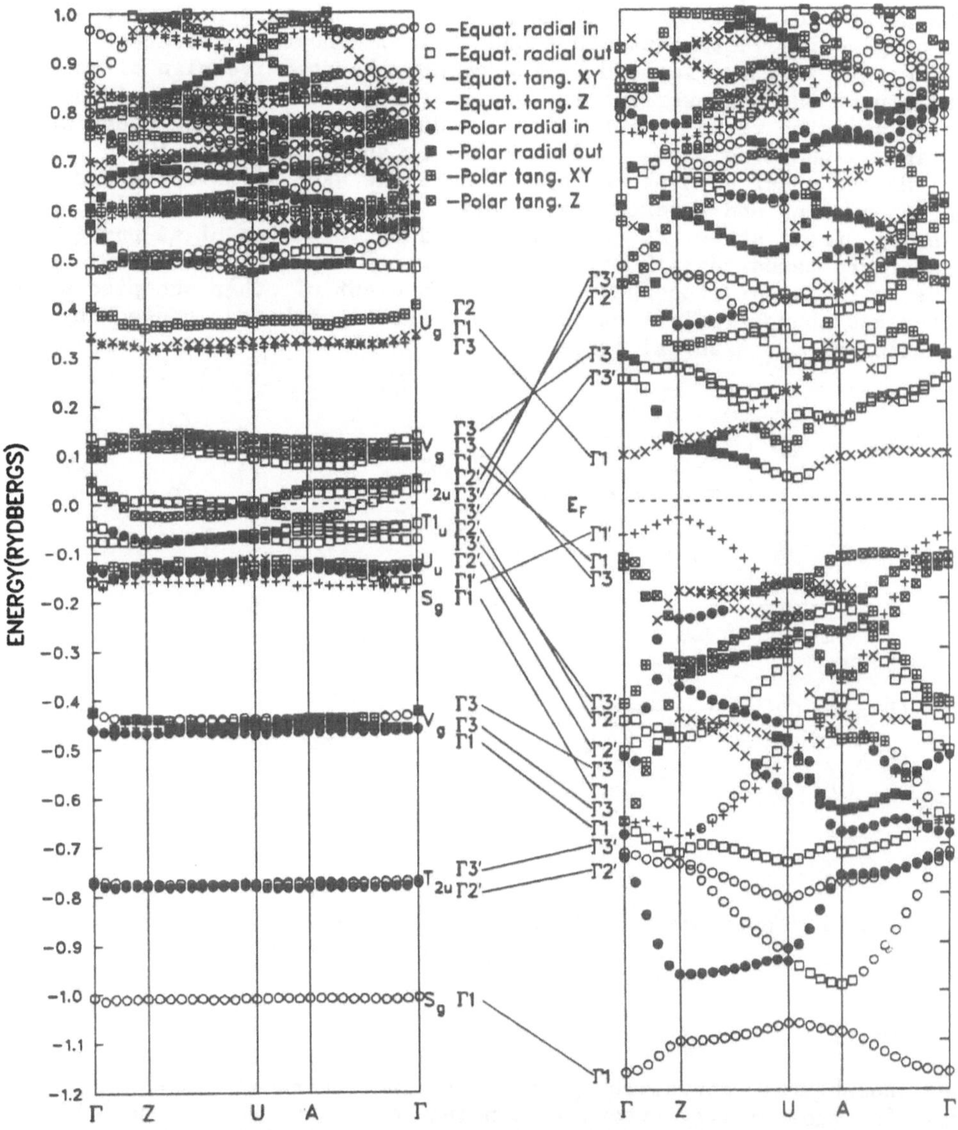

Figure 8. Bands for expanded alpha boron. Left panel ex.=1.4, right panel ex.=1.0.

The nodal surface of Γ2' does not fall between interacting atoms since they are all at the same z-level on different icosahedra and all have the same phase. These two bands hold four electrons and are largely localized in the exterior region between the equatorial atoms. This is the origin of the external three centered triangular or delta bond(two of which are shown dotted) containing four electrons per icosahedra.

The two $\Gamma 3'$ states and the $\Gamma 2'$ state at or above the Fermi energy have polar radial out or polar tang. Z character and hence point towards the neighboring icosahedra above and below. However because of the inversion symmetry implied by the prime, the positive sign of the wave function at the top overlaps the negative sign at the bottom of the one above and vice versa maintaining a node(antibond) causing these states to rise as they begin to overlap. This is shown in Figure 10.

On the other hand the $\Gamma 3$ and $\Gamma 1$ have the same signs above and below and can overlap and interact favorably. This leads to their energy lowering from above E_F to below and the creation of an energy gap. These three bands contain six electrons which form six external polar bonds with other icosahedra. The movement of other occupied states represents charge redistribution within the icosahedra caused by the formation of the external bonds.

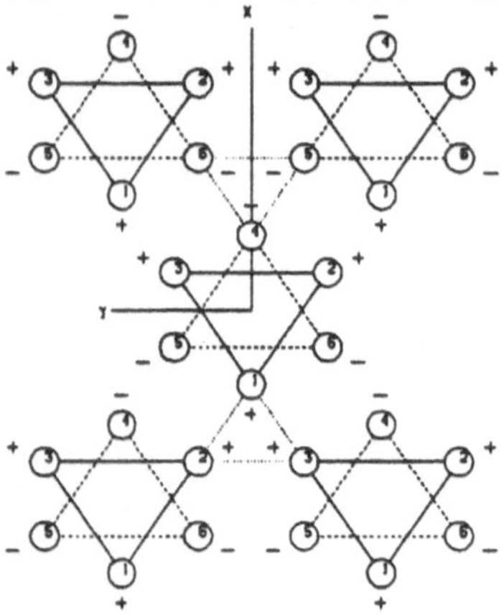

Figure 9. $\Gamma 2'$ equatorial delta bond.

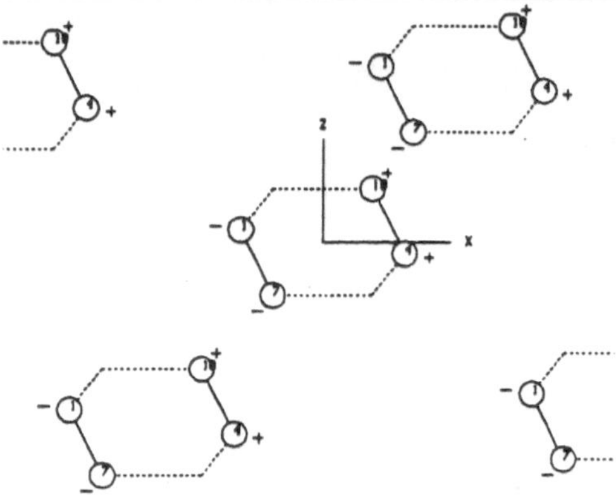

Figure 10. $\Gamma 3'$ polar sigma antibonding state.

4. Summary

The icosahedral structure gives rise to thirteen internal bonding orbitals and twelve external bonding orbitals [2]. The thirteen internal orbitals are filled in <u>both</u> the isolated and interacting regimes. The highest occupied orbital(HOMO) and the lowest unocuppied oribital(LUMO) are degenerate in the case of the isolated icosahedra and do not lead to a closed shell(gap). These orbitals and their neighbors in energy are external in nature but because of their symmetry do not bond with neighboring icosahedra and become empty in α-boron. External polar orbitals which do have the ability to form bonds with neighboring icosahedra form strong bonds with a consequent lowering below E_F and filling with electrons. The emptying of five states containing six electrons and the filling of three bands with six electrons allows the formation of a gap which is observed experimentally for α-boron. The location of the charge in the orbitals and its symmetry(relative phases) determines whether favorable(unfavorable) interactions, bonds(antibonds) will form with energy lowering(raising) in this system.

5. Acknowledgements

I wish to acknowledge helpful discussions with T. A. Green. This work was supported by the Division of Materials Sciences, Office of Basic Energy Research, U. S. Department of Energy contract DE-AC-04-DP00789.

6. References

1. Lipscomb, William N. (1981) 'Borides and boranes', J. Less Common Metals 82, 1-20

2. Longuett-Higgins, H. C. and Roberts, M. de V. (1955) 'The electronic structure of an icosahedron of boron atoms', Proc. Roy. Soc. A230, 110-9.

3. Bambakidis, G. and Wagner, R. P. (1981) 'Electronic structure and binding energy of the icosahedral boron cluster B_{12}', J. Phys. Chem. Solids 42, 1023-25.

4. Solov'ev, N. E., Averbakh, E. M. and Ugai, Ya. A. (1984) 'Electronic energy levels of the icosahedron in rhombohedral α-boron', Sov. Phys. Solid State 26, 114-6.

5. van Schilfgaarde, M. and Harrison, W. A. (1985) 'Electronic structure of boron', J. Phys. Chem. Solids 46, 1093-1100.

6. Howard, I. A., Beckel, C. L. and Emin, D. (1987) 'Bipolarons in boron icosahedra', Phys. Rev. 35, 2929-33.

7. Green, T. A., Switendick, A. C. and Emin, D. (1988) 'Ab initio self consistent field calculations of borane icosahedra with zero, one or two substituted carbon atoms', J. Chem. Phys. 89, 6815-22.

8. Perrot, F. (1981) 'First approach to the band structure of α-rhombohedral boron' , Phys. Rev. 23, 2004-10.

9. Schöttke, H. (1983) 'The electronic band structure of α-rhombohedral boron', J. Less Common Metals 91, 159-165.

10. Armstrong, D. R., Bolland, J. and Perkins, P. G. (1984) 'The electronic structure of α-B_{12}, $B_{12}P_2$, and $Ba_{12}As_2$', Theoret. Chim. Acta(Berlin) 64, 501-514.

11. Bullett, D. W. (1983) 'Electronic structure studies of boron and boron-rich borides', in D. Emin, T. Aselage, C. L. Beckel, I. A.

Howard and C. Wood (eds.), Boron Rich Solids, American Institute of Physics Conference Proceedings 140, New York, pp-249-59.

12. Lee, S., Bylander, D. M. and Kleinman, L. (1990) 'Bands and bonds of B_{12}', Phys. Rev. B42, 1316-20.

13. Li, D., Xu, Y-N. and Ching, W. Y. (1991) 'Electronic structures, total energies and optical properties of α-rhombohedral B_{12} and α-tetragonal B_{50} crystals', preprint.

14. Koelling, D. D., and Arbman, G. O. (1975) 'Use of energy derivative of the radial solution in an augmented plane wave method: application to copper', J. Phys. F:Metal Phys. 5, 2041-2054.

15. Hedin, L. and Lundqvist, B. I. (1971) 'Explicit local exchange-correlation potentials', J. Phys. C: Solid State 4, 2064-2083.

16. Mattheiss, L. F., Wood, J. H. and Switendick, A. C. (1968) 'A procedure for calculating electronic energy bands using symmetrized augmented plane waves', in B. Alder, S. Fernbach and M. Rotenberg (eds.), Methods in Computational Physics, Volume 8, Academic Press, New York, pp. 63-147.

17. Switendick, A. C. and Morosin, B. (1991) 'Electronic charge density and bonding in α-boron: An experimental-theoretical comparison' in D. Emin, T. L. Aselage, A. C. Switendick, B. Morosin and C. L. Beckel (eds.), Boron Rich Solids, American Institute of Physics Conference Proceedings 231, New York pp. 205-211.

18. Jansen, H. J. F. and Freeman, A. J. (1984) 'Total-energy full-potential linearized augmented-plane-wave method for bulk solids: electronic and structural properties of tungsten', Phys. Rev. B 30, 561-9.

19. Weyrich, K. H., Brey, L. and Christensen, N. E. (1988) 'Full-potential linear-muffin-tin-orbital calculation of phonon frequencies in semiconductors', Phys. Rev. B 38, 1392-6.

20. Morosin, B. Mullendore, A. W., Emin, D. and Slack, G. A. (1986) 'Rhombohedral crystal structure of compounds containing boron-rich icosahedra', in D. Emin, T. Aselage, C. L. Beckel, I. A. Howard and C. Wood (eds.), Boron Rich Solids, American Institute of Physics Conference Proceedings 140, New York, pp. 249-59.

21. Cotton, F. A. (1971) Chemical Applications of Group Theory, Wiley-Interscience, New York.

22. Bouckaert, L. P., Smoluchowski, R., Wigner , E. (1936) 'Theory of Brillouin zones and symmetry properties of wave functions in crystals', Phys. Rev. 50, 58-67.

INVESTIGATION BY THE GUTZWILLER APPROACH OF THE ^3He AND ^4He LIQUID CLUSTERS

R.J. TARENTO[1], P. JOYES[2], J. VAN DE WALLE[2]

[1] CNRS Laboratoire de Physique des Matériaux, 1 Place Aristide Briand, 92195 Meudon Cédex, France.

[2] Laboratoire de Physique des Solides,Bât. 510, Université Paris-Sud, 91405 Orsay Cédex (France)

ABSTRACT

The size stability of ^3He and ^4He clusters has been investigated by using the Hubbard Hamiltonian. The correlations have been treated for ^3He_n with the classical Gutzwiller method, and for ^4He_n we have proposed an extension of the Gutzwiller method to the Boson systems.

1. Introduction

Theoretical descriptions of He_n aggregates can be divided into three classes. Firstly a purely quantum mechanical technique where the solution of a n-body hamiltonian is searched by a variational Monte-Carlo calculation [1]. Secondly, a mean field method which involves simpler calculations where the energy of the helium cluster is expressed as a functional of $\rho(r)$ ane $\tau(r)$, respectively, the density and the kinetic energy density inside the cluster [2,3]. One of the main results of these studies is the existence of a critical size N_c($N_c \sim$ 20-40 [4] above which ^3He_n aggregates begin to be stable while ^4He_n aggregates are stable for any size.

A third kind of method, based on the Hubbard hamiltonian can be developed for studying He_n clusters. This technique has already been applied with success to bulk [5,6]. In the Vollhardt's paper [6] the Gutzwiller technique is used for solving the Hubbard hamiltonian and the dependence with pressure of various response functions (suceptibility, compressibility) are derived. In part II, devoted to ^3He_n aggregates, we will follow the main lines of the Vollhardt procedure, in part III we study ^4He_n aggregates and extend the Gutzwiller technique to the case of bosons.

P. Jena et al. (eds.), Physics and Chemistry of Finite Systems: From Clusters to Crystals, Vol. I, 691–696.

2. ^3He_n aggregate.

2.1 THE CLASSICAL GUTZWILLER METHOD FOR FERMIONS [6].

For a strongly correlated Fermi liquid like ^3He, below the degeneracy temperature, the system is governed by two features : the fermion character of the atoms and the hard-core repulsion between the atoms. Such a system could be described by the classical Hubbard hamiltonian for a lattice :

$$H = \sum_{i,j,\sigma} \beta_{ij}\, c^+_{i\sigma}\, c_{j\sigma} + U \sum_i n_{i\uparrow} n_{i\downarrow} \tag{1}$$

β_{ij} is the kinetic energy due to the hopping between nearest neighbour sites i and j : in the following it will be assumed a constant β. U is the correlation energy, $c^+_{i\sigma}$ ($c_{i\sigma}$) are the creation (annihalation) operators for fermions at the site i with spin σ, $n_{\uparrow(\downarrow)} = c^+_{\uparrow(\downarrow)}\, c_{\uparrow(\downarrow)}$ is the density operator for particle of spin $\uparrow(\downarrow)$. The resolution of the model Hubbard hamiltonian is unknown, owing to the correlation term. By applying a variational procedure (the Gutzwiller method) an approximative, but accurate solution can be obtained. Let us summarize the ideas of this method. Let n, $N\uparrow$, $N\downarrow$ and D be the number of lattice points, up spins, down spins, and doubly occupied sites respectively and $n\uparrow = \frac{N\uparrow}{n}$, $n\downarrow = \frac{N\uparrow}{n}$ and $d = \frac{D}{n}$. The state of the correlated system $|\psi>$ is derived from the state $|\psi_0>$ of the uncorrelated system (U = 0) by reducing the number of doubly occupied sites. So

$$|\psi> = \prod_i (1 - (1-d)\, n_{i\uparrow} n_{i\downarrow} > |\psi_0> \tag{2}$$

d is a variational parameter ($0 \leq d \leq 1$). The energy is

$$<\bar{E}> = \frac{<\psi|H|\psi>}{<\psi|\psi>} = q_\uparrow(d,n_\uparrow,n_\downarrow)\bar{\epsilon}_\uparrow + q_\downarrow(d,n_\uparrow,n_\downarrow)\bar{\epsilon}_\downarrow + Ud \tag{3}$$

with

$$q_\sigma = \frac{\{[(n_\sigma - d)(1-n_\sigma - n_{-\sigma} + d)]^{1/2} + [n_{-\sigma} - d)d]^{1/2}\}^{1/2}}{n_\sigma(1-n_\sigma)} \tag{4}$$

and $\bar{\epsilon}_\sigma = \frac{1}{n} \sum_{|k|<k_{F\sigma}}$ are calculated in the uncorrelated case. The ground state is obtained by minimizing (3) with respect to d.

2.2 STABILITY OF ^3He_n AGGREGATE.

In the present contribution, we restrict our calculation to the spinless case (i.e. : $N_\uparrow = N_\downarrow$). Therefore $\bar{\epsilon}_\uparrow = \bar{\epsilon}_\downarrow$ and let $\bar{\epsilon}$ be the one-electron energy per atom

$\overline{\epsilon} = \overline{\epsilon}_{\uparrow} + \overline{\epsilon}_{\downarrow}$. So the ground state energy is given by

$$\overline{E} = \overline{\epsilon}\left(1 - \frac{U}{8|\overline{\epsilon}|}\right)^2 \qquad (6)$$

The one electron energy $\overline{\epsilon}$ of liquid ^3He cluster has been derived from a classical procedure based on the modification of the discrete electronic spectra of a solid cluster by broadening the line spectra by a convolution with a Gaussian. Let $\{E_i\}$ be the set of discrete energies therefore the modified density of state for the liquids is

$$\rho(E) = \sum_i \frac{1}{\sqrt{2\pi}\ \beta} e^{-(E-E_i)^2/2\beta^2} \qquad (7)$$

The subscript i runs over the states and the width of the Gaussian is taken to be β. The energies E_i are the ones of a solid cluster compactly stacked. The ^3He_n (n = 38, 30, 20, 14) have been studied. The geometrical structure has been derived from the truncated dodecahedron. Our β parameter is $\beta = -10^{-4}$ eV [5]. The value of U is fixed in such a way that we obtain the same energy as in the Stingari-Treiner calculation [4] for $^3He_{38}$. We obtain $U/|\beta| = 8$. The present results are given in table 1. It can be seen that the stability of $^3He_{22}$ is very low. We have obtained a positive energy for $^3He_{14}$ (non stable) and by interpolation we have derived the critical N_C value at about $N_C \sim 21$. This result is of the same order of magnitude as in other studies.

n	$10^5\ \overline{E}$
38	- 1.6
30	- 1.87 10^{-1}
22	- 6.1 10^{-3}

Table 1. Energy per atom, $\overline{E}$, (in eV) of ^3He_n aggregates in the non magnetic state.

3. ^4He_n aggregate : Gutzwiller method extension to bosons.

A liquid ^4He cluster is controlled by the boson character and the hard-core repulsion between the atoms. Such a system is yet described by the classical Hubbard hamiltonian (eq. 1). But now $c^+_{i\sigma}(c_{i\sigma})$ are the creation (annihilation) operators for bosons at the site i with spin σ. The solution of eq.(1) for Boson is unknown. We propose here a derivation of the Gutzwiller method for Boson, our method is easily applicable only for cluster (i.e. for the bulk case we are confronted to a counting problem as it will be seen below). The Boson problem is analogous to the fermion case with a band degeneracy different from a s band. We have proposed an extension of the Gutzwiller case for such Fermion systems for Pr [7] and Al [8] cluster. Let us summarize the method. The method is based on the modification of the wave function Φ_0 of the system for U = 0, by introducing a variational parameter $\eta(\eta \leq 1)$ so that the component of Φ_0 that are not singly occupied are defavorized. Therefore in the Gutzwiller framework, the energy is the minimum of $E(\eta)$

$$E(\eta) = \tilde{q}(\eta)\, E^H + \tilde{\chi}(\eta)\, U \quad (8)$$

E^H is the Hückel energy (i.e. E^H is the energy of the system without the correlation term). The quantities $\tilde{q}(\eta)$ and $\tilde{\chi}(\eta)$ are written classically as

$$\tilde{q}(\eta) = q(\eta)/N(\eta)$$
$$\tilde{\chi}(\eta) = \chi(\eta)/N(\eta) \quad (9)$$

where $N(\eta)/q(\eta)$ and $\chi(\eta)$ have the following physical meanings : $N(\eta)$ the normalization of the variational wave function ; $q(\eta)$ the modification of the hopping term and $\chi(\eta)$ the modification of the correlation terms. Let us consider a system made up of N atoms. Let S be the set of configurations obtained by distributing N atoms on N sites. For each configuration we count the number α of atom pairs on the same site, then we write $N(\eta) = \sum_{S} \eta^{2\alpha}$ (10)

which can be recast into $N(\eta) = \sum_{\alpha \in G} N_1(\alpha)\, \eta^{2\alpha}$ (11)

where G is the set of α values. The $q(\eta)$ term is obtained by $q(\eta) = \sum_{S,S'} \eta^{\alpha+\alpha'}$ where the summation is made on all the couples of configurations, which differ from one another by the hopping of one boson. Moreover we have

$$\chi(\eta) = \sum_{S} \alpha\, \eta^{2\alpha} = \sum_{\alpha \in G} \alpha\, N_1(\alpha)\, \eta^{2\alpha} \quad (12)$$

The function $q(\eta)$ and $N(\eta)$ specific to ^4He_n (n = 2, 3, 4, 5, 6, 8) are reported in table 2. The set of parameters (β and U) have been taken the same as for the ^3He case. From the ^4He_n energy calculation (table 3) we conclude that the ^4He_n are stable for any n, at T = O K. The second remark is that, the formation energy being small (smaller than in the other calculations [1 - 4]) we expect that ^4He_n aggregates will be stable at very low temperature. This is confirmed by Gspann experiments [9].

4. Conclusion

The calculations on He_n cluster with the Gutzwiller method have confirmed the trend obtained with other methods : i.e. a critical size for ^3He_n, and a stability of He_n at T = 0 K for any n.

n = 2

$$q(\eta) = 6\eta$$
$$N(\eta) = 4 + 2\eta^2$$

n = 3

$$q(\eta) = 24\eta^4 + 32\eta^2 + 30$$
$$N(\eta) = 6\eta^6 + 72\eta^2 + 12$$

n = 4

$$q(\eta) = 36 + 48\eta + 42\eta^2 + 24\eta^3 + 42\eta^4 + 12\eta^5 + 18\eta^6 + 18\eta^9$$
$$N(\eta) = 60 + 99\eta^2 + 18\eta^4 + 54\ \eta^6 + 9\eta^{12}$$

n = 5

$$q(\eta) = 112 + 160\eta + 192\eta^2 + 168\eta^3 + 212\eta^4 + 72\eta^5 + 192\eta^6 + 72\eta^7 + 12\eta^8 + 104\eta^9 + 24\eta^{10} + 48\eta^{12} + 32\eta^{16}$$

$$N(\eta) = 192 + 432\eta^2 + 216\eta^4 + 320\eta^6 + 96\eta^8 + 128\eta^{12} + 16\eta^{20}$$

n = 6

$$q(\eta) = 320 + 480\eta + 720\eta^2 + 760\eta^3 + 950\eta^4 + 460\eta^5 + 1040\eta^6 + 640\eta^7 + 380\eta^8 + 390\eta + 320\eta^{10} + 160\eta^{11} + 340\eta^{12} + 200\eta^{12} + 200\eta^{13} + 210\eta^{16} + 40\eta^{17} + 100\eta^{20} + 50\eta^{25}$$

$$N(\eta) = 560 + 1575\eta^2 + 1350\eta^4 + 1500\eta^6 + 1200\eta^8 + 900\eta^{12} + 200\eta^{14} + 250\eta^{20} + 25\eta^{30}$$

n = 8

$$q(\eta) = 2240 + 3584\eta + 7392\eta^2 + 9324\eta^3 + 13230\eta^4 + 10794\eta^5 + 18508\eta^6 + 14224\eta^7 + 18312\eta^8 + 13062\eta^9 + 12138\eta^{10} + 8064\eta^{11} + 12264\eta^{12} + 9408\eta^{13} + 6720\eta^{14} + 5124\eta^{15} + 5796\eta^{16} + 1764\eta^{17} + 4116\eta^{18} + 420\eta^{19} + 4158\eta^{20} + 3108\eta^{21} + 1974\eta^{22} + 504\eta^{24} + 1736\eta^{25} + 504\eta^{26} + 504\eta^{27} + 84\eta^{29} + 1554\eta^{30} + 504\eta^{31} + 602\eta^{36} + 84\eta^{37} + 294\eta^{42} + 98\eta^{98}$$

$$N(\eta) = 4032 + 15631\eta^2 + 23961\eta^4 + 28224\eta^6 + 37142\eta^8 + 14700\eta^{10} + 21756\eta^{12} + 20727\eta^{14} + 2205\eta^{16} + 5292\eta^{18} + 7840\eta^{20} + 5292\eta^{22} + 294\eta^{24} + 588\eta^{26} + 3185\eta^{30} + 588\eta^{32} + 686\eta^{42} + 49\eta^{56}$$

Table 2 - Polynomial expression $N(\eta)$ and $q(\eta)$ versus the number of ^{4}He atoms.

n	$10^6\ \bar{E}$ (eV)
2	– 3.25
3	– 4.15
4	– 5.75
5	– 6.20
6	– 7.10
8	– 7.50

Table 3 - Energy per atom, $\overline{E}$, (in eV) of ^4He_n aggregates $\overline{E}$.

References

1. D.S. Lewart, V.R. Pandharipande, S.C. Pieper, Phys. Rev. B 37 (1988) 4950.
2. D.M. Brink and S. Stingari, Z.Phys. D 15 (1990) 257.
3. L.C. Balbas, A. Mañanes, M. Membrado, A.F. Pacheco and J. Sañudo, J. Chem. Phys. 94 (1991) 7335.
4. S. Stingari and J. Treiner, J. Chem. Phys. 87 (1987) 5021.
5. G. Montambaux, M. Héritier and P. Lederer, J. of Low Temp. Phys. 47 (1982) 39.
6. D. Vollhardt, Rev. Mod. Phys. 56 (1984) 99.
7. R.J. Tarento and P. Joyes, Phys. Rev. B41 (1990) 4547.
8. R.J. Tarento and P. Joyes, J. of Chem. Phys. (1991) to appear.
9. J. Gspann and R. Rico "Physics and chemistry of small clusters", edited by J. Jena, B.K. Rao and S.N. Khanna, ASI Series, vol. B158 (1987) 199.

A PRELIMINARY STUDY OF ENERGY BANDS OF 2D PERIODIC QUANTUM DOT ARRAYS

XU, Guangzhao, JIANG, Ping and XU, Zhizhong
Physics Department
Fudan University
Shanghai 200433
P.R.China

ABSTRACT. The energy bands of two types of 2D quantum dot array are calculated. One of them is a 2D square lattice consisting of circular quantum dots. Tight binding method is applied to calculate the energy band for this model. The other is a 2D Kronig-Penney model, whose electronic structure is obtained by solving the Schrœdinger equation

1. Introduction

In recent years there has been growing interest in the electronic properties of quantum-dot system. These structures can be formed by means of confining the quasi-two-dimensional electron gas (2DEG) near the interface of a semiconductor heterojunction or in a narrow quantum well, so that the electron motion is constrained in all three dimensions. Experimentally, a quantum dot could be defined on the 2DEG by field effect[1-4]. A plausible example would be that a circular quantum dot could be made by depositing a metallic gate with a circular hole (with radius of the order of 10 nm) on a heterostructure and applying a negative voltage on the gate so that the underneath electrons would be confined in a circular potential well just below the hole. In this way people can make many many quantum dots on a single substrate. If these dots are arranged periodically and are so closely distributed the coupling between the neighbor dots may be strong enough to make the periodic array behave like a 2D crystal. In this paper, the energy bands of some of these semiconductor 2D nanostructure arrays are shown. From the calculated results, we may get some ideas about in what manner and what degree the arrays of the quantum dots tailor the electronic properties. The calculations are conducted for $GaAs/Al_xGa_{1-x}As$ heterojunction. Thus we have the electron effective mass $m^*=0.067m_e$ which is the effective mass of GaAs, where m_e is the mass of the free electron.

2. Model and Result

2.1. 2D Array of Circular Quantum Dots -- Type 1.

P. Jena et al. (eds.), Physics and Chemistry of Finite Systems: From Clusters to Crystals, Vol. I, 697–702.

The 2D array of circular quantum dots is supposed to be formed by the method mentioned above. Figure 1. shows the scheme of this model. Tight binding method is applied to calculate the energy band for such an array. At first, we must obtain the electronic energy levels and wave functions of a single quantum dot which acts just like a single artificial atom.

For simplicity, we assume the potential of such a circular quantum dot is of square-well type:

$$V(r)=\begin{cases} V_0 & r>R \\ 0 & r\leq R \end{cases} \tag{1}$$

where R is the radius of the dot. Solving the Schroedinger equation for a single electron with effective mass m* in such a potential gives us the electronic energy levels and wave functions.

Based on these energy levels and wave functions, the tight binding method can be followed straightforwardly. In our calculation, only the overlap of the wave functions for the same energy level of the nearest quantum dots are taken into account. However, the mixing of the degenerate states must be considered.

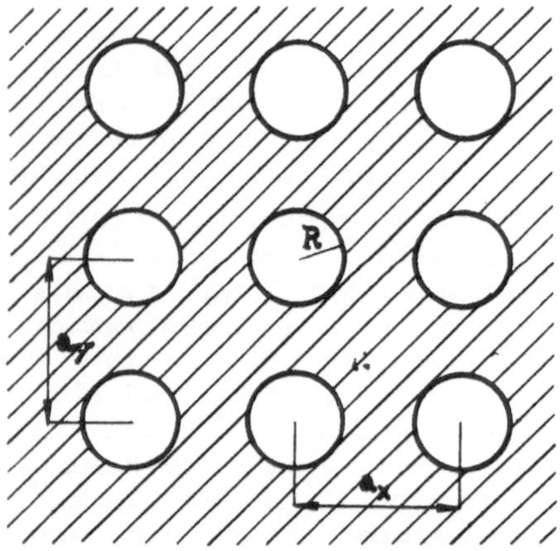

Figure 1. Scheme of Type 1. Shaded area -- barrier region

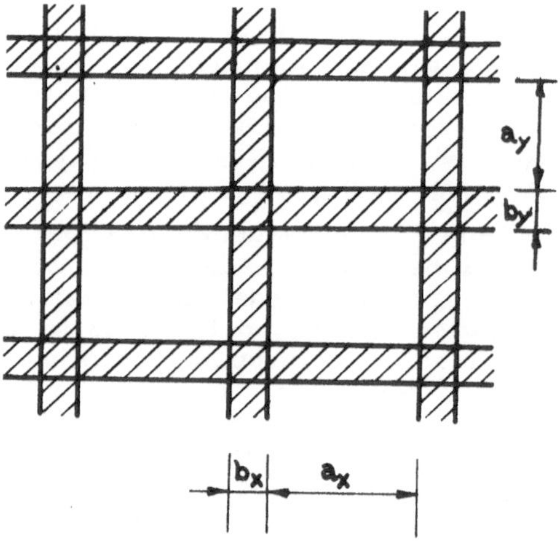

Figure 2. Scheme of Type 2. Shaded area --barrier region

It has been shown that the confining potential is expected to be more likely parabolic[5][6]. So for a more realistic approximation, we also use the parabolic potential for each quantum dot:

$$V(r)=\begin{cases} \frac{1}{2}m^*\omega^2R^2 & r>R \\ \frac{1}{2}m^*\omega^2r^2 & r\leq R \end{cases} \tag{2}$$

It is worthwhile noting that although the potential of equation (2) is of muffin-tin type, i.e. the potential of the barrier region (r>R) is constant, as an approximation, however, we still use the energy levels and wave functions of a 2D harmonic oscillator for the quantum dot. This approximation is more reliable for the lower energy levels.

The parameters in the calculation are chosen as follows: For the square-well potential, the depth of potential well V_o=100mev, the radius of quantum dots R=10nm. The 2D arrays are of square lattice and two lattice constants a=30nm and a=25nm are considered respectively.

In such a quantum dot there are four energy levels of bound states, two of which are 2-fold degenerate. The energy levels expand to six energy bands when the quantum dots are periodically arranged in square lattice. The width of bands is of the order of 1mev. Higher energy level expands to wider energy band. The maximum width of energy band is 1.7mev in the case of a=30nm and 2.9mev of a=25nm. No mixing occurs between the different energy bands.

For parabolic potential we chose the barrier height $\frac{1}{2}m^*\omega^2R^2$=100mev, radius of quantum dots R=10nm and the same lattice as that mentioned above. Then we get $\hbar\omega$=22.8mev. There are five energy levels in the well. The energy bands are a little bit wider than that of the square-well potential, but the width is also of the same order of magnitude and there is also no band mixing.

It seems from the calculation that the band effects in such systems of 2D quantum dots are not quite strong. It is expected that reducing the lattice constant would widen the energy bands.

2.2. 2D Kronig-Penney Model -- Type 2.

The schematic configuration of a 2D Kronig-Penney model considered is shown in Figure 2. It can be made by fabricating a metallic orthogonal lattice as a gate on a semiconductor heterojunction and applying a negative voltage on the gate. The 2D Kronig-Penney periodic potential can be written:

$$V=V_x(x)+V_y(y)-V_{xy}(x,y) \tag{3}$$

where both $V_x(x)$ and $V_y(y)$ are 1D Kronig-Penney potential:

$$V_x(x)=\begin{cases} U_0 & n(a_x+b_x)<x<n(a_x+b_x)+b_x \\ 0 & n(a_x+b_x)+b_x<x<(n+1)(a_x+b_x) \end{cases} \tag{4}$$

$$V_y(y)=\begin{cases} U_0 & m(a_y+b_y)<y<m(a_y+b_y)+b_y \\ 0 & m(a_y+b_y)+b_y<y<(m+1)(a_y+b_y) \end{cases} \tag{5}$$

$$m,n=0,\pm1,\pm2,\cdots\cdots$$

where a_x and a_y are well widths ,and b_x and b_y are barrier widths in the x direction and y direction respectively. For the barrier crossing region we add the last term $-V_{xy}(x,y)$ to keep the potential uniform in the whole barrier area:

$$V_{xy}(x,y)=\begin{cases} U_0 & n(a_x+b_x)<x<n(a_x+b_x)+b_x \text{ and } m(a_y+b_y)<y<m(a_y+b_y)+b_y \\ 0 & \text{otherwise} \end{cases} \tag{6}$$

Solving the Schroedinger equation for this periodic potential yields the band structure of the model. The last term $-V_{xy}(x,y)$ is treated as a perturbation in our calculation. The 0-order wave function $\Psi_0(x,y)$ satisfies:

$$\left(-\frac{\hbar^2}{2m^*}\left(\frac{d^2X}{dx^2}+\frac{d^2Y}{dy^2}\right)+V_x(x)+V_y(y)\right)\Psi_0(x,y)=E_0\Psi_0(x,y) \qquad (7)$$

Assuming $\Psi_0(x,y)=X(x)Y(y)$, the equation can be separated into two independent 1D Kronig-Penney equations[7]:

$$\left(-\frac{\hbar^2}{2m^*}\frac{d^2X}{dx^2}+V_x(x)\right)=E_xX(x) \qquad (8)$$

$$\left(-\frac{\hbar^2}{2m^*}\frac{d^2Y}{dy^2}+V_y(y)\right)=E_yY(y) \qquad (9)$$

The energy up to the 1-order perturbation

$$E=E_0-\frac{\langle\Psi_0|V_{xy}(x,y)|\Psi_0\rangle}{\langle\Psi_0|\Psi_0\rangle} \qquad (10)$$

where $E_0=E_x+E_y$. By using the standard Kronig-Penney approach and taking the perturbation into account we have obtained the dispersion E(k) for this model.

The parameters for this model are chosen as: barrier height U_0=100mev, the widths of potential well and potential barrier: a_x=30nm, b_x=6nm; a_y=40nm, b_y=6nm for the first case, a rectangular lattice and $a_x=a_y$=30nm, $b_x=b_y$=10nm for the second case, a square lattice, respectively.

Figure 3 and Figure 4 show the energy band structure of our 2D Kronig-Penney model. Figure 3 is for the first case and figure 4 for the second case, of which (a) is the bands along $k_y=0$ in the Brillioun zone while (b) alone the diagonal of the Brillioun zone. In Figure 3(a) and Figure 4(a) the abscissa k is shown in $\pi/(a_x+b_x)$, while in Figure 3(b) and Figure 4(b) it is shown in $\pi[(1/(a_x+b_x)^2+1/(a_y+b_y)^2]^{1/2}$. It can be seen in these figures that there is mixing between the energy bands and the band structure of this model is much more complicated in comparison with that of Type 1. Such a 2D nanostructure system can be considered more like an artificial 2D crystal.

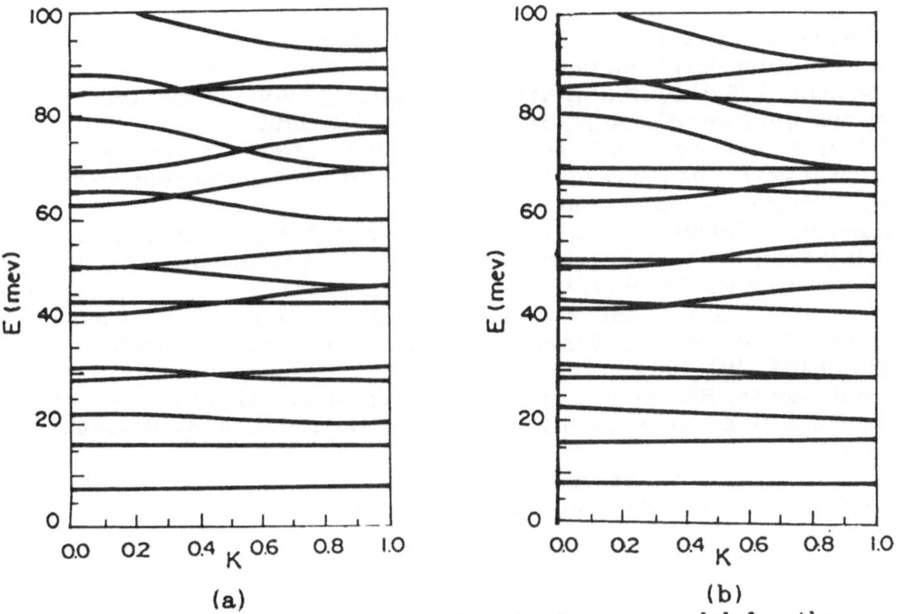

Figure 3. Energy bands of the Kronig-Penney model for the rectangular lattice

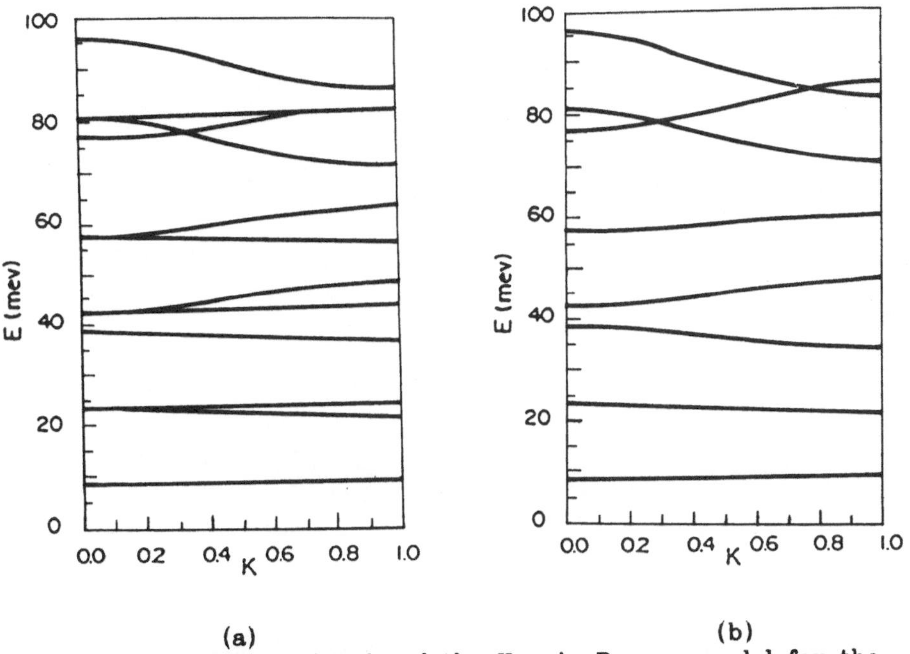

Figure 4. Energy bands of the Kronig-Penney model for the square lattice

References

1.Ch.Sikorski and U.Merkt. (1989), 'Spectroscopy of electronic state in IbSb quantum dots', Phys.Rev.lett.62, 2164-2167

2.J.Alsmeier, E.Batke, and J.P.Kotthaus. (1990), 'Voltage-tunable quantum dots on silicon', Phys.Rev.B41, 1699-1702

3.C.T.Liu, K.Nakamura, D.C.Tsui, K.Ismail, D.A.Antoniadis, and Henry I.Smith. (1989) 'Magneto-optics of a quasi-zero-dimensional electron gas', Appl.Phys.Lett.55, 168-170

4.W.Hansen, T.P.Smith,III, K.Y.Lee, J.A.Brum, C.M.Knoedler, J.M.Hong, and D.P.Kern. (1989) 'Zeeman Bifurcation of quantum-dot spectra', Phys.Rev.Lett.62, 2168-2171

5.S.E.Laux, D.J.Frank, and Frank Stern. (1988), 'Quasi-One dimensional electron states in a split-gate GaAs/AlGaAs heterostructure', Surf.Sci.196, 101-106

6.Arvind Kumar, Steven E.Laux, and Frank Stern. (1990), 'Electron states in a GaAs quantum dot in a magnetic field', Phys.Rev.B42, 5166-5175

7.R.del.Kronig and W.G.Penney. (1931), 'Quantum mechanics of electrons in crystal lattices', Proc.Roy.Soc. (London) A130, 499-513

COMPOUND VERSUS PURE METAL MICROCLUSTERS: HOW THE PHOTOABSORPTION PROFILES RELATE TO THE ABUNDANCE SPECTRA

C. Yannouleas, P. Jena, and S.N. Khanna
Physics Department
Virginia Commonwealth University
Richmond, VA 23284-2000

ABSTRACT. We report a strong connection between the abundance spectra and the photoabsorption profiles of heteroatomic metal clusters of composition YA_x, where A is an alkali metal and Y is either a divalent or a monovalent impurity atom (Y=A encompasses the limiting case of pure clusters). This trend is established within the framework of a two-step-background jellium model by allowing for a continuous variation of the density parameter representing the impurity. For divalent impurities, parameters substantially away from the bulk values are required to reproduce the observations. Parallel molecular calculations are being carried out to elucidate this behavior in depth.

INTRODUCTION

Because of the similarities between nuclear and metal-clusters physics (e.g., cf. Refs. [[1]-[2]]), the nuclear matrix-RPA (random phase approximation) was adapted to the case of spherical, closed-shell, homonuclear clusters in order to describe their optical properties [3]. In this approach, the collective motion arises as a linear superposition over many single-particle transitions from the occupied to the unoccupied orbitals of the effective average field that confines the delocalized valence electrons. For specific mass numbers and species, like neutral Na_{20} and Na_{40}, this method predicted multipeak photoabsorption profiles more complex than the simple one-peak profile expected [4] from the Mie theory for the charge oscillations of a classical metal sphere. Such profiles have recently been observed experimentally [[4]-[5]]. They represent quantum size effects due to the discreteness of the single-particle levels. They are sensitive to the position of these levels, and thus to the corresponding effective potential that characterizes the delocalized electrons (for the fragmentation of the

P. Jena et al. (eds.), Physics and Chemistry of Finite Systems: From Clusters to Crystals, Vol. I, 703–708.

optical response in Cs_8 and $Cs_{10}O$, cf. Refs. [[6]-[7]]; for the absence of fragmentation in cationic potassium clusters K_9^+ and K_{21}^+, cf. Refs. [[8]-[9]]; for a molecular *ab initio* interpretation of photoabsorption spectra cf. Ref. [10]).

In this paper, we focus on the optical response of the compound metallic clusters $ZnNa_8$ and NaK_{19}. Since the impurity atoms represent an important perturbation that results in unequal shifts in the single-particle levels with respect to the homologous pure clusters, strong modifications of the photoabsorption profiles are to be expected. Furthermore, we show that these modifications have a direct counterpart in the appearance of additional magic numbers (or the disappearance of magic numbers that are prominent in pure clusters), and even more in detail in the precise modifications reported earlier in the abundance spectra [[11]-[12]].

TWO-STEP JELLIUM BACKGROUND

The influence of the impurity atom can be treated in the spirit of a modified jellium approximation. For a compound cluster containing the heteroatom (impurity, I) at the center of the otherwise homonuclear (H) cluster, the positive background n_+ must be modified as follows:

$$n_+ = n_0^I\,\theta(R_I - r) + n_0^H\,\theta(r - R_I)\theta(R_H - r)\ , \tag{1}$$

where $\theta(R-r) = 0$ for $r > R$ and 1 for $r < R$; n_0^I and n_0^H are positive-ion densities. A first choice for these densities could be the corresponding bulk values. Although preserving the overall trend in the abundance spectra, such a choice, however, leads to significant discrepancies with the experimental observation in the case of divalent impurities [13]. Best known is the case of the series MgK_x, where this procedure predicts strong abundances for both MgK_6 and MgK_8 [14], even though only MgK_8 exhibits a maximum in the experimental abundances [11]. In the present work, we will adopt a more flexible approach. Indeed, there is no apparent reason why the density n_0^I corresponding to the divalent impurity should be the same as the bulk density. In the following, we will thus assign to it a set of varying values, while retaining the bulk value for the alkali atoms.

EFFECTIVE POTENTIALS - SHIFT OF THE 2s ELECTRONIC LEVEL - SECOND DIFFERENCES

A conspicuous effect of the heteroatom at the center is the dramatic change it can cause in the effective central potential that binds the delocalized electrons compared to the case of pure clusters. Specifically, for $n_0^I > n_0^H$, a depression centered at the origin develops in the potential. This depression increases with increasing n_0^I (decreasing Wigner-Seitz radius r_s^I, cf. Fig. 1.), and results in unequal shifts in the single-particle levels. The bound single-particle levels that are most influenced are the s orbitals, while the p and d orbitals remain comparatively unchanged. This is a

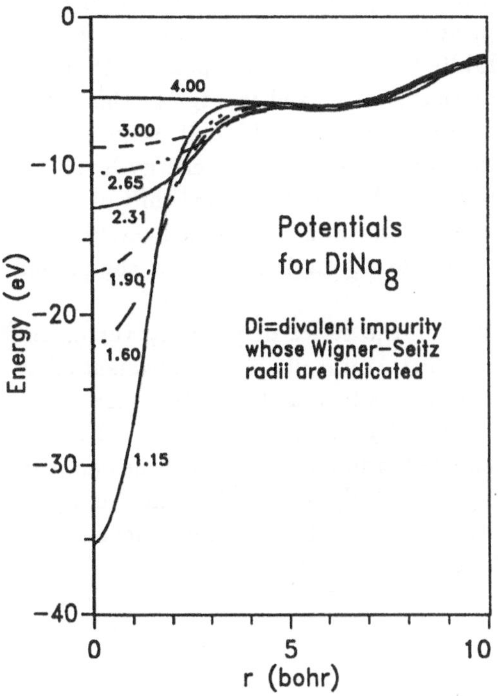

Fig. 1.

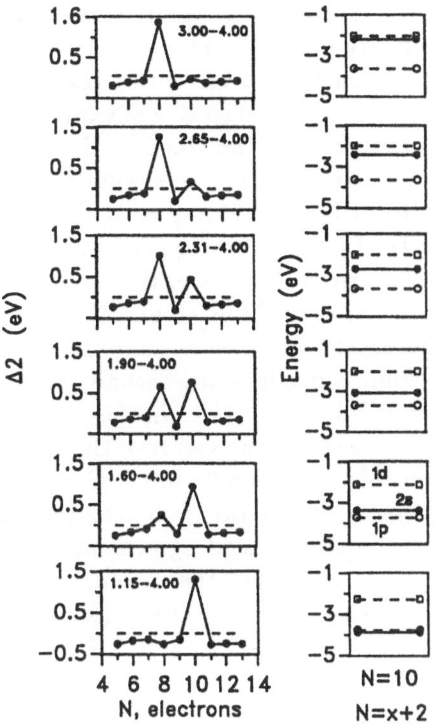

Fig. 2.

Fig. 1. Potentials for $DiNa_8$. The Wigner-Seitz radius of Na is 4.00 a.u.
Fig. 2. Second differences Δ_2 and single-particle levels for the series $DiNa_x$.

consequence of the $\hbar^2 l(l+1)/2mr^2$ repulsive centrifugal potential that insulates the higher angular momenta from the narrow depression. The deeper the depression, the deeper the 2s state is pushed down. As a result, it can be shifted below the 1d level, and a new magic number for N=10 electrons appears for the series DiA_x, where Di denotes a divalent impurity (N=x+2). As shown in Fig. 2 (right column), this has happened already for $r_s^I = 3.00$ in the series $DiNa_x$. In this case, the compound aggregate is spherical, like the case of Na_8, and in sharp contrast to the prolate Na_{10}. The competition of the peaks between N=8 electrons and N=10 electrons in the abundance spectra, however, is determined by the relative position of the 2s state with respect to both the 1d and 1p levels. Fig. 2 displays the evolution of the second differences Δ_2 in the total energies of the $DiNa_x$ aggregates, when the Wigner-Seitz radius of the central impurity varies from 3.00 a.u. to 1.15 a.u. A lower r_s^I favors the N=10 peak. For $r_s^I = 1.15$ the N=8 peak has disappeared, in agreement with the experimental data on $ZnNa_8$ [11] (for comparison, the bulk Wigner-Seitz radius of Zn is 2.31). For $MgNa_8$, the experiment exhibits two equally sized peaks [11]. From an inspection of the trend in the second differences displayed in Fig. 2, this latter compound seems to correspond to $r_s^I = 1.90$ (for comparison, the bulk Wigner-Seitz radius of Mg is 2.65). Fig. 2 also displays for N=10 electrons the energy levels for the 1p, 1d and 2s single-particle states (right column). The downward shift of the 2s level between the 1d and the 1p levels is well noticeable and is in agreement with the competition of the N=8 and N=10 peaks in the corresponding plots for the second differences.

Fig. 3 displays the second differences Δ_2 for the series of pure K_x versus the compound series NaK_x. In this case, since the central impurity is a monovalent alkali atom, we retain the values of the bulk for r_s^I, namely 4.00 a.u. for sodium (for potassium this value is 4.86 a.u.). Going from K_x to NaK_x, the peak for N=18 electrons disappears, while the N=8 peak remains unaffected, and no N=10 peak does appear. This trend is in good agreement with the experimentally observed abundances [12]. Fig. 3 also displays the correponding positions of the 1s, 1p, 1d, and 2s levels both for N=8 and N=20 electrons (N=x+1, in the case of NaK_x).

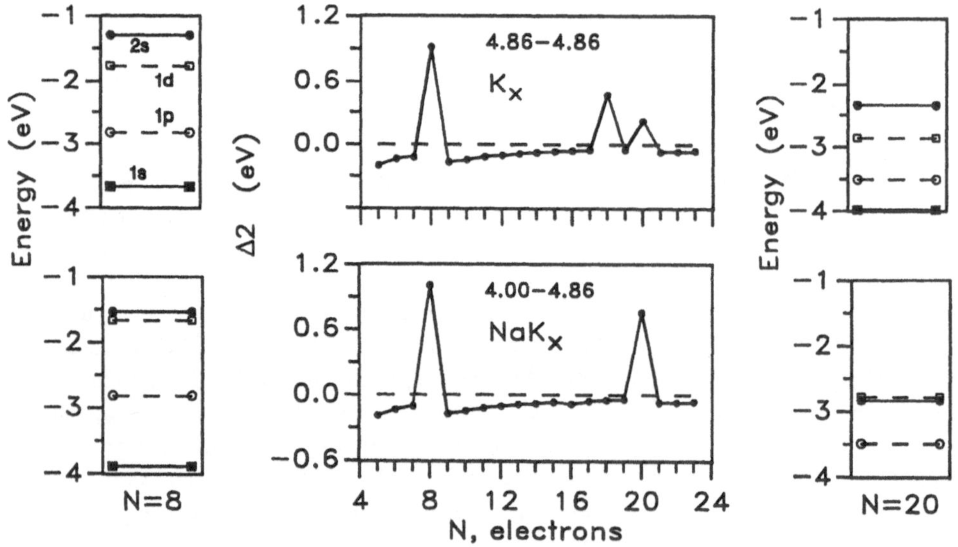

Fig. 3. Second differences Δ_2 and single-particle levels for the series K_x and NaK_x. The Wigner-Seitz radii for Na and K are 4.00 and 4.86 a.u., respectively.

OPTICAL RESPONSE - PHOTOABSORPTION

It has been found [3] that the split-plasmon that characterizes the photoabsorption profiles of homonuclear Na_{20} and K_{20} clusters can be treated as a two-level model due to a degeneracy between the collective plasmon and a particular 2s → 3p particle-hole transition. In the case of the compound NaK_{19}, the impurity at the center shifts the 2s level downwards, but barely affects the 3p level, while, at the same time, the position of the plasmon remains unchanged. As a result, it is to be expected that the degree of the splitting in the photoabsorption spectrum will be affected by the presence of the Na impurity. Indeed, matrix-RPA calculations for K_{20} and NaK_{19} are displayed in Fig. 4. In the case of K_{20}, the plasmon is split into two components, while in the case of NaK_{19} there appears one dominant line with 80% of the Thomas-Reiche-Kuhn sum rule.

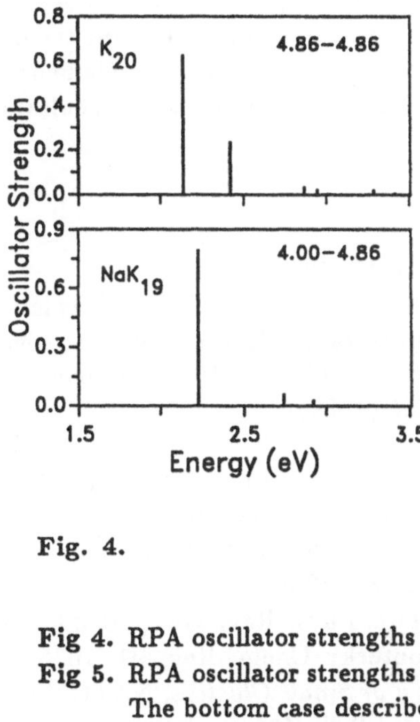

Fig. 4.

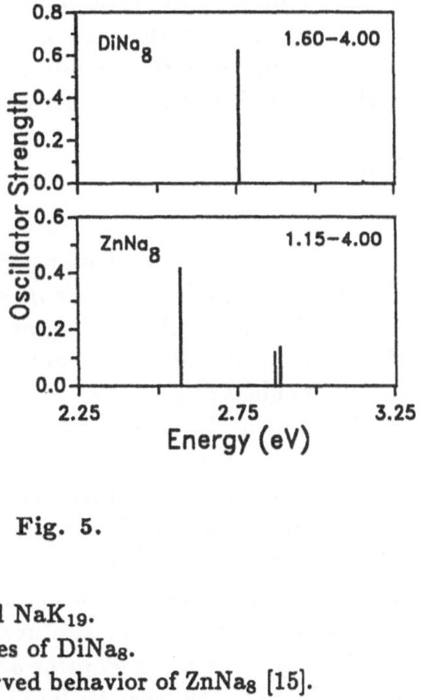

Fig. 5.

Fig 4. RPA oscillator strengths for K_{20} and NaK_{19}.
Fig 5. RPA oscillator strengths for two cases of $DiNa_8$.
The bottom case describes the observed behavior of $ZnNa_8$ [15].

Another typical example of the influence of the impurity upon the photoabsorption profiles is offered by $ZnNa_8$. The photoabsorption cross section of this cluster has recently been measured by Kappes et al [15]. A double-peak has been observed: the higher component lies at 2.97 eV and carries a smaller amount of the strength than the lower component at 2.63 eV. Fig. 5 displays the RPA response for the system $DiNa_8$ for two different values of r_s^I. For $r_s^I = 1.60$, one single line at 2.76

eV carries 62% of the TRK sum rule. In the case of $r_s^I = 1.15$, the single line is fragmented into three smaller lines. Two of these lines at higher energies (at 2.87 eV) are closely spaced and will appear as one peak after the broadening that accounts for the experimentally observed widths. Between the two of them, they carry 26% of the total strength, while the stronger line at 2.57 eV carries 42% of the strength. The variations in the photoabsorption profiles displayed in Fig. 5 depend again upon the extent of the shift sustained by the s levels. In particular, the fragmentation at the value $r_s^I = 1.15$ is due to a degeneracy that occurs between the plasmon and the 2s $\rightarrow$ 2p and 1p $\rightarrow$ 3s particle-hole transitions. A very strong shift downwards supplied by the value $r_s^I = 1.15$ is needed for this degeneracy to develop. As discussed in the earlier paragraph, the same shift accounts sufficiently well for the trend in the observed abundances of the series $ZnNa_x$.

We conclude that a two-step jellium background with a Wigner-Seitz radius for the central region substantially different from the bulk value can describe sufficiently well both the optical properties and the abundances of compound clusters possessing a divalent impurity. In the case of an alkali impurity, use of the bulk value seems to suffice. Molecular *ab initio* calculations [16] are in progress to elucidate this behavior in depth. We wish to thank F. Reuse for helping us with the molecular calculations.

REFERENCES

[1] M.L. Cohen and W.D. Knight, Physics Today **43**, 42 (1990).
[2] P. Jena et al, in *VII Int. Conf. on Recent Progress in Many-Body Theories*, eds. C.E. Campbell et al, (Minneapolis, Minnesota, 26-31 Aug. 1991) in press.
[3] C. Yannouleas and R.A. Broglia, Phys. Rev. **A44** (1 Nov. 1991) in press; C. Yannouleas et al, Phys. Rev. Lett. **63**, 255 (1989).
[4] K. Selby et al, Phys. Rev. **B43**, 4565 (1991).
[5] S. Pollack, C.R.C. Wang, and M.M. Kappes, J. Chem. Phys. **94**, 2496 (1991).
[6] H. Fallgren and T.P. Martin, Chem. Phys. Lett. **168**, 233 (1990).
[7] C. Yannouleas and R.A. Broglia, Europhys. Lett. **15**, 843 (1991).
[8] C. Brechignac et al, Chem. Phys. Lett. **164**, 433 (1989).
[9] C. Yannouleas, J.M. Pacheco, and R.A. Broglia, Phys. Rev. **B41**, 6088 (1990).
[10] V. Bonacic-Koutecky, P. Fantucci, and J. Koutecky, Chem. Rev. **91**, 1035 (1991).
[11] M.M. Kappes et al, in *Physics and Chemistry of Small Clusters*, NATO ASI Series **B**, Vol. 158, eds. P. Jena, B.K. Rao, and S.N. Khanna (Plenum, New York, 1987) p. 145; M.M. Kappes et al, Chem. Phys. Lett. **119**, 11 (1985).
[12] W.A. de Heer et al, Solid State Phys. **40**, 93 (1987).
[13] C. Baladron and J.A. Alonso, Physica **B154**, 73 (1988).
[14] C. Baladron, J.A. Alonso, and M.P. Iniguez, J. Phys. **F17**, L197 (1987).
[15] M.M. Kappes et al, to be published.
[16] B.K. Rao, S.N. Khanna, and P. Jena, Phys. Rev. **B36**, 953 (1987); B.K. Rao and P. Jena, Phys. Rev. **B37**, 2867 (1988).

MAGNETISM OF FINITE SYSTEMS: FROM CLUSTERS TO CRYSTALS

S. N. Khanna and P. Jena
Physics Department
Virginia Commonwealth University
Richmond, VA 23284-2000

ABSTRACT. Using self-consistent molecular orbital theory as well as ab-initio tight-binding formalism, we have investigated the magnetic moments of atomically engineered materials as a function of their size, dimensionality, symmetry and topology. The magnetic moments of magnetic materials can be enhanced by reducing their size as well as dimension. Similarly, atoms of otherwise non-magnetic materials are found to be magnetic in lower dimension. In small isolated clusters, the effect of externally applied magnetic field and temperature can have substantial effect on the measured magnetic moment.

I. Introduction

The study of magnetism has attracted renewed interest in recent years. The primary reason for this is that through innovative experimental techniques, novel materials with atomic dimension can be produced. For example, molecular beam epitaxy techniques allow one to deposit controlled layers of various elements on a substrate[1]. This includes not only thin films of the same material but also modulated layers consisting of bi-metallic elements. Using supersonic jet expansion, atomic clusters up to thousands of atoms have been produced[2] and their magnetic properties analyzed through Stern-Gerlach type experiments[3-6]. Non-magnetic elements such as V have been substituted into polymers and are found to be magnetic even at room temperature[7].

These experiments have provided some of the most startling results in magnetism. (1) The magnetic moment of atoms confined to linear chains (1-dimension, 1-D) are larger than those confined to a plane (2-D) and the later carries a larger moment than the atoms confined in the bulk (3-D). (2) The magnetic moment per atom in isolated

P. Jena et al. (eds.), Physics and Chemistry of Finite Systems: From Clusters to Crystals, Vol. I, 709–720.

clusters are higher than in bulk, but their values are very sensitive to temperature and external magnetic field. (3) The atoms in a cluster of fixed size with high symmetry carry larger moment than those with reduced symmetry. (4) While V-atoms confined to a linear chain or plane are magnetic, they cease to be magnetic when confined to a cluster mimicking the bulk phase.

All these varied experiments can be understood by realizing that a single factor that dominates magnetism is the immediate environment of the magnetic atom under investigation. If the magnetic atom has a higher coordination number with other magnetic atoms, it will tend to have a lower moment. Similarly, if the symmetry is reduced the moment too will be reduced. We have arrived at the above understanding by carrying out selfconsistent calculations of the electronic structure of Fe, Co, Ni, and V confined to linear chains, surfaces, layers and bulk. The magnetic moment was studied by analyzing the spin density of states.

The ground state magnetic properties can be significantly affected if the atoms are confined to small clusters and subjected to external magnetic field and temperature[3-6]. Isolated clusters can have additional degrees of freedom namely rotation and vibration. The coupling of the rotational axis with intrinsic magnetic moment of the system can give rise to interesting temperature and field dependencies. The effective magnetic moment per atom as measured in Stern-Gerlach experiments can therefore appear to be less than the moments calculated in the ground state of the system. At very high temperatures, the spins can also melt leading to further reduction in the moment. In the following, we provide a brief description of our theory[7] and compare the results with experiments[3-6].

II. Theory

The first principle study of electronic structure and magnetic properties of systems ranging from clusters to crystals has evolved along two main directions. For small atomic clusters, it is conventional to use the ab-initio self-consistent-field linear combination of atomic-orbitals molecular-orbital (SCF-LCAO-MO) method[8]. The simplest among these is the local density approximation in the density functional formulation[9-10]. This scheme is usually restricted to study clusters consisting of less than 50 atoms. For infinite systems, (chains, surfaces and bulk) with periodic boundary conditions the conventional band structure based on linearized augmented plane wave (LAPW) is the most appropriate method[11]. For systems consisting of defects or clusters larger than 50 atoms, no first principles method is available. Approximations in the calculation are therefore necessary to study electronic structure and magnetic properties of clusters in the nanometer scale or systems with no periodic boundary conditions (e.g. finite chains).

We have used the SCF-LCAO-MO theory based upon the local density approximation and the discrete variational method[10] to study clusters of less than 50 atoms. For larger clusters we have developed a new tight-binding theory[12-13] where the parameters (the overlap and the hopping matrix elements) are computed from first principles by carrying out LCAO-MO studies on dimers with varying interatomic distances. This tight-binding method is then applied to larger systems with or without defects and with or without high symmetry for systems ranging from clusters to crystals. We have compared the predictions of this model for magnetic moments of Fe, Co, and Ni confined to infinite chain, plane, slabs, and bulk with those obtained from LAPW technique. The quantitative agreement between our results and that of LAPW gives us confidence that the ab-initio tight-binding (ATB) model indeed can provide quantitative results on the magnetic moments. This method has been described in detail elsewhere[12] and the reader is referred to this paper for details. In the following we only give salient feature of our results.

III. Results and Discussions

We discuss the effect of dimensionality, symmetry and size on magnetic moment separately. These results will then be coordinated in to one comprehensive picture where the magnetism can be understood by the local coordination of the probe atom. We will finally present the effects of applied field and temperature on the magnetic moment.

A. Effect of Dimensionality

In Table 1 we present the magnetic moment of V, Fe, Co, and Ni confined to linear chain, plane, and bulk arrangements. The interatomic spacings in each of these configurations were held at the equilibrium lattice spacings. The magnetic moments were calculated using the ATB formalism mentioned above. We note that the moments per atom increase as the dimensionality decreases. In the case of V, the bulk phase is non magnetic, however, in linear chains, the moments are ferromagnetically coupled[14].

As the dimensionality of the system decreases, the probe atom has fewer nearest neighbors. As a result the electron density of states become narrower as the overlap between the neighboring sites decreases. Narrowing of the density of states give rise to a higher magnetic moment.

TABLE 1. The magnetic moment of V, Fe, Co. and Ni confined to linear chain, plane, and bulk arrangements[12]. (The atomic moments are based on local spin density calculations.)

	ATOM	CHAIN	PLANE	BULK
Fe	4.0	3.21	3.15	2.53
Co	3.0	2.23	1.85	1.69
Ni	2.0	1.02	0.76	0.59
V	5.0	4.0	--	0

B. Effect of Symmetry

Clusters can exist in various geometrical forms. The clusters characterized by high symmetry would tend to have degenerate electron levels compared to those with reduced symmetry. Since degenerate electronic states give rise to high spin configurations because of Hund's rule coupling, it is expected that a high symmetry cluster would carry a larger magnetic moment per atom than the corresponding cluster with reduced symmetry. This is indeed the case as was shown by Dunlap[15] for Fe_{13} cluster having icosahedric, as well as cuboctahedric structures. He had observed that the number of unpaired spins in icosahedric Fe cluster was larger than the corresponding cuboctahedric cluster with the same radial bond distance.

We had earlier shown[16] that clusters of non-magnetic elements can exhibit similar properties. For example, in Na_4 cluster confined to a tetrahedral or square geometry, the p_x, p_y states are degenerate and due to Hund's rule coupling, favor a spin 1 ground state. However, the energy of the system can also be lowered by distorting Na_4 in to a rhombus form due to Jahn Teller effect. This distortion left the degeneracy of the p-state and consequently prefers a spin zero ground state. Thus if the cluster is continually distorted from a square planar structure to that of a rhombus structure, the magnetic ground state in the former will transform to a non magnetic ground state in the later. This magnetic crossover as a function of geometrical distortion is a unique feature of clusters and could find, sometime in the future, practical applications.

C. Effect of Size

We have calculated[13] magnetic moments of Fe and Ni clusters consisting of up to 43 atoms and confined to their respective bulk geometries. The results on the magnetic moments of atoms in various shells as a function of size are given in Table 2. We note that the moments on the innermost atom is lower than its immediate neighbor. As the cluster size increases, the inner moments decrease and asymptotically approach the bulk value. The moment on the outermost shell, however, is almost independent of the cluster size. The above result were obtained using the SCF-LCAO-MO-DVM technique. Due to the computer limitations, it is impossible to carry forth this calculation indefinitely. Thus in this context when does a cluster become a crystal becomes hard to answer and approximations in the theory are needed.

TABLE 2. Spin moments of atoms (in μ_B) in various coordination shells (A,B,C,D) in the Ni_n and Fe_n clusters. A corresponds to central atom, and B,C,D, correspond to atoms in the first, second and third coordination shells respectively.

Cluster	A	B	C	D
Ni_{13}	0.89	1.37		
Ni_{19}	0.86	1.02	1.17	
Ni_{43}	0.60	0.82	0.91	1.06
Fe_{15}	2.77	3.23	3.48	
Fe_{27}	2.47	2.41	2.90	3.96

To address this issue we have used the ATB technique and calculated the magnetic moment of Fe, Co, and Ni confined to linear chains as a function of number of atoms in the chain[17]. Calculations were done for chains up to 20,000 atoms. The results are plotted in Fig. 1. There are two interesting aspects to note here. First the average moment per atom reaches the asymptotic value when the chain has about 19 atoms or the central atom has nine shells. This is different from the case in bulk where an atom surrounded by three shells already has the bulk moment. Second, the moments oscillate with size in small chains. In Ni in particular, there is an odd-even alteration in the moment; chains with odd number of atoms have higher moments than those with even number of atoms. This quantum size effect originates entirely from the electronic structure of the system. In the chains with odd number of atoms, a non bonding orbital appears in the middle of the band whereas for chains with even number of atoms there are only bonding and antibonding states. The need to fill the non-bonding orbital with electrons from the minority band gives rise to a larger moment.

All the above results can be understood in a concise way by realizing that the effect of size and/or dimension is to change the number of nearest neighbor atoms around the probe site[12]. This coordination number increases as the dimension and/or size increases. For example, the coordination number of the probe site in a linear chain, square lattice, and simple cubic lattice are respectively 2,

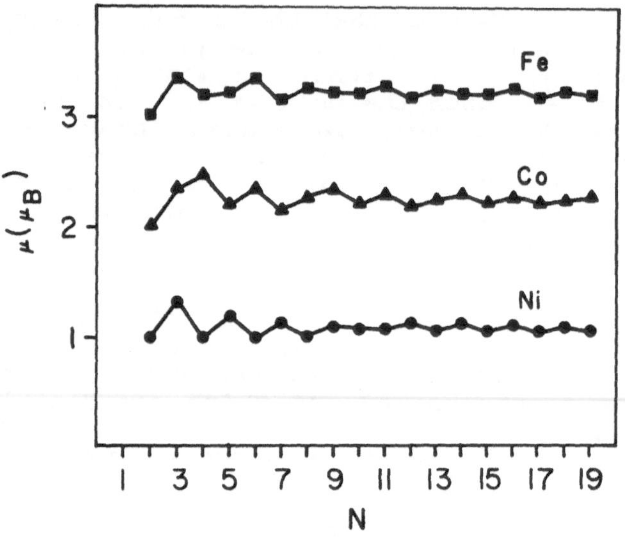

Figure 1. Magnetic moment μ in Fe, Co, and Ni chains as a function of number of atoms in the chain.

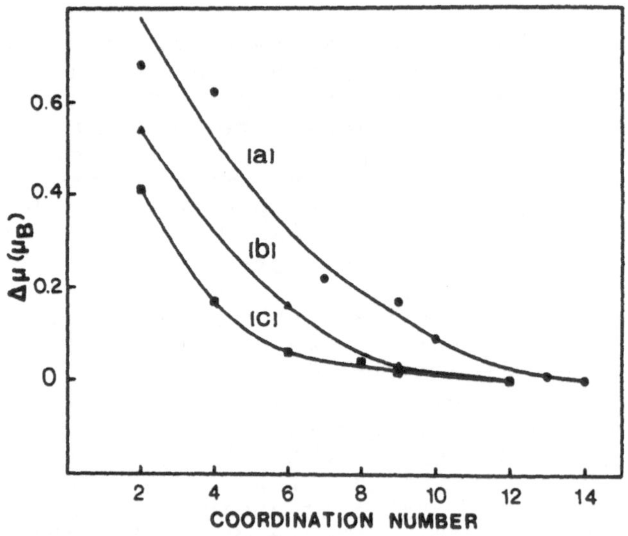

Figure 2. Deviations from the bulk magnetic moment, $\Delta\mu$, in Fe, Co, and Ni as a function of coordination number in various structures).

4, and 8. With increasing coordination number, the electron overlap increases and with increasing overlap, the magnetic moment decreases. Thus from the results given above and from calculations on imperfect systems (vacancies and voids), an universal curve on the change of the magnetic moment as a function of coordination number can be drawn[12]. This is plotted in Fig. 2.

D. Experiments on Isolated Clusters

The above analysis clearly suggests that the average magnetic moment per atom in a real cluster produced during supersonic expansion has to be larger than the bulk. While earlier experiment[3] tend to agree with this view, several later experiments[4-6] using Stern-Gerlach magnet proved otherwise. The moments per atom in Fe, Co, and Ni were found to be almost order of magnitude less than the corresponding bulk value. We explain in the following that this apparent contradiction is not due to incorrect theory described earlier but due to the effect of temperature on the measured magnetic moment.

The magnetic behavior of a ferromagnetic metal can be understood by regarding that the metal consists of ferromagnetic domains each containing at least several hundred atoms. The atomic moments within a domain are ferromagnetically aligned and different domains may be aligned in different directions. As the size of the system is reduced below a critical domain size, as is the case in clusters containing several hundred atoms, the exchange forces dominate and the moments are aligned with respect to each other[18]. A cluster thus behaves as a single magnet with a moment of $N\mu$ where N is the number of atoms in the cluster and μ is the moment per atom. The orientation of the moment $N\mu$ depends on the geometry of the cluster. The lowest energy configurations correspond to $N\mu$ along the easy axes which are usually the symmetry axis. The quantity which dominates the magnetic behavior is then the anisotropy energy, E_{ani}, required to change the orientation of $N\mu$ from one easy axis to another. It has contributions coming from cluster volume, stress, surface, and shape anisotropy. For 2.5 nm Fe particles in a SiO_2 matrix, the anisotropy energy per unit volume, K, has been measured[19] to be ~ 1.7×10^6 J/m^3. This value has to be compared to a corresponding K value of 1×10^4 J/m^3 in bulk Fe and indicates that K is enhanced over the bulk at small sizes. Using the above measured K one can calculate E_{ani} for small clusters. For example, a cluster containing 100 atoms has an E_{ani} of 0.0125 eV. The thermal behavior of the clusters at any given temperature is then determined by a competition of the thermal energy k_BT and the anisotropy energy. In particular, the thermal relaxation time is given by[20]

$$\tau = \tau_o e^{kV/k_B T} \tag{1}$$

where $\tau_o \sim 10^{-10}S$. The observation time τ for the experiments on clusters is ~ 100 μs. Using Eq. (1), this corresponds to a temperature, T_B, of around 10K for clusters having 100 atoms. The temperature T_B is called the blocking temperature. When the cluster temperature is below T_B, the moment $N\mu$ is locked around the easy axis. On the other hand, for clusters with temperatures above T_B, the moment $N\mu$ can change from one direction to another. The resulting paramagnetic behavior is called superparamagnetism. (We would like to point out that the changes in the direction of $N\mu$ require coupling to other contributions to angular momentum. The details of this mechanism are discussed elsewhere[21].) Note that the recent Stern-Gerlach experiments on Fe_n and Co_n clusters were carried out with cluster tempertures ranging from 70K--300K. These temperatures are much above T_B calculated above and therefore these clusters undergo superparamagnetic relaxation. The effective moment per atom along the direction of the field, H, is then given by the classical expression[18],

$$\mu_{eff} = \mu(\coth(\frac{N\mu H}{k_B T}) - \frac{k_B T}{N\mu H}) \tag{2}$$

where k_B is the Boltzmann constant. For small fields, μ_{eff} depends linearly on H and N, and is given by

$$\mu_{eff} = \frac{N\mu^2 H}{3k_B T} \tag{3}$$

We first discuss the case of Fe_n clusters. In order to calculate μ_{eff} using Eq. (2) and (3) one needs to know $N\mu$, the total spin moment of the cluster. In the absence of a knowledge about μ in clusters, we assumed that μ in clusters has the same value as in bulk namely μ_{Fe} = 2.2 μ_B. In Fig. 3, we show the μ_{eff} calculated using Eq. (2) for clusters containing 64, 130, and 240 atoms. These signs correspond to the average sizes for which de Heer et al. have measured μ_{eff} experimentally. Also shown are the measured[4] μ_{eff} corresponding to clusters having 56-72, 120-140, and 225-256 atoms. In all the comparisons the temperature T is the only parameter. We found that the theoretical curves reproduce the experimental data at T = 135K for 64 atom clusters, T = 146K for 130-atom clusters and T = 230K for 240-atom clusters. The theoretical curves are in good agreement with experiment supporting the superparamagnetic model.

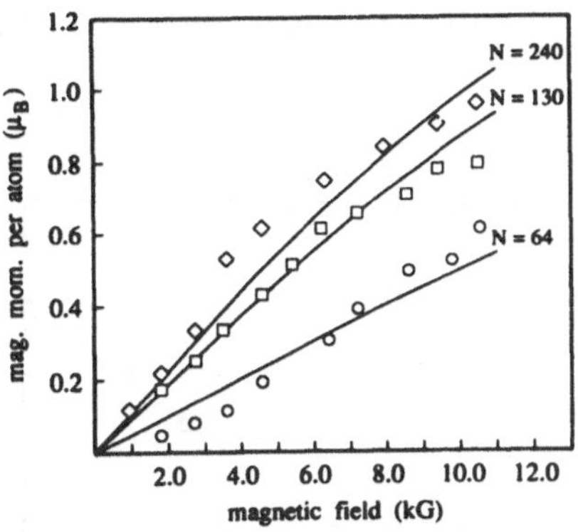

Figure 3. Magnetic moment per atom as a function of magnetic field for Fe_N clusters having 64, 130 and 240 atoms calculated using Eq. (2) and tempertures of 135, 146 and 230 K, respectively. Continuous curves are the theoretical values and , , and are the experimental results[4] corresponding to clusters having 56-72, 120-140 and 225-256 atoms, respectively.

We now discuss the results on Co_n clusters[5]. There are two important aspects to these measurements. Firstly, Bloomfield and coworkers[5] have carried out experiments as a function of the residency time of clusters. The residency time corresponds to the duration in which the clusters equilibrate with the carrier gas and by increasing residence time, one can bring cluster temperatures closer to source temperatures. Secondly they have measured μ_{eff} as a function of cluster size and this provides an independent check on the model. We first start with measurements on 115-atom Co clusters as a function of residency time. They found that the measured μ_{eff} increases with the residency time and saturates to a constant value of 0.402 μ_B at H = 0.307 T and temperature T = 83 K. These measured values allow one to determine μ in clusters using Eq. (2). One obtains μ = 2.08 μ_B. A comparison of this value with the bulk moment[12] of 1.72 μ_B shows that the moment in small clusters is indeed enhanced over the bulk value (by about 20%) as expected theoretically. Bucher et al[5]. have also measured μ_{eff} in Co clusters having between 20 and 110 atoms for residency times of 850 μs and source temperature 82 ± 1 K. Since the measurement as a function of residency time indicates that it takes about 1.6 ms for the clusters to attain source temperatures, these clusters have temperatures higher than the source temperatures. Employing Eq. (1) and μ_{eff} for Co_{85} clusters determined above, the cluster temperature can be estimated to be 277 K. Assuming this temperature for all clusters and μ = 2.08 μ_B, μ_{eff} vs N can be calculated. This is shown in Fig. 4, along with the experimental values. One finds a remarkable agreement between the calculated and the experimental data, establishing, once again, the validity of the theoretical model.

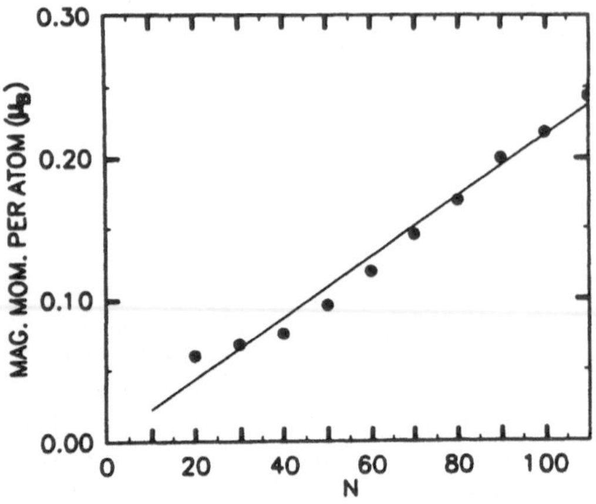

Figure 4. Variation of the magnetic moment per atom as a function of the number of atoms in Co clusters for T = 170 K and H = 0.53 T. Continuous curve is the theoretical values and solid circles are the experimental results (5).

To summarize, we have shown that the observed magnetic behavior of the transition metal clusters can be explained by assuming that they undergo superparamagnetic relaxation. We have also shown that the μ in small Co_n clusters is indeed higher than bulk as expected theoretically. We would like to add that Eq. (2) is valid only when $T >> T_B$. This will not be the case if $T < T_B$. In this case, the moment $N\mu$ will be fixed along the easy axis and the cluster would behave like a gyroscope[22]. This will be discussed in a different paper[21].

This work was supported in part by a grant from the Army Research Office (DAAL-03-89-K-0015). One of us (S.N.K.) is grateful to L. Bloomfield and J.P. Bucher for discussions.

IV. REFERENCES

1. E. Kaldis, Ed., "Crystal Growth of Electronic Materials", (North-Holland, Amsterdam, 1985).

2. "Physics and Chemistry of Small Clusters", edited by P. Jena, B.K. Rao, and S.N. Khanna, (Plenum Press, NY, 1987).

3. D.M. Cox, D.J. Trevor, R.L. Whetten, E.A. Rohlfing and A. Kaldor, Phys. Rev. B32, 7290 (1985).

4. W.A. de Herr, P. Milani, and A. Chatelain, Phys. Rev. Lett. 65, 488 (1990).

5. J.P. Bucher, D.C. Douglass, and L.A. Bloomfield, Phys. Rev. Lett. 66, 3052 (1991); D.C. Douglass, J.P. Bucher, D.B. Haynes, and L.A. Bloomfield (this proceedings).

6. R.L. Whetten (Private Communication).

7. Ivan Amato, Science, p. 1379 (June 1991).

8. J.C. Slater, "The Self-Consistent Field for Molecules and Solids" (McGraw-Hill, New York, 1974).

9. P. Hohenberg and W. Kohn, Phys. Rev. B136, B864 (1964); W. Kohn and L.J. Sham, ibid. 140, A1133 (1965).

10. D.E. Ellis, Int. J. Quantum Chem. S2, 35 (1968); D.E. Ellis and G.S. Painter, Phys. Rev. B2, 2887 (1970).

11. E. Wimmer, H. Krakauer, M. Weinert and A.J. Freeman, Phys. Rev. B24, 864 (1981).

12. F. Liu, M.R. Press, S.N. Khanna, and P. Jena, Phys. Rev. B39, 6914, (1989).

13. M.R. Press, F. Liu, S.N. Khanna, and P. Jena, Phys. Rev. B40, 399 (1989).

14. F. Liu, S.N. Khanna, and P. Jena, Phys. Rev. B43, 8179 (1991).

15. B.I. Dunlap, Phys. Rev. A41, 5691 (1990).

16. S.N. Khanna, B.K. Rao, P. Jena, and J.L. Martins, in "Physics and Chemistry of Small Clusters", Ed. P. Jena, B.K. Rao, and S.N. Khanna (Plenum Press, 1987) p. 435.

17. F. Liu, S.N. Khanna and P. Jena, Phys. Rev. B42, 976 (1990).

18. S.N. Khanna and S. Linderoth, Phys. Rev. Lett. 67, 742 (1991); J. Merikoski, J. Timonen, M. Manninen and P. Jena, Phys. Rev. Lett. 66, 938 (1991). (These authors assumed cluster temperatures larger than Tc, the real temperatures are much below Tc); P.J. Jensen, S. Mukherjee and K.H. Bannemann (this conference proceedings).

19. G. Xiao, S.H. Liou, A. Levy, J.N. Taylor, and C.L. Chien, Phys. Rev. B34, 7573 (1986).

20. I.S. Jacobs and C.P. Bear, in "Magnetism", edited by G.T. Rado and H. Suhl (Academic Press, New York, 1963).

21. S.N. Khanna (to be published).

22. D.C. Douglass, J.P. Bucher, and L. Bloomfield (to be published).

MAGNETISM OF FREE TRANSITION METAL AND RARE EARTH CLUSTERS

J.P. BUCHER
Department of Physics, University of Virginia
Charlottesville, VA 22901, Virginia, USA
Present address : Institut de Physique Expérimentale, EPFL,
PHB – Ecublens, CH – 1015 Lausanne, Switzerland

ABSTRACT. Current magnetic deflection experiments on transition metal and rare earth clusters as well as their connection to theory are discussed. It is found that magnetic moments of Co and Fe clusters are well described by a Langevin function, hinting strongly at the superparamagnetic nature of these particles. Rare earth metals like Gd and Tb exhibit a variety of behaviors some being well described by considering the particles as spinning tops with a body fixed magnetic moment. Informations on magnetic moments, T_{rot}, and magnetic anisotropies are obtained.

1. Introduction

Until recently, transition metal clusters have only been studied for their stability and their chemical activity. In the past two years however, studies of magnetic properties of transition metal[1-6] and rare earth[7] clusters have developed into a new direction of research. Unlike the paramagnetic clusters in which the net magnetic moment per atom depends strongly on the parity of the number of valence electrons, and generally decreases as $1/N$ (where N is the number of atoms), the clusters of ferromagnetic materials have huge magnetic moment per atom that do not depend strongly on N for big clusters. Magnetic field gradient deflection measurements performed on these clusters provide precious informations on their magnetic properties.

Ferromagnetic ordering and magnetic phase transitions can be understood in terms of fluctuations with many scales of length.[8] Therefore, interesting effects are expected when the correlation length of the spin system is no longer commensurate with the linear dimensions of the particle. For instance in magnetically ordered materials, there is a size (typically 20 nm) below which the grains lose their spontaneous magnetization at temperatures well below the bulk Curie temperature. This effect is called superparamagnetism and is due to the fact that the giant (single domain) magnetic moments of the particles begin to fluctuate in direction under thermal activation. Since the magnetic anisotropy of an isolated particle is proportional to its volume, room temperature superparamagnetism is observed in ultrafine particles only.[9]

Recent developments in molecular beam technology offer the unique opportunity to study free clusters of even smaller sizes (typically 2 to 500 atoms) in which the atomic character becomes apparent. The evolution of their magnetic properties can be followed very precisely by stepwise addition of atoms. In this regime, additional effects due to the surface, the low dimensionality,[10] and the symmetry[11] play important roles. Transition metal and rare earth clusters are expected to show quite different behaviors owing to the

P. Jena et al. (eds.), Physics and Chemistry of Finite Systems: From Clusters to Crystals, Vol. I, 721–732.

different origin of their magnetism but also because of the different relative importance of exchange and anisotropy energy in both systems.

The experimental technique is similar to the one which lead Stern and Gerlach to discover space quantization in atoms.[12] It involves the passage of the clusters through an inhomogenous magnetic field. Space quantization is not observed in ferromagnetic clusters because laboratory fields are too small to decouple individual spins. One essentially observes the behavior of the cluster as one single object.

2. Experimental techniques

All facilities in use[2-7] are of the type described by Cox et al.[1] This technique couples the conventional Stern-Gerlach deflection scheme with laser vaporisation cluster source (LVCS) technology and time-of-flight mass spectrometry (TOFMS).

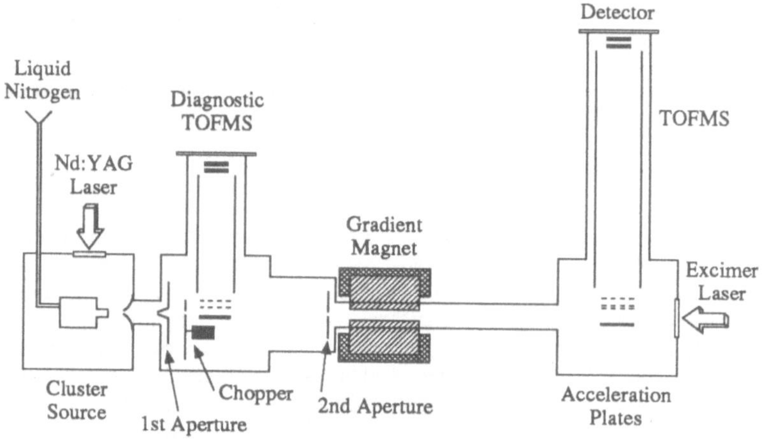

Fig. 1. Diagram of the experimental apparatus used at the University of Virginia.[4,5,7]

A schematic diagram of the experimental apparatus used at the University of Virginia (UVA)[4,5,7] is shown in Fig. 1. The metal clusters are grown in a helium-filled source from the pulsed laser vaporisation of a sample material. Unlike other LVCSs, the source used at UVA retains the clusters in a cavity during the pump-out time of a helium gas pulse backed by a pressure of several bars. During this time (approximately 0.7 ms) the clusters stop growing and thermalize with the source cavity. The helium-cluster mixture then undergoes a free-jet expansion into the vacuum, producing a supersonic molecular beam. The beam is collimated by two slits, 0.4 mm wide and 2.5 mm high and passes through a 250 mm long gradient magnet assembly, where the clusters are deflected according to the projection of their magnetic moments onto the field axis. The clusters travel from the magnet to the last vacuum chamber through a flight tube and are finally ionized 2.770 m downstream from the source by a narrowly collimated pulsed ultraviolet light (193 nm) from an ArF excimer laser. The cluster ions are analyzed by means of a time-of-flight mass spectrometer (TOFMS). The collimated laser beam (0.5 mm wide) is scanned accross the cluster beam in order to determine the deflection profile. Due to deflections of several millimeters observed for clusters of magnetic materials, the drift space of the TOFMS is made bigger than in conventional spectrometers.

2.1. SOURCE CONCEPTION AND OPERATIONAL MODE

The success of a magnetic experiment depends critically on the careful control of all experimental parameters. Temperature effects on the magnetic properties of clusters have been recognized. They are not just limited to the temperature dependence of the order parameter since other excitations not present in the bulk come into play. In order to observe new effects and insure their unequivocal interpretation, a control of the cluster temperature is still a prerequisite.

Liquid nitrogen has been used to cool the nozzle[2,3] or even the whole source assembly.[4,5,7] The advantage of reducing the translational velocity of the carrier gas by lowering its preexpansion temperature appears clearly in the increased sensitivity. Clusters that spend more time in the magnet deflect more. However, the fact that the internal center of mass temperature of the carrier gas pulse can be made very low, a few 100 mK,[13] does not necessarily ensure a low vibrational temperature of the seeded clusters since the two systems interact only for a very short time.

Indeed, it was found that the supersonic expansion is quite ineffective in cooling down the vibrational degree of freedom of the clusters.[5] As a result, a completely different strategy was adopted at UVA. The clusters are allowed to cool down in the source cavity during a residence time τ_{res}. Intrasource cooling of the clusters involves a succession of events. At one end of the chain one finds the source walls, at the other the clusters lattices and ultimately the spins. The heat exchange is mediated by the helium atoms. The collisional process produces an exponential evolution of the clusters temperature towards the source equilibrium temperature with an estimated time constant of about 1 ms.[5] Therefore, from a practical point of view, after 1 ms the clusters are essentially in equilibrium with the source. Well below 1 ms, it is likely that the clusters are considerably warmer than the source.

2.2. TIME DELAYS AND VELOCITY

In magnetic measurements with pulsed molecular beam techniques, three different stages can be distinguished. (i) vaporization and cluster confinement. (ii) flight time through the apparatus and the magnet. (iii) mass and deflection measurement.

To help separate different events in time, the apparatus used at UVA has a rotating chopper wheel located just after the cluster source. The velocity selection is achieved during the second stage since only clusters that pass through the chopper when it opens and arrive at the acceleration region of the TOFMS when the ionizing laser fires are detected. The velocity v_x thus obtained is important in determining the magnetic moment. The time at which the detected clusters left the source can be extrapolated back and the residence time τ_{res} of the clusters in the source can be determined. For given source conditions, the residence time controls the vibrational temperature of the clusters. The longer τ_{res}, the closer the vibrational temperature T_{vib} will be to the source temperature.

2.3. EFFECTIVE MAGNETIC MOMENT AND INSTRUMENTAL BROADENING

Transition metal clusters, typically Fe, Co and Ni were found to deflect exclusively towards increasing magnetic field. They produce a relatively narrow, well defined deflection profile. Therefore the following expression relating the deflection to the effective magnetic moment per atom can be used:

$$d = (G/m\, v_x^2)\, B\, \mu_{eff}(N, B, T) \ , \tag{1}$$

where v_x is the axial velocity of the cluster, m is the atomic mass, B is the magnetic field in the z-direction and G is a geometrical factor, constant for a given experimental configuration. The symbols also appear in Fig. 2. μ_{eff} must be considered as a time average value of the projection of the magnetic moment onto the field axis, it depends implicitely on B, T and the number N of atoms in the cluster. Actually, most ferromagnetic clusters exhibit a μ_{eff} that is proportional to the applied field.

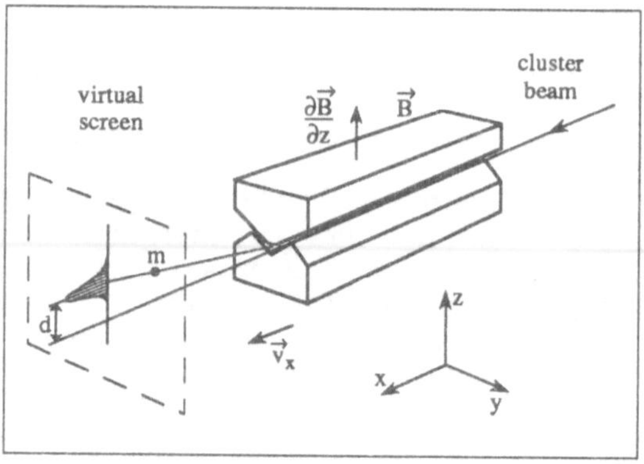

Fig. 2. Perspective view of the quadrupole sector pole faces used at UVA. This configuration offers the optimal field gradient homogeneity. Observable parameters are also represented and correspond to Eq. (1).

Since a gradient field is used to deflect the clusters, Eq. (1) is only valid for small deflections. When precision measurements are required, B can no longer be considered constant along the trajectory. This correction may be significant for large deflections since the process is amplified: the more the clusters deflect, the higher is the magnetic field and the bigger the projection of the magnetic moment onto the field axis. Due to the finite width of the beam, the same effect may produce an instrumental broadening of the deflection profile. This contribution is easily calculated in the case of the quadrupole magnet (Fig. 2), that has an almost perfectly uniform field gradient near the hyperbolic pole face but may lead to instrumental distorsions that are hard to estimate in the less favorable case of the generally used "two wire" magnet.[14]

2.4. ENSEMBLE AVERAGE VERSUS SINGLE PARTICLE EXPERIMENT

Magnetic measurements on small particles are not new. They have been practiced for decades on macroscopic samples, mostly particles embeded in a solid matrix or suspended in a liquid.[9,15-17] A force or an induction voltage measurement[18] on such a sample provides an ensemble average of the magnetic moment over a huge amount of clusters. Apart from a few exceptions,[16] the particles have a size distribution and a surface state that is not well characterized. In this context, the Stern Gerlach type of experiment can be classified into the category of force measurements as for example the Faraday method.[18] It is the instantaneous force $F_z = \mu_z \, \partial B/\partial z$ acting on a particle with magnetic moment μ in a field gradient that produces the curved trajectory. In a cluster beam experiment however, each cluster size can be studied separately since the whole mass spectrum is available from the cluster source. No ensemble averaging takes place, the only averaging is on the time the particles spend in the magnet. In a way, each cluster is a "new experiment" and produces its own signal at the detector. There is a profile of the deflected beam that contains informations on the strength of the coupling between the magnetic moment and the lattice, and on different states of the same particle that coexist in the beam.

3. Experimental results

Investigations on clusters of magnetic materials have been performed at Exxon (iron), EPFL-Lausanne (iron), UVA (cobalt and gadolinium) and UCLA (cobalt and iron). Since the purpose of this paper is essentially to present a comparative study of magnetic properties of transition metal and rare earth clusters we will focus on the study of Co and Gd clusters performed at UVA. It is also the only one which allows quantitative comparision with theoretical predictions. Some statements of comparison with the other studies will be given in the last paragraph of this section.

3.1. COBALT CLUSTERS

The measured magnetic moment of cobalt clusters increases linearly as a function of the magnetic field as is seen in Fig. 3 for three different cluster sizes. A linear increase of the effective magnetic moment is also observed as a function of cluster size as is shown in Fig. 4 for two different source temperatures. Due to the small amount of clusters remaining in the source after 1 ms or so, equilibrium measurements are difficult. Therefore, most Co data have been taken at shorter residence times. The real cluster temperature is then interpolated from a small number of "equilibrium" measurements.

Interestingly, the effective magnetic moment is found to increase with decreasing cluster temperature. In order to demonstrate the characteristic scaling behavior of the effective magnetic moment with N, B, and T, source temperature equilibrium measurements (open symbols) have been reported in Fig. 5 as a function of NB/kT. Effective magnetic moment versus field data[5] have also been reported (full symbols). One obtains a straight line passing through the origin. The proportionality factor must have the units of the square of a magnetic moment. It turns out that the value of the magnetic moment μ thus obtained, is of the order of the bulk magnetic moment per atom. These

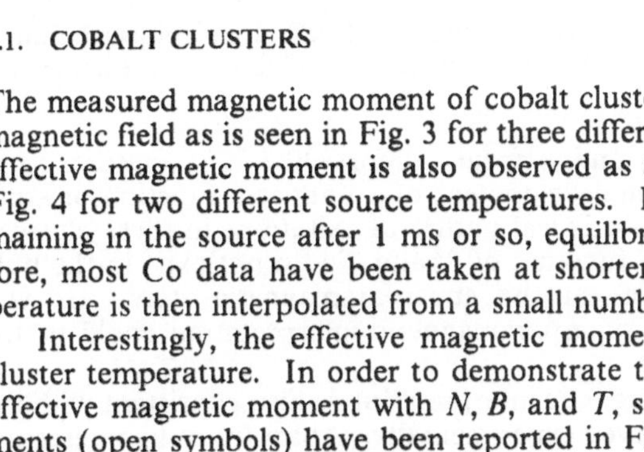

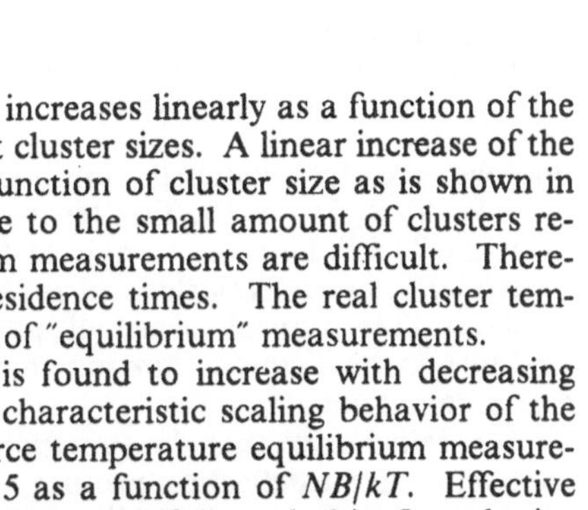

Fig. 3. Measured magnetic moment per atom as a function of magnetic field.[5]

Fig. 4. Measured magnetic moment per atom as a function of cluster size.[5]

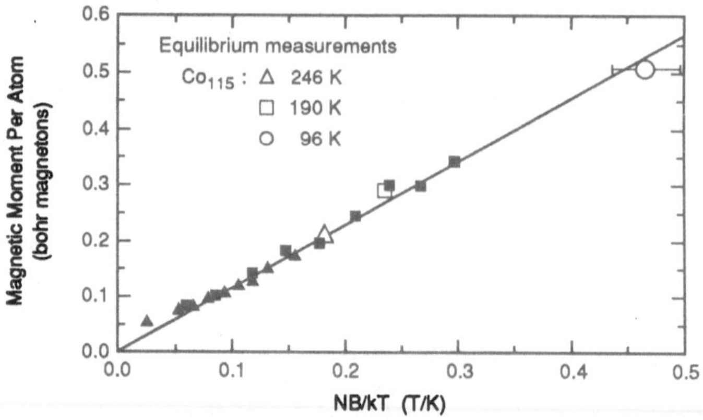

Fig. 5. Measured magnetic moment per atom as a function of NB/kT. Data points are from effective magnetic moment versus field data.[5] Source temperature equilibrium measurements are indicated by the open symbols.

findings are tremendously important and appear almost like a signature since $\mu_{eff}/\mu = N\mu B / 3kT$ is the first term of the development of a Langevin function of argument $N\mu B / kT$, hinting strongly at the statistical nature of exitations in these clusters. Particularly, linear variations of μ_{eff} as a function of N and T^{-1} do not exist in the bulk solid and are typical of small particles in the superparamagnetic state. It was Khanna[19] who first emphazised that a superparamagnetic interpretation could fit the experimental data.

It is remarkable that the superparamagnetic interpretation is also consistent with the relatively narrow deflection profiles observed experimentally. Apart from the interesting appearance of two magnetically resolved isomers for clusters between Co_{55} and Co_{66};[5] the statistical time averaging produces nearly the same deflection for all clusters of the same size. While cobalt clusters are well described by the superparamagnetic assumption within the sizes and the temperatures accessible experimentally (the lowest accessible temperature being 77 K), it is legitimate to ask if there is any limit to the superparamagnetic behavior. The answer is provided by the rare earth clusters.

3.2. GADOLINIUM CLUSTERS

The monotonic behavior of the magnetic moment as a function of size observed in magnetic transition metal clusters, is essentially a result of the relatively low magnetic anisotropy (much lower than the exchange energy) compared to the thermal energy kT. The magnetic anisotropy energy of rare earths is much bigger in absolute value than the one of ferromagnetic transition metals, and it is the competition between the exchange energy and the crystal field anisotropy energy that dictates most of the magnetic properties of these metals.[20] The study of the magnetic structure of Gd clusters[7] presented in this review is the first study of this type performed on rare earth clusters. Previously, Eu, Yb and Sm clusters have only been studied for their mass aboundances.[21]

While Co clusters deflect towards high fields by leaving their initial deflection profile unchanged, Gd clusters on the contrary exhibit a variety of behaviors. One category of cluster sizes immediately modify the width of their deflection profile at comparatively low field. These clusters are caracterized by an extreme spatial spreading. The second category of clusters that appear as magnetic "magic numbers" in Fig. 6 only show a moderate deflection and do not spread. A careful analysis shows that Gd_{22}, Gd_{30}, and Gd_{33} behave very much like superparamagnetic clusters. They show the right behavior as a function of B and T.

The spreading of the deflection profiles deserves particular attention since it is the shape of the profile rather than the average deflection that contains most of the information. As we will show later, the spreading behavior appears to be due to clusters with magnetic moments locked to their lattices. Figure 7 shows the deflection profile for Gd_{21} at four value of the magnetic field. These profiles are representative of all the non-superparamagnetic clusters. At very small applied field, the clusters spread almost symmetrically to both strong and weak field. As the field increases, the profile begins to shift toward strong field. Data of Fig. 6 and 7 were taken with τ_{res} long enough to insure that the clusters had T_{vib} = 105 K.

As the source temperature is increased, more clusters behave superparamagnetically. However, some clusters still exhibit a spreading behavior at 300 K, and it is only above 600 K that all Gd clusters are found to be superparamagnetic.[7] These clusters still deflect although the Curie temperature of bulk Gd is reached just around room temperature.

3.3. CLUSTERS OF OTHER MATERIALS

While effective magnetic moments of Fe and Co clusters are reported to show linear variations as a function of cluster size and magnetic field,[6] only the linear dependence in the field is observed by the authors of ref. 2 for Fe. Both studies only provide qual-

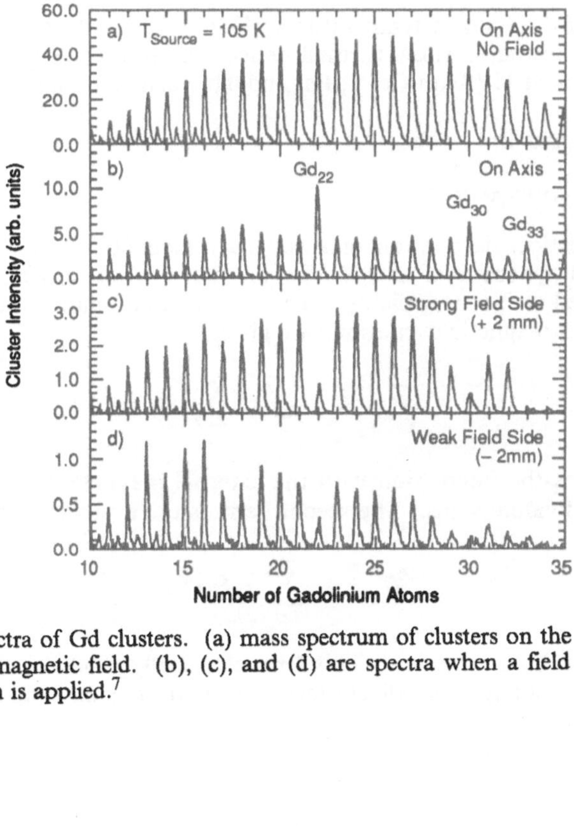

Fig. 6. Mass spectra of Gd clusters. (a) mass spectrum of clusters on the undeflected beam axis without any magnetic field. (b), (c), and (d) are spectra when a field of 0.132 T and a gradient of 50 T/m is applied.[7]

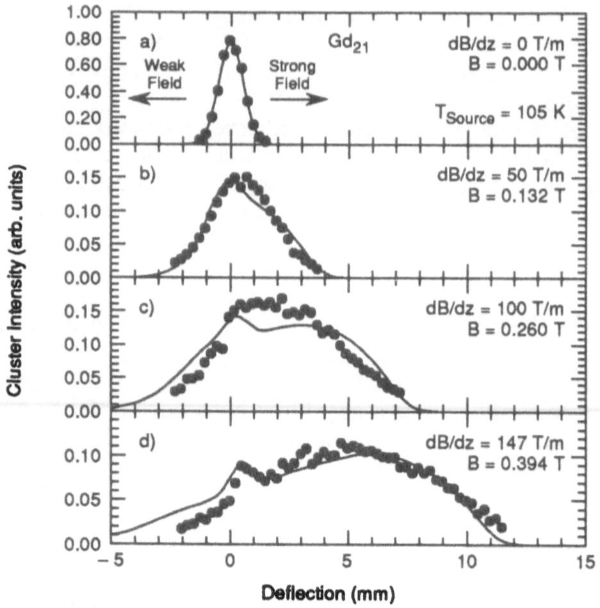

Fig. 7. Deflection profile of Gd_{21} as a function of magnetic field. The solid lines are the predictions of the rotating particles with body-fixed magnetic moment using T_{rot} = 5.8 K and μ = 2.51 μ_B per atom.[7]

itative informations on cluster temperatures. Due to the μ^2 dependence of μ_{eff} (Eq. (4)), Ni clusters are expected to show a small deflection, about ten times smaller than Fe. Investigations done at UVA on Cr, V, Pd, and Tb clusters are discussed elsewhere.[22]

4. Interpretation and discussion

4.1. STATISTICAL AVERAGING

The simplest statistical description of a superparamagnetic cluster that takes into account the internal dynamics of individual spins, but neglects anisotropy effects, is provided by a localized Heisenberg picture. The total Hamiltonian is the sum of two terms. The truly superparamagnetic excitations are given by

$$\mathcal{H}_{ze} = -g\mu_B B \sum_i S_i^z . \tag{2}$$

This term describes the interaction with the external magnetic field. The S_i^z are the z-components of individual spins. The second term is the exchange interaction

$$\mathcal{H}_{xc} = \sum_{i<j} J_{ij} \mathbf{S}_i \cdot \mathbf{S}_j , \tag{3}$$

equally observed in bulk materials. This term essentially produces a small temperature dependent variation of μ in small clusters. In analysing data on superparamagnetic systems, it has been generally admitted that both effects may be treated independently and that there is some kind of superposition principle.

A statistical mechanical treatment of the above problem[23] shows that it is equivalent to consider a giant magnetic moment of magnitude $\mu' = N\mu$ that fluctuates rapidly in direction under thermal agitation. The probability of a cluster having a magnetic energy $U_{mag} = -\boldsymbol{\mu}'\cdot\mathbf{B}$ is then proportional to the factor $e^{-U_{mag}/kT}$. The clusters explore most of the Boltzmann distribution since it takes them a few hundred μs to fly through the magnet, a time long compared to the thermal fluctuations. Averaging over time leads to an effective magnetic moment governed by the Langevin function

$$\mu_{eff} = \mu\left[coth\left(\frac{N\mu B}{kT}\right) - \frac{kT}{N\mu B}\right]. \tag{4}$$

A bulk dependence of μ of the type $(1 - T/T_c)^\beta$ has been assumed quite generally. This assumption is only true for big particles, since for a finite system like a cluster, fluctuations in μ are relatively important.[23]

When $T \ll T_c^{\,bulk}$, μ can be assumed independent of temperature and a detailed study of magnetic properties of free clusters becomes possible. As an example, Co_{115} clusters brought into thermal equilibrium with a source temperature of 83 K prior to the free jet expansion show an intrinsic magnetic moment per atom $\mu = (2.08 \pm 0.20)\ \mu_B$. This value is enhanced over the bulk value of 1.72 μ_B. Enhancements of magnetic moments have been predicted theoretically for low-dimensional systems.[10,24] First principle cluster calculations, although limited to $N<15$ show similar trends.[25]

As is clear from the experiment, the fact that the magnetic moments exchange angular momentum with the particles lattice through crystal anisotropy couplings does not invalidate the simple picture presented above (at least not within sizes and temperatures accessible experimentally). When the coupling between the magnetic moment and the lattice is increased, a departure from the statistical behavior is expected. The phenomenon becomes difficult to describe and a rigorous theoretical treatment must include consideration of relativistic effects. So far, only simple models assuming uniaxial anisotropy could be solved numerically.[26,27] While the first model[26] predicts correctly the trivial limit of superparamagnetism, the predictions reported in ref. 27, in the strong coupling limit, do not seem to agree with experimental results. It is also unclear from these calculations how the coupling with the lattice would affect the shape of the deflection profile. In the high temperature limit, "spin melting" occurs[28] and the magnitude of μ itself is affected.

4.2. LINE SHAPE INFORMATION

The rotational degree of freedom is essential to explain the results on Gd. Particularly the initial stage of profile broadening (Fig. 7b), is typical for a situation in which all orientations of the magnetic moment would still be equally probable in the presence of an applied magnetic field. The clusters are pulled to strong and weak fields by a force $F_z = \mu' \cos\theta\, \partial B/\partial z$ in a way very similar to what was expected in the event of a classical result of the Stern Gerlach experiment on atoms.[29] Such a situation is realised in a small particle with body-fixed magnetic moment when the rotational energy $U_{rot} = I\omega^2/2$ is much bigger than the magnetic energy $U_{mag} = \mu' B$. Due to the high rotational temperature, the body-fixed magnetic moment will not align in the field but will produce a net magnetic moment along the axis of rotation that will point in every possible direction, different for each cluster. As long as the isotropic distribution is not perturbed significantly by the applied magnetic field, the profile will broaden symmetrically. Its width will be proportional to the magnitude of the magnetic moment $\mu\, B$.

This way of measuring the magnetic moment of any locked moment particle in low magnetic fields has not been fully exploited yet.

The locked moment reasoning developed above can be extended to the case where $U_{rot} \ll U_{mag}$. This is the textbook problem of a physical pendulum. The particle oscillates in and out of alignment with the direction of the magnetic field. Solving the differential equation leads to elliptic integrals. In the intermediate regime, where U_{rot} is comparable to U_{mag}, the complexity of the behavior together with the fact that a great many different initial conditions leading to different values of the deflection must be considered, imposes a numerical solution of the problem.

Taking the rotational temperature T_{rot} and the intrinsic magnetic moment μ as free parameters, equation of motions have been integrated for 10^7 different initial orientations of the magnetic moment vector and the angular moment vector in space.[7] All four profiles in Fig. 7 have been fitted with only one value of $T_{rot} = 5.8\ K$ and one value of $\mu = 2.51\ \mu_B$ per atom. The agreement between experiment and simulation is rather good. As a matter of fact, the separation between superparamagnetic and locked moment behavior is not so clear cut for all cluster sizes and some of the spreading cluster profiles must be explained by allowing a small percentage of superparamagnetic component.[7] Other clusters with a spreading component are in similar agreement with theoretical profiles based on the same rotational temperature and approximately the same magnetic moment per atom.

4.3. MAGNETOCRYSTALLINE ANISOTROPY

Crystal field anisotropy is the mechanism that locks the magnetic moment to the lattice. If we consider the case of a single domain particle, the crystal anisotropy energy E can be expressed in the following way

$$E = b_i\alpha_i + b_{ij}\alpha_i\alpha_j + b_{ijk}\alpha_i\alpha_j\alpha_k + \textit{higher order terms}\,, \tag{5}$$

where the tensor notation is used. The b-coefficients are 1st, 2nd, and 3rd rank tensors and are physical constants of the crystal. The α's are the direction cosines of the magnetization vector with respect to the crystal axes. Since the b-coefficients must conform to the crystal symmetry, the topography of the anisotropy energy surface only depends on the respective values (magnitudes and signs) of a small number of b's.

The spin-orbit coupling is generally believed to be responsible for the strong magnetic anisotropy observed in rare earth metals. The fact that Gd is an S-state (and therefore has no first order term in the spin-orbit coupling) is irrelevant and only poses a problem when E is expressed in terms of symmetric combinations of spherical harmonics with well defined J-multiplets.[30] Experimentally, Gd is found to have a quite appreciable anisotropy (of the same order than iron), that is well described by an expression like Eq. (5).[31]

Equilibrium easy directions of the magnetic moment correspond to the local minima of E. For the sake of simplicity, lets consider a uniaxial problem where a local minimum is well described by the stiffness $S = \partial^2E/\partial\theta^2$ at $\theta = 0$ with which the magnetic moment is bound to an easy direction. When $S \rightarrow \infty$, the magnetic moment will be trapped in a sharp potential minimum and the cluster will exhibit locked moment behavior. When S is finite the magnetic moment is allowed to explore a larger solid angle under thermal activation. When $S \rightarrow 0$, the moment is loosely coupled to the lattice. Such a cluster may already exhibit a behavior that resembles superparamagnetism although its magnetic moment is not completely unlocked from the lattice.

In clusters, the symmetries are not restricted to the ones dictated by the Bravais lattices and a variety of lower symmetries than in the bulk are expected. All situations described above may be found separately or even coexist, giving rise to the dramatic size dependence observed experimentally. While Gd_{21} and Gd_{23} have their magnetic moment strongly locked to their lattice, Gd_{22} appears to be superparamagnetic. The fact that Tb, an adjacent element of Gd, shows a similar behavior [22] is not trivial and may be related to the process of valence fluctuations observed in rare earth clusters.[32] As a result, the 4f multiplet structures of Tb and Gd may closely resemble one another.

Whether a given cluster has locked moment or not depends critically on temperature. As the cluster temperature increases, a moment that was tightly confined in a potential minimum at low temperature may begin to explore larger regions of solid angle. If the temperature is further increased, the potential barrier between two minima may be overcomed, the moment may eventually unlock from the lattice or tunnel through the barrier and the cluster will exhibit superparamagnetism.

The observed magnetic moments per atom are substantially lower than the ones expected from the atomic ground state multiplet. This may be due to the fact that individual magnetic moments are not necessarily aligned but may be at an angle with the easy axis of magnetization. This is also well known from the bulk rare earths; for instance in Gd, the ferromagnetic moment moves away from the c-axis with a maximum deviation of 65° at 180 K.[33] The effect may be accentuated in clusters, where the single ion behavior is very sensitive to the surroundings and where long range indirect interactions mediated by conduction electrons, the so called RKKY interactions,[34] can no longer develop normally.

5. Conclusion

The newly developed source concept allows for the first time the production of beams of clusters with a controled cluster temperature. This opens up the possibility of studying in great detail the magnetic properties of clusters as a function of size, temperature and magnetic field.

A consistent although not yet complete picture has emerged from existing results. Experimentally, two extreme situations can be distinguished. (i) Superparamagnetic relaxation produces a narrow deflection profile due to statistical averaging. (ii) Locked moment behavior which produces a characteristic spreading of the deflection profile. Not only have both behaviors been identified in agreement with theoretical predictions, but transitions from one behavior to the other under thermal activation has been observed. The exact temperature at which this happens, however, does not have a direct physical meaning, since it is difficult to formulate a free cluster equivalent of the concept of blocking temperature.[9]

In contrast, the fact that Gd clusters remain magnetic (they deflect) at temperatures well above their bulk Curie temperature has direct implications on the temperature dependence of the order parameter μ. While in a superparamagnetic particle μ_{eff} strongly depends on the magnetic field, μ can be expected independent of B as long as B is small compared to the exchange field. Therefore, the observation that μ still has a finite value at temperatures of at least twice T_c^{bulk}, is a direct contribution to our knowledge of the magnetism of isolated zero-dimensional systems.

The fact that Co clusters show a characteristic statistical behavior, which manifests itself by the Langevin dependence, implies that the magnetic moments relax. The fact that internal relaxation takes place in very small isolated systems has far reaching implications. Among others, it means that clusters made of a few tens of atoms possess their own heat bath and most importantly that the concept of temperature still has a meaning in such small systems. Molecular beam studies on cluster magnetism will no doubt contribute to test the validity of fundamental theories and calculations.

REFERENCES

1. D.M. Cox, D.J. Trevor, R.L. Whetten, E.A. Rohlfing, and A. Kaldor, Phys. Rev. B **32**, 7290 (1985).
2. W.A. de Heer, P. Milani, and A. Châtelain, Phys. Rev. Lett. **65**, 488 (1990).
3. W.A. de Heer, P. Milani, and A. Châtelain, Z. Phys. D **19**, 241 (1991).
4. J.P. Bucher, D.C. Douglass, P. Xia, B. Haynes, and L.A. Bloomfield, Z. Phys. D **19**, 251 (1991).
5. J.P. Bucher, D.C. Douglass and L.A. Bloomfield, Phys. Rev. Lett. **66**, 3052 (1991).
6. J.L. Persson and R.L. Whetten, unpublished.
7. D.C. Douglass, J.P. Bucher, and L.A. Bloomfield, to be published.
8. K.G. Wilson, Scientific American, August 1979.
9. I.S. Jacobs and C.P. Bean in *Magnetism*, edited by G.T. Rado and H. Suhl (Academic, New York, 1963), Vol. III, p. 272.
10. F. Liu, M.R. Press, S.N. Khanna, and P. Jena, Phys. Rev. **B 39**, 6914 (1989).
11. B.I. Dunlap, Phys. Rev. A **41**, 5691 (1990).
12. W. Gerlach and O. Stern, Z. Phys. **8**, 110 (1922); **9**, 349 (1922).
13. J.P. Bucher, D.C. Douglass, P. Xia, and L.A. Bloomfield, Rev. Sci. Instrum. **61**, 2374 (1990).
14. For a discussion, see D. McColm, Rev. Sci. Instrum. **37**, 1115 (1966).
15. J. Korecki and K. Krop, Surf. Sci. **106**, 444 (1981).
16. G. Xiao, S.H. Liou, A. Levy, J.N. Taylor, and C.L. Chien, Phys. Rev. B **34**, 7573 (1986).
17. T. Furubayashi and I. Nakatani, Solid State Commun. **74**, 821 (1990).
18. H. Zijlstra in *Experimental Methods in Magnetism*, edited by E.P. Wohlfarth (North Holland, Amsterdam, 1967), chap. 3.
19. S.N. Khanna and S. Linderoth, Phys. Rev. Lett. **67**, 742 (1991).
20. A.R. Mackintosh, Physics Today, June 1977, p. 23; Europhys. News **19**, 41 (1988).
21. C. Brechignac, Ph. Cahuzac, F. Carlier, M. de Frutos, A. Masson, and J. Ph. Roux, Z. Phys. D **19**, 195 (1991).
22. D.C. Douglass, J.P. Bucher, D.B. Haynes, and L.A. Bloomfield, this conference.
23. J.P. Bucher and L.A. Bloomfield, Phys. Rev. B, in press.
24. G.M. Pastor, J. Dorantes-Davila, and K.H. Bennemann, Phys. Rev. B **40**, 7642 (1989).
25. C.Y. Yang, K.H. Johnson, D.R. Salahub, J. Kaspar, and R.P. Messmer, Phys. Rev. B **24**, 5673 (1981); A. Rosen and T.T. Rantala, Z. Phys. D **3**, 205 (1986).
26. P.J. Jensen, S. Mukherjee, and K.H. Bennemann, Z. Phys. D, to be published.
27. P. Ballone, P. Milani, and W.A. de Heer, to be published.
28. J. Merikoski, J. Timonen, M. Manninen, and P. Jena, Phys. Rev. Lett. **66**, 938 (1991).
29. O. Stern, Z. Phys. **7**, 249 (1921).
30. R.R. Birss in *Symmetry and Magnetism* edited by E.P. Wohlfarth (North Holland, Amsterdam, 1964), chap. 5.
31. R.J. Elliott in *Magnetic Properties of Rare Earth-Metals*, edited by R.J. Elliott (Plenum, London, 1972), chap. 1.
32. M.G. Mason, S.T. Lee, G.A. Apai, R.F. Davis, D.A. Shirley, A. Franciosi, and J.H. Weaver, Phys. Rev. Lett. **47**, 730 (1981); W. Niemann, M. Lübcke, W. Malzfeldt, P. Rabe, and R. Haensel, J. Magn. Magn. Mater. **47 & 48**, 462 (1985).
33. S.K. Sinha in *Handbook of Physics and Chemistry of Rare Earths*, edited by K.A. Gschneidner and L.R. Eyring (North Holland, Amsterdam, 1982), Vol. 1, chap. 7.
34. T. Kasuya in *Magnetism*, edited by G.T. Rado and H. Suhl (Academic Press, New York, 1966), Vol. IIB.

BAND STRUCTURE EFFECTS AND MAGNETISM IN QUASICRYSTALS

F.CYROT-LACKMANN, C.BERGER, A.GOZLAN, T.KLEIN, D.MAYOU
CNRS - Laboratoire d'Etudes des Propriétés Electroniques des Solides
BP166 - 38042 GRENOBLE CEDEX 09 - FRANCE
Associated to Université Joseph Fourier, Grenoble.

ABSTRACT : Two different types of quasicrystals have been studied, both formed by alloying sp elements with a small proportion of transition elements. The metastable AlMn(Si) icosahedral and decagonal phases present a weakly magnetic behaviour increasing sharply with the Mn concentration. Transport measurements are also similar to those observed in amorphous materials in the same composition range.

On the other hand, stable AlCuFe and AlPdMn phases of high strutural quality present peculiar bahaviors such as very high resistivities associated to a reduced density of states at the Fermi level and a diamagnetic susceptibility. The temperature and magnetic field dependences of the conductivity can be described by quantum interference effects including a strong electron-electron interaction contribution to the magnetoconductivity as already observed in other systems close to a metal-insulator transition. This behavior can be interpreted as a peculiar band structure effect due to a large number of Bragg diffractions and a strong interaction between the Fermi surface and the pseudo Brillouin zone. A model which takes into account these diffractions and their effects on sp and d states via the sp-d coupling is proposed. A pseudogap in both sp and d D.O.S. is obtained in agreement with ab initio LMTO calculations. This can lead to very special features in the electronic and magnetic properties of the quasicrystals.

1 INTRODUCTION

Many metastable and even stable quasicrystals (QC) are formed by alloying sp elements with a small proportion of transition elements. We present here transport and magnetic properties of three kinds of such quasicrystalline samples : AlMn(Si), AlCuFe and AlPdMn. The AlMn(Si) system is of peculiar interest since one can obtain in the same compositional range amorphous (a-), icosahedral (i-) and decagonal (T-) phases[1,2]. Contrary to the icosahedral phase which is a 3D quasicrystal, the decagonal phase possesses one axes of periodicity perpendicular to quasicrystalline planes and is thus often presented as an intermediate phase between the icosahedral and crystalline phases. However these metastable systems can only be obtained by out of equilibrium methods such as melt spinning and possess thus a great amount of structural defects. On the other hand, the stable AlCuFe and AlPdMn systems allow to obtain very high structural quality quasicrystals[3] (very low density of defects, absence of phason strains) and are thus of great interest to study the specific properties of the quasicrystalline structure.

P. Jena et al. (eds.), Physics and Chemistry of Finite Systems: From Clusters to Crystals, Vol. I, 733–742.

The most salient feature is that the stable QC show very high resistivity values[4] (up to 10000 μΩcm at 4K in $Al_{62.5}Cu_{25}Fe_{12.5}$) associated to a reduced density of states at the Fermi level and a diamagnetic behaviour[4]. On the contrary the resistivity in the metastable AlMn(Si) i- and T-phases are lower[1,2] (a few hundred μΩcm) and of the same order of magnitude as that of amorphous samples of the same composition ; the magnetic properties are also different showing a low fraction of magnetic moments increasing with the Mn concentration and a spin glass transition at low temperature[5]. These electronic properties are remarkable and show that these alloys can be classified in two different regimes : one is a simple extension of solid solution of transition metals in a sp matrix (AlMn(Si)) and the other is a Hume Rothery type system (AlCuFe,AlPdMn). Indeed, it appears that the diffraction of electrons by Bragg planes is important in the stable QC and induces for example strong variation of the density of states as compared to the free electron model. We will present a model, starting from an Anderson Hamiltonian (used to describe a transition metal impurity in a sp matrix), that allows to discuss some effects concerning band structure and magnetism.

2 HIGH RESISTIVITY VALUES

Master ingots of the different systems were prepared by melting high purity elements under argon atmosphere. Thin ribbons of quasicrystalline phases were then prepared by melt spinning. A subsequent annealing is sometimes necessary to remove additionnal crystalline phases. The purity and stuctural quality were confirmed by powder X-ray diffraction patterns. The resistivity was measured using the classical four probes technic between 300 K and 350 mK .

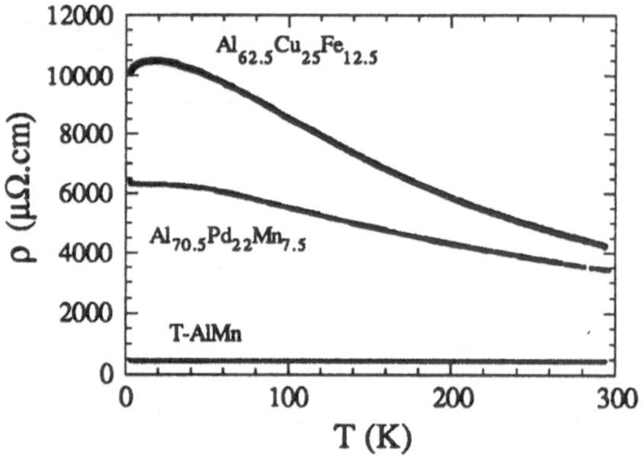

Figure 1 : resistivity as a fonction of temperature for $Al_{62.5}Cu_{25}Fe_{12.5}$, $Al_{70.5}Pd_{22}Mn_{7.5}$ and T-AlMn.

Figure 1 presents the temperature dependence of the resistivity of i-AlCuFe, i-AlPdMn and T-AlMn. At 4K, resistivities of about 400 μΩcm have been measured in both amorphous and decagonal AlMn phases and ρ_{4K} = 200 μΩcm in i-AlMn(Si). These resistivities are associated to a small negative temperature dependence ($\Delta\rho/\rho\Delta T = 2.10^{-4}\ K^{-1}$) in agreement with the Mooij criterium. However, these metastable systems possess a great amount of structural defects and it is interesting to notice that an approximant of high structural quality can be obtained in $Al_{72.5}Mn_{17.5}Si_{10}$, this approximant (α-AlM(Si)) has then a very high resistivity value[6] : $\rho_{4K} = 6700$ μΩcm which is of the same order of magnitude than those observed in our stable quasicrystals of high structural quality : $\rho_{4K} = 6500$ μΩcm[7] in $Al_{70.5}Pd_{22}Mn_{7.5}$ and $\rho_{4K} = 10000$ μΩcm[4] in $Al_{62.5}Cu_{25}Fe_{12.5}$. These values are associated with a reduced density of states at the Fermi level[4,22] which could be due to peculiar Bragg like diffractions by similarity to Hume Rothery rules. Moreover in these systems the resistivity varies strongly with a slight change in nominal composition : $\rho_{4K} = 4300$ μΩcm in $Al_{63}Cu_{25}Fe_{12}$ and $\rho_{4K} = 10000$ μΩcm in $Al_{62.5}Cu_{25}Fe_{12.5}$ for instance.The temperature dependence of the conductivity can also be well described by quantum interference effects[8] (weak localisation and electron-electron interaction effects) as usually observed in amorphous systems of much lower resistivities[9]or heavily doped semiconductors close to the metal - insulator transition[10,11].

Moreover we have shown in a previous paper[4] that in AlCuFe and AlCuRu the conductivity at room temperature and the conductivity at 4K are strongly correlated with each other. This correlation shows that our nearly perfect QC are close to a metal - insulator transition[21] where the zero temperature conductivity should vanish for a characteristic finite value at the room temperature. This correlation can be extended to $Al_{70.5}Pd_{22}Mn_{7.5}$ and α-AlMn(Si) as shown in Figure 2.

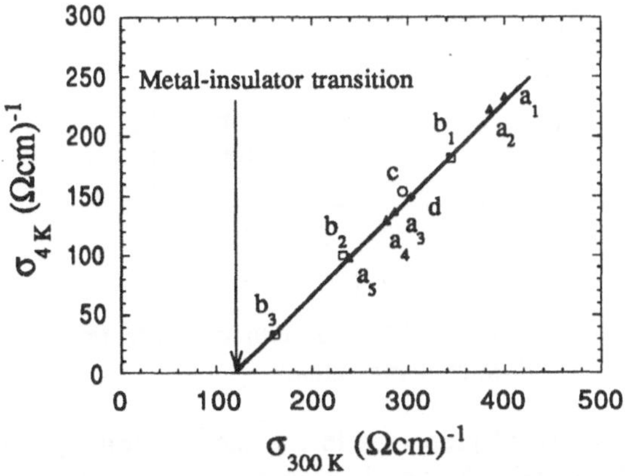

figure 2 : conductivity σ_{4K} as a function of σ_{300K} for our i-AlCuFe (a), i-AlCuRu from ref. 23 (b), i-AlPdMn (c) and α-AlMnSi from ref. 6 (d).

3 MAGNETIC PROPERTIES

3.1. THE AlMn(Si) SYSTEM

The study of the i- and T-AlMn(Si) phases has suggested the existence of two different classes of Mn atoms, one being magnetic and the other non-magnetic. The presence of magnetic moments in this system is confirmed by a.c. susceptibility measurements (in a field of 1 Oe) showing a Curie-Weiss law and a spin glass maximum at a temperature increasing with the Mn concentration[2,5].

The magnetoresistance (MR) a-, i-, T-AlMn phases are also dominated by magnetic contributions at low temperature. Indeed, these systems have a negative MR at low temperature which increases up to a positive value as the temperature increases . This negative MR could be understood by electron scattering on magnetic sites. Another way of investigating the scattering of electron by magnetic moments is by studying the Hall effect. Indeed in this case, an anomalous contribution is added to the ordinary Hall effect and the Hall resistivity (ρ_H) could be related to the magnetization (M) by the relation[12] $\rho_H = R_0H + 4\pi R_S M$ where R_0 and R_S are the ordinary and anomalous Hall coefficients respectively. ρ_H is then not linear with H but presents curvatures that strenghen with decreasing temperature. Figure 3 presents ρ_H / H versus M / H at different temperatures with H varying from 0 to 7 T for T-AlMn. This curve shows that R_S is almost temperature independent and about 10^4 times largerhan R_0. This large anomalous Hall effects could be due to a strong spin orbit scattering[13].

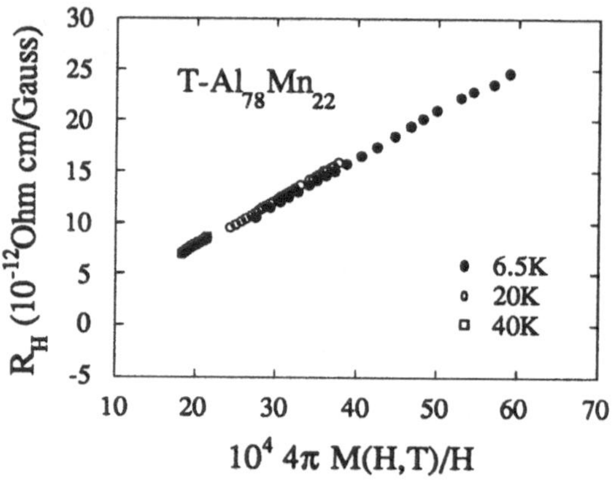

Figure 3 : linear dependence of ρ_H / H upon M(H,T) / H with M varying from 0 to 7 Tesla and at different temperatures.

In conclusion, all the measured properties in i, a and T AlMn(Si) samples show the presence of a low fraction of magnetic Mn sites that increases with the Mn concentration. This behavior has to be opposed to that of α(AlMnSi) and orthorhombic Al_6Mn approximants of the i -phase which have no magnetic moments [14].

3.2. THE AlPdMn AND AlCuFe SYSTEMS

The magnetic susceptibility of AlCuFe and AlPdMn have been measured using a SQUID susceptometer between 1 and 2 Tesla . In AlPdMn, both paramagnetic and diamagnetic behaviors have been observed depending on the Mn concentration. Indeed, the samples become more and more diamagnetic as the Mn concentration decreases (Figure 4). Such a diamagnetic behavior has already been observed in $Al_{63}Cu_{25}Fe_{12}$ The origin of that diamagnetism is still an open question but could be due to a reduced effective mass (m*) leading to a strong Landau diamagnetism as observed in semiconductors ($\chi_{Landau} / \chi_{Pauli} = -1/3\ (m/m^*)^2$). It is interesting to notice that the most resistive AlPdMn sample is also diamagnetic. Moreover the resistivity of our samples increases as their structural quality is improved. The paramagnetic contributions might thus be due to structural defects and the diamagnetic behavior might then be characteristic of the perfect quasicrystalline structure.

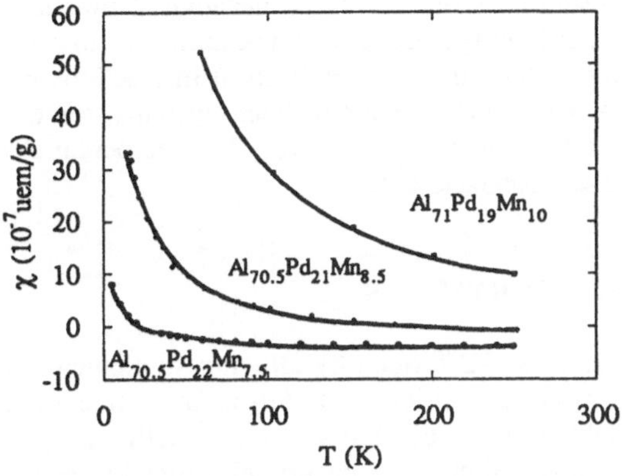

Figure 4 : variation of the high field (1-3 Tesla) magnetic susceptibility with T for i-AlPdMn of various compositions.

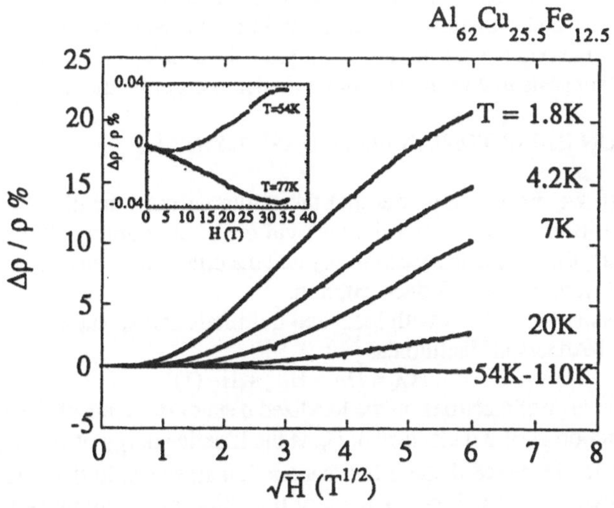

Figure 5: Magnetoresistance as a function $\sqrt{H}$ of for Al62Cu25.5Fe12.5. In the insert : the variation with T of the magnetoresistance at 54K and 77K shows a minimum.

The magnetic field dependence of the conductivity of AlCuFe could then be described by non magnetic quantum interference effects including strong electron-electron interaction contributions as already observed in other systems close to the metal insulator transition such as heavily doped semiconductors. Indeed, as seen on figure 5, at low temperature the MR is positive up to 35 Tesla and $\Delta\rho/\rho \propto \sqrt{H}$ in a large field range. Furthermore at 54K we can observe a non classical feature with the MR being negative at low field and positive at higher field (insert of Figure 5). This striking behavior could be attributed to competing effects between weak localisation (negative MR) and electron-electron interactions (positive MR). Such a behaviour has never been observed in amorphous systems for which electron-electron interactions can usually be neglected (except at very low temperature) but has already been observed in heavily doped semiconductors[11].

4. MODEL FOR QUASICRYSTALS

We have studied various QC formed by alloying sp element swith transition metals showing very different electronic properties, depending on their structural quality. The results on the thermodynamically stable samples of high structural quality suggest that they are Hume Rothery alloys for which diffraction of conduction electrons by Bragg planes is strong and determinant for their properties[15,16,17]. Indeed, the condition for having Hume Rothery alloys is that the differences in electronegativities and atomic radii between the constituents are small and this is well obeyed in these compounds[17]. The magnetic and transport properties as well as the small density of states at the Fermi level are different from what would be expected by considering them as extension of solid solution of transition metals. In particular, for AlCuFe, iron is not magnetic and probably not Kondo and the D.O.S. at the Fermi level is small. This contradicts the virtual bound state model which predicts an enhancement of the D.O.S. due to the d resonance, and suggests that we need another model to describe these systems.

4.1. ANDERSON HAMILTONIAN FOR QUASICRYSTALS

In this part, we propose a model that takes into account the diffraction of electrons by Bragg planes and its effect on sp and d states via the sp-d coupling. We want to discuss the electronic structure, the magnetic susceptibility and the criterion for the appearance of magnetism which is different from the usual Stoner criterion.

When considering systems with localized d orbitals and sp states, a widely used starting point is the Friedel-Anderson Hamiltonian[18] :

$$H_A = H_d + H_{dc} + H_c \quad (1).$$

H_d describes the energy of electrons on the localized d states (in terms of the on site energy of the d level and interaction U of 2 d electrons), H_c is the kinetic energy of free electrons and H_{dc} is the hopping between localized d states and conduction sp states. In the Q.C. studied here there are no, or very few, nearest neighbour pairs of transition metal atoms and the direct hopping between d orbitals can be neglected. In quasicrystals for which the diffraction of conduction electrons by Bragg planes is important, it is necessary to add a term that describes it. Thus we propose a hamiltonian for quasicrystals and more generally for Hume-Rothery phases containing transition element in a sp matrix with the form :

$$H = H_A + H_B$$

$$H_B = \sum_j V(\vec{K}_j) \cos(\vec{K}_j . r + \varphi_j) \qquad (2)$$

($V(\vec{K}_j) = V(-\vec{K}_j)$ is real and positive and φ_j is a phase factor)

H_B describes the diffraction by Bragg planes. Formula (2) amounts to describe the pseudopotential in a local approximation but this approximation could easily be relaxed without changing the following discussion. $V(\vec{K})$ in equation (2) is the self consistent potential in the Hume-Rothery phase. In the following we shall use reasonnable estimates of this pseudopotential to discuss physical properties. The $V(\vec{K})$ which are expected to be important for physical properties are such that $K \sim 2\,k_F$ and modify the electronic structure near the Fermi energy.

The H_B part of the hamiltonien obviously breaks the translationnal invariance of the Anderson part H_A with respect to the position of the transition atom. Thus we see that the electronic structure depends on the position of the transition metal atom . A simple physical image for the effect of the position of the transition metal atoms is the following. Due to the diffraction by Bragg planes the proper states of the conduction band (without coupling to the d orbitals) are no longer plane waves. For example, if we consider a plane wave $|\vec{k}\rangle$ with $\vec{k}$ lying near the Bragg plane associated to $\vec{K}$, then the important coupling (due to $V(\vec{K})$) is with $|\vec{k}-\vec{K}\rangle$ and gives rise to a bonding state $|\varphi_B(\vec{r})|^2 \approx \sin^2(\frac{\vec{K}.\vec{r}+\varphi}{2})$ and an antibonding one $|\varphi_{AB}(\vec{r})|^2 \approx \cos^2(\frac{\vec{K}.\vec{r}+\varphi}{2})$.

Depending on its position $\vec{r}$, the localized d orbital will be more coupled to the bonding ($\varphi_B(\vec{r})^2 > \varphi_{AB}(\vec{r})^2$) or antibonding state ($\varphi_{AB}(\vec{r})^2 > \varphi_B(\vec{r})^2$). This is the reason for the dependence of the D.O.S. on the position of transition atoms in our model.

We now consider more specifically systems with a sufficiently small concentration of transition elements. In these systems the electronic structure can be discussed by neglecting interferences between transition atoms. In other words each transition atom is treated as an impurity. In reference 19 we discuss the effect of the coherent diffraction of sp states by d orbitals.

4.2. MAGNETIC SUSCEPTIBILITY AND CRITERION FOR MAGNETISM

Within an Hartree-Fock approximation we calculate the paramagnetic susceptibility. We find that :

$$\chi = \chi_0 + \Delta\chi$$

with $\chi_0 = \mu_B{}^2 N_T(E_F)$

and $\Delta\chi = \mu_B{}^2 N_D(E_F) \left[\frac{U\, n_d(E_F)}{1-U|f_d'(E_d)|}\right]$

χ_0 is the Pauli susceptibility for independent electrons with total D.O.S. $N_T(E_F)$ and $\Delta\chi$ is the reinforcement due to the d orbitals. $Nd(E_F)$ is the partial d D.O.S. at the Fermi level, $n_d(E_F)$ is the partial d D.O.S. per spin ($n_d(E_F) = 1/2\, Nd(E_F)$), $f_d(E_d)$ is te total number of the electron per spin and orbital when the energy of the d orbital is E_D, and $f_d'(E_d)$ is the derivative with respect to E_D .The contribution of the impurity diverges for $U > U_C$ where $U_C = 1/|f_d'(E_d)|$ and corresponds to the appearance of magnetic moments. Without diffraction by Bragg planes one has $|f_d'(E_d)| = n_d E_F)$ and we recover the usual Stoner criterion. However, even for moderate value of the pseudopotential $V(\vec{r})$ we find that $|f'_d(E_d)|$ is very different from $n_d(E_F)$ and thus the Stoner criterion is not applicable in these systems. We note that in the non magnetic state the susceptibility $\Delta\chi$ is favored by the term $1 / (1\text{-}U|\ f_d'(E_d)|)$; however $n_d(E_d)$ tends to be small if the Fermi level lies in a pseudo gap and this could lead to a small value of $\Delta\chi$ consistent with the diamagnetic behaviour observed in many samples. This simple Hartree-Fock analysis suggests that magnetic properties can be much affected by Hume-Rothery effect and clearly further investigations are needed.

4.3. DENSITY OF STATES

We have performed density of states calculations with Hamiltonian (2) in the impurity approximation[19]. In this calculation we considered the effects of N equivalent Bragg planes (N=42) associated to N reciprocal vectors $\vec{K}$. We calculated the proper states of the conduction band (before coupling to the d states) within a classical two bands approximation which retains the coupling of the plane wave $|\vec{k}\rangle$ with only one state $|\vec{k} - \vec{K}\rangle$ (the one which has the closest kinetic energy). The parameter $V(\vec{K})$ was chosen such that the density of conduction states (before coupling) is divided by 2 as compared to the free electron value in the pseudogap where we put the Fermi energy). In Figure 6 we show the partial d D.O.S. (per eV) for ten degenerated orbitals containing 8 electrons which simulates the case of cobalt and the variation of density of states of the conduction band due to its coupling with the impurity. The d D.O.S. is different from the Lorentzian shape of the virtual bound states model and the variation of the sp D.O.S. is important. Qualitatively this result looks very much like that of ab initio calculation for AlMn by Fujiwara[20] or Al_9Co_2 by G.Trambly et al[19] (figure 7).

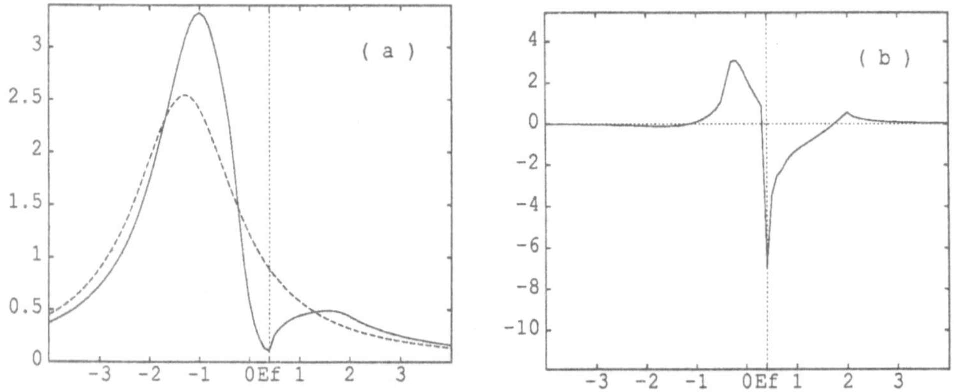

Figure 6 : Model calculation of the d density of states (a) and the variation of sp D.O.S. due to the impurity (b) for an atom containing 8 d electrons and coupled mainly to antibonding states. In (a) (- - -) represents the virtual bound state result and (—) the Q.C. result. Energie are in eV and the D.O.S. are in states/eV.

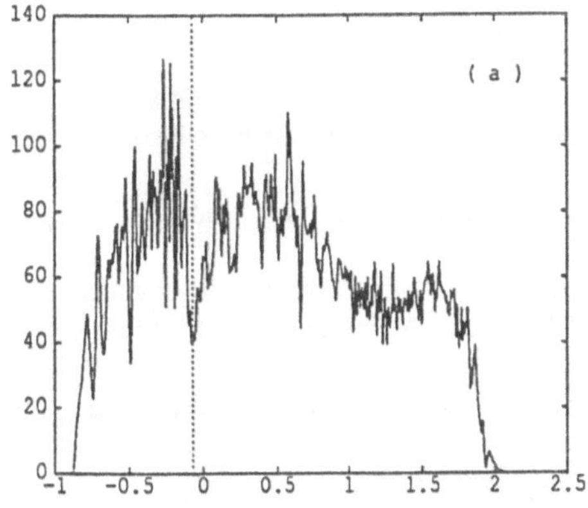

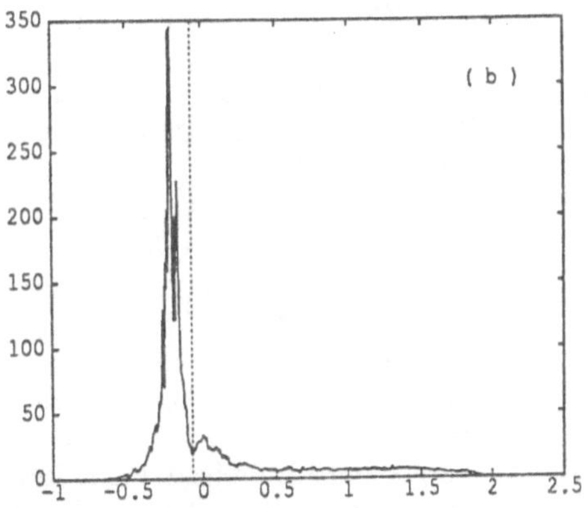

Figure 7 : Total sp states D.O.S. (a) and Co d D.O.S. (b) for Al_9Co_2 from an initio calculations. Energies are in Rydberg and D.O.S. in states/Ryd. unit cell.

5. CONCLUSION

To conclude we recall that the AlMn(Si), AlCuFe, AlCuCo and AlPdMn quasicrystals have remarkable transport and magnetic properties. In particuliar stable AlCuFe and AlPdMn have high resistivities, small density of states at the Fermi level and are often diamagnetic. Despite the presence of Mn, Fe or Cr atoms, very few magnetic moments are observed.

The Hume-Rothery-Jones effect is probably essential to understand these properties and we have proposed an Anderson hamiltonien that takes it into account. This model gives a description of the d and sp states and shows a strong departure from virtual bound state in agreement with ab initio calculations. Further studies are under investigation in particular for understanding the magnetic properties

We would like to thank Dr. Friedel for fruitful discussions,D.Nguyen Manh and G.Trambly for the numerical calculations, G.Fillion , M.Bouvier and G.Sulpice for the SQUID measurements , J.J.Prejean for the a.c. susceptibility measurements and P.Lanco for the study of the AlPdMn system. We also thank G.Fourcaudot and J.C.Grieco for technical assistance. T.Klein acknowledges the C.E.A. - C.E.R.E.M. for financial fellowship.

REFERENCES

[1] C. Berger, G. Fourcaudot, A. Gozlan, F. Cyrot-Lackmann, and J.C. Lasjaunias, 10th General Conference on Condensed Matter EPS (Lisbon, April 1990), Physica Scripta., T35, 90, 1991.

[2] A. Gozlan, C. Berger G. Fourcaudot, R. Omari, J.C. Lasjaunias, and J.J. Préjean, Phys. Rev. B, 44, 575, 1991.

[3] Y. Calvayrac, A. Quivy, M. Bessiere, S. Lefebvre, M. Cornier-Quiquandon, and D. Gratias, J. Phys., 51, 417, 1990 - A.P. Tsaï, A Inoue, and T. Masumoto, Phil. Mag. Lett., 62, 95, 1990.

[4] T. Klein, C. Berger, D. Mayou, and F. Cyrot-Lackmann, Phys. Rev. Lett., 66, 2907, 1991.

[5] C. Berger and J.J. Préjean, Phys. Rev. Lett., 64, 1769, 1990.

[6] S.J. Poon, submitted to Advances in Physics.

[7] P. Lanco, T. Klein, C. Berger, G. Fourcaudot, J.C. Grieco and F. Cyrot-Lackmann, submitted to Europhysics Letters.

[8] A.L. Alt'shuler, A.G. Aronov in Electron-electron Interactions in Disordered Systems (Eds. Al Efros, M. Pollack 1985, Elsever Science Publishers) Chap 1, and references therein.

[9] M.A. Howson, B.L. Gallagher, Physics Reports, 170, 265, 1988.

[10] C.M. Hurd, The Hall effect in Metal and Alloys (Plenum, New York, 1972).

[11] G.A. Thomas, A. Kawabata, Y. Ootuka, S. Katsumoto, S. Kobayashi, and W. Sasaki, Phys. Rev. B.,26, 2113, 1982.

[12] T.F. Rosenbaum, R.F. Milligan, M.A. Paalanen, G.A. Thomas, R.N. Bhatt, and W. Lin, Phys. Rev. B., 27, 7509, 1983.

[13] C.M. Hurd, Contemp. Phys., 16, 517, 1975.

[14] J.J. Hauser, H.S. Chen, G.P. Espinosa, and J.V. Waszczak, Phys. Rev. B., 34, 4674, 1986.

[15] J. Friedel, Helv. Phys. Acta, 61, 538, 1988.

[16] A.P. Smith and N.W. Ashcroft, Phys. Rev. Lett., 59, 1365, 1987.

[17] P.A. Bancel and P.A. Heiney, Phys. Rev. B., 33, 7917, 1986.

[18] P.W. Anderson, Phys. Rev., 124, 41, 1961.

[19] G. Trambly, D. Nguyen Manh, and D. Mayou, to be published.

[20] T. Fujiwara, Phys. Rev. B., 40, 942, 1989 - T. Fujiwara and I. Yokokawa, Phys. Rev. Lett., 66, 333, 1991.

[21] S. Sanquer, R. Tourbot and B. Boucher, Europhys.Lett. 7, 635, 1988.

[22] M. Mori, S. Matsuo, T. Ishimasa, T. Matsuura, K. Kamiya, H. Inokuchi and T. Matsukawa, J.Phys. Condens. Matter, 3, 767, 1991.

[23] B.D. Biggs,S.J.Poon and N.R. Murirathan, Phys.Rev.Lett. 65,2700, 1990.

MAGNETISM OF SMALL PARTICLES

S. GANGOPADHYAY, L. YIPING and G. C. HADJIPANAYIS
Department of Physics, University of Delaware, Newark, Delaware 19716.

C. M. SORENSEN and K. J. KLABUNDE*
*Department Of Physics, *Department Of Chemistry, Kansas State University, Manhattan, Kansas 66506.*

V. PAPAEFTHYMIOU and A. KOSTIKAS*
Department of Physics, University of Ioannina, Greece.
**Institute of Material Science, NCSR Demokritos, Athens, Greece.*

ABSTRACT. The magnetic and structural properties of fine Fe particles and Fe-B particles prepared by vapor deposition and $NaBH_4$ reduction respectively, have been studied. Fe particles in the range of 50-300 Å were obtained by varying the argon pressure during evaporation in the range of 0.5-30 torr. The Fe-B powders were prepared by reducing $FeCl_3$ with $NaBH_4$ in aqueous solution. Particle sizes in the range of 300-600 Å were obtained. The X-ray patterns indicate that powders with B content higher than 20 % are amorphous and those with lower than 20 % are crystalline. The dependence of magnetization and coercivity on particle size and temperature have been studied, using SQUID magnetometry and Mössbauer spectroscopy. The magnetic and structural data suggest a core/shell type of structure, where the core consists of metallic Fe (and/or Fe-B) and the shell is composed of Fe-oxides. The values of coercivity obtained (1200 Oe) are two orders of magnitude higher than those in bulk Fe and Fe-B metallic glasses.

1. Introduction

Metals and alloys in the nanocrystalline form exhibit properties which are drastically different from the bulk. This opens up an area of research for nano-structured materials that is very challenging scientifically and technologically. Fine particles are widely used for magnetic recording media, ferro-fluids, catalysts, medical diagnostics, drug delivery systems and pigments in paints and ceramics [1-3]. Enhanced magnetic properties of fine ferromagnetic particles make them very promising candidates for high density magnetic recording media.

Fine Fe particles (~ 200 Å) have been reported to have coercivities as high as two orders of magnitude higher than bulk Fe and their saturation magnetization can be varied between 20 - 90 % (of the bulk value) depending on particle size [4]. Particles in their fine form (few 100's of Å) are very pyrophoric and hence require a controlled

P. Jena et al. (eds.), Physics and Chemistry of Finite Systems: From Clusters to Crystals, Vol. I, 743–752.

surface passivation. The magnetic properties of the particles are strongly dependent on the form and constituents of the surface layers which constitute a major volume fraction in ultra-fine particles (UFPs). Thus a greater understanding and control of the surface layer in UFPs would result in optimum magnetic properties.

Ultrafine transition metal boron particles can be synthesized by reduction of metal salts with $NaBH_4$ in aqueous solution. A number of studies have dealt with the magnetic and structural properties of these fine particles [5-7]. The chemical reduction technique is very easy to use and it is very important scientifically and technologically. So far this technique has been used as granted to prepare boron containing particles. Very few studies if any exist to study the chemistry of the process. Our group is presently studying the detailed chemistry of the reduction process as it progresses to form the final product, in order to be able to get a better control over the latter part.

In this paper we will briefly review our recent studies on the structural and magnetic properties of fine Fe and Fe-B particles made by vapor deposition and $NaBH_4$ reduction technique, respectively. The size and temperature dependence of magnetic properties have been studied. A shell/core model [8] has been used to explain the magnetic properties and their temperature dependence.

2. Experimental

2.1. VAPOR DEPOSITION

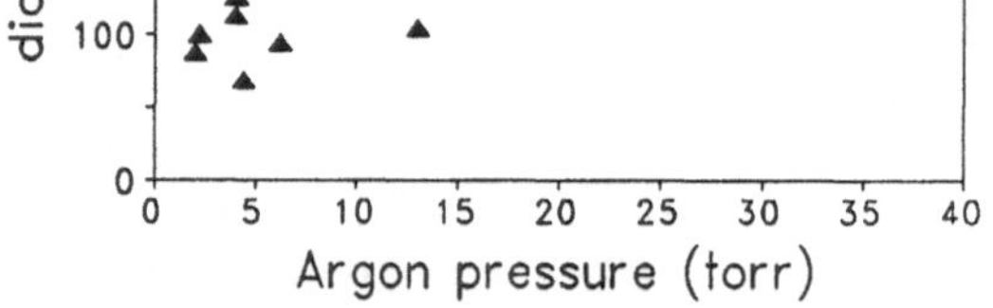

Figure 1. The particle size increases as the argon pressure during the evaporation increases.

In the vapor deposition technique the material is evaporated in the presence of an inert gas. The metal vapor condenses due to collisions with the inert gas molecules to form clusters of atoms which subsequently grow to form particles. These particles then agglomerate together to form chains/dendrites as they travel towards a cold substrate, due to gas convection currents and thermophoresis. The resulting sample is in the form of black soot or "smoke", which can be easily scraped off the substrate using a glass slide.The particles formed by the condensation technique are very pure and clean chemically. The particle size can be effectively controlled by various means including, the molecular weight of the inert gas, evaporation source temperature, inert gas pressure and the substrate temperature.

The joule heating element used in this study was an alumina-coated tungsten crucible. The distance between the crucible and the water cooled Cu substrate was optimized in order to maximize the yield. Argon gas was used to provide the inert atmosphere during evaporation. The dynamic pressure of argon used during evaporation varied between 1 and 30 torr. The particles obtained were of the size of 50-300 Å. The increase in particle diameter as the gas pressure increased is shown in Figure 1. With increasing gas pressure the mean free path of metal vapor is decreased, because of more collisions, leading to formation of larger particles.

The metal particles made by this technique are highly pyrophoric and require surface passivation which, as it will be shown later, plays an important role in the magnetic behavior of the whole particle. An argon-air mixture was introduced into the chamber for this purpose after evaporation of particles (volume ratio of argon to O_2 being about 300). The extent of oxidation could be varied by carefully changing the O_2 to argon ratio in the chamber after the evaporation. Samples with virtually no oxide could be prepared by sandwiching the metal particles between two thin films of Ag, with the whole assembly deposited on a Kapton substrate.

2.2. $NaBH_4$ REDUCTION

Powdered Fe-B samples were prepared by reducing aqueous solutions of iron salts with $NaBH_4$. The composition and structure of the powders were found to depend on the method of sample preparation. Crystalline Fe-B particles can be prepared using a "Y junction" through which aqueous solutions of $FeCl_3$ and $NaBH_4$ are mixed in the presence of an external magnetic field of 1 kOe. The maximum amount of boron that was found in the Fe-B powders was only 15 at% when the "Y junction" technique was used. However, if the $NaBH_4$ solution is added into a well stirred iron salt solution, a high boron-content powder is obtained which is also found to be amorphous (approximately 40 at% B). The black precipitates were separated and repeatedly washed with distilled water and subsequently in acetone, and then dried in a N_2 atmosphere chamber.

3. Magnetic and structural properties

3.1. VAPOR DEPOSITION

The enhanced and interesting magnetic properties of nanostructured materials are due to finite size effects, which makes the study of structure morphology and microstructure very crucial. Particles size and morphology in this study was determined using a transmission electron microscope. Different areas in the bright field micrographs were enlarged and sampled to arrive at the median particle size. Particle size distribution was found to follow a log-normal distribution. Small particles were more uniform in their size distribution ascompared to the bigger particle samples.

Particles formed were roughly spherical in the case of Fe and Co. Small deviations from sphericity were observed as the particle size increased. In the case of Ni particles, however, complete sphericity was observed up to about 150 Å. In most of the particles the observed images showed a distinct contrast with the center of particle being darker than its periphery. This could be due to the presence of an oxide shell around a metallic core brought in by the intentional passivation. The core/shell structure was also evident from dark field micrographs. A typical transmission electron micrograph (TEM) of an Fe sample is shown in Figure 2.

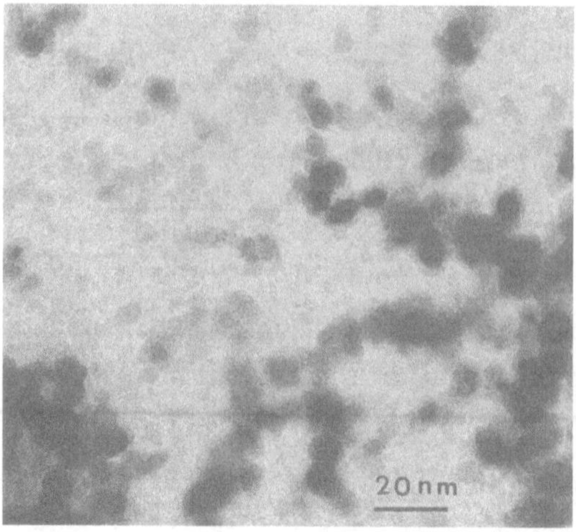

Figure 2. Bright field micrograph of fine Fe particles with a median particle size of 113 Å.

The sharpness of the metal diffraction lines in both X-ray and selected area diffraction patterns increased as the particle size increased. The lines due to the oxides were always broad and diffuse, which could be due either to their amorphous or poly-crystalline form. The oxides found in the case of Fe particles were Fe_3O_4 and/or γ-Fe_2O_3. CoO and NiO were found in the case of Co and Ni, respectively. X-ray photoelectron spectroscopy (XPS) was also used to determine the type of oxides. The amount of Fe_3O_4 was always found to be much higher than γ-Fe_2O_3.

Saturation magnetization as a function of particle diameter is shown in Figure 3. In the case of Fe particles the magnetization values varied between 25 and 190 emu/g as the particle size increased from 60 to 200 Å, respectively. The general decrease of magnetization with decreasing particle diameter can be attributed to the increased volume fraction of oxides in smaller particles. The reduction of magnetization below the oxide value will be discussed in section 4.

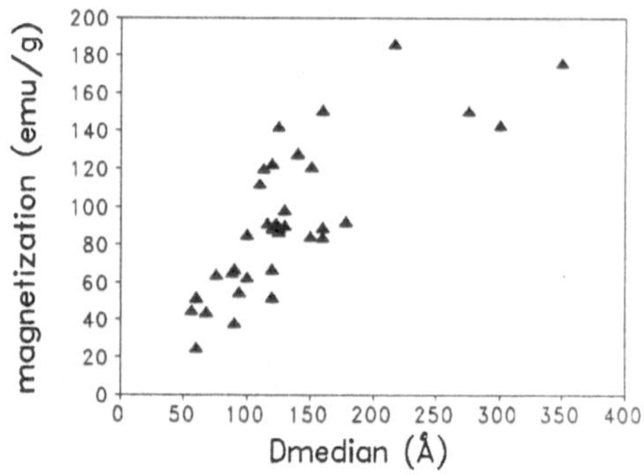

Figure 3. Size dependence of saturation magnetization at 10 K.

Coercivity as function of size is shown in Figure 4 at room temperature. The smaller particles are superparamagnetic even below room temperature. The particle size obtained in Fe particles was always below the single domain size, hence an increase in coercivity (due to thermal effects) with particle size was observed [9]. Temperature dependence of coercivity for different particle sizes is shown in Figure 5. The main feature of this graph is the crossover temperature at about 40 K, where the particles show an opposite size dependence of coercivity above and below this temperature.

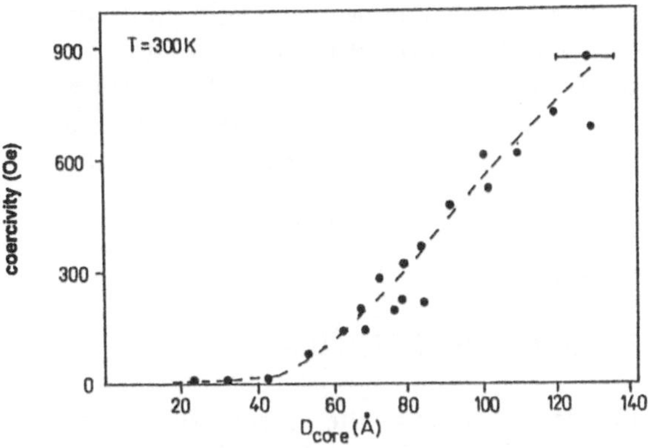

Figure 4. Size dependence of coercivity for Fe particles at room temperature. Particles below 40 Å core diameter show superparamagnetism.

The high field magnetization data were fit to the law of approach to saturation [10], to determine the anisotropy constant K using the expression

$$M=M_s[1-A/H-B/H^2-C/H^3]+\chi H \qquad (1)$$

where the first term (A/H) is related to the magnetization process in lower fields. The coefficients B and C are functions of the magnetocrystalline anisotropy constant K_1 and saturation magnetization M_s and χ is the high field susceptibility. Furthermore, the H_c vs $T^{1/2}$ law [9] was also used to estimate the value of "K". Both of these techniques gave values of K of the order of 10^6 ergs/cc, which is an order of magnitude higher than the bulk anisotropy value for Fe (4.8 x 10^5 ergs/cc).

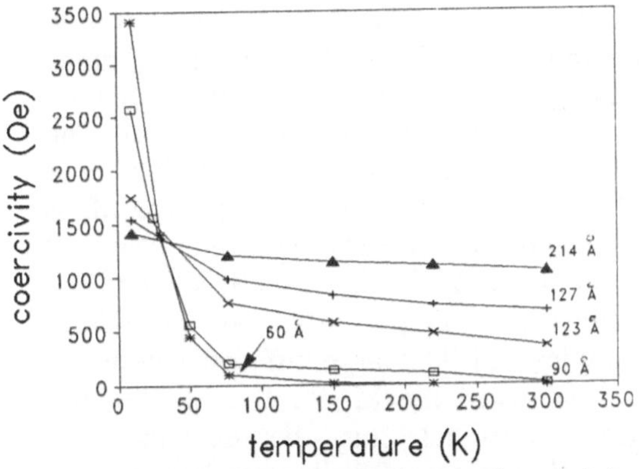

Figure 5. Temperature dependence of coercivity for Fe particles with different size. Note the crossover temperature of curves at about 30 - 40 K.

Mössbauer spectra on fine Fe particles [11] with median diameter 114 Å are shown in Figure 6. The spectra at the three temperatures are characteristic of mixtures of α-Fe with its oxides Fe_3O_4 and/or γ-Fe_2O_3. The presence of the sharp sextet as the temperature increases from 4.2 to 300 K indicates the ferromagnetic nature of α-Fe. However the oxide spectra evolve from a magnetically split spectrum to a very broad quadrupole doublet as the temperature increases from 85 to 300 K. This broadness of the oxide spectra at room temperature indicates that there exists a magnetic interaction between the superparamagnetic oxide shell and the ferromagnetic Fe core [12].

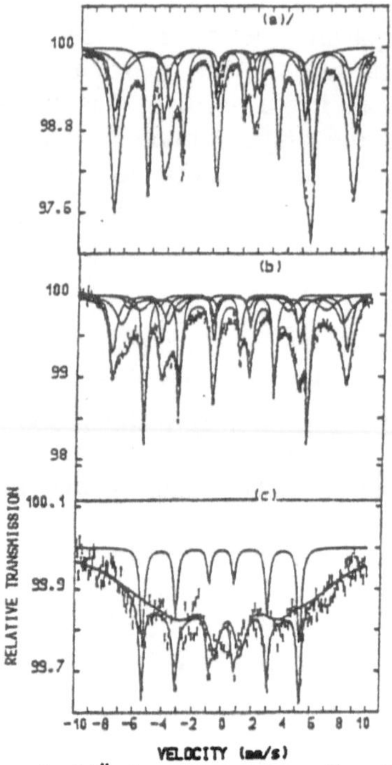

Figure 6. Mössbauer spectra of a 114 Å sized Fe sample at: (a) 4.2 K (b) 85 K (c) 300 K

3.2. $NaBH_4$ REDUCTION

Figure 7 shows the x-ray diffraction patterns for Fe-B powders with different boron concentrations. When the boron content is higher than 20 at%, the powders were found to be amorphous; otherwise they were crystalline with a bcc structure and with slightly broad peaks due to particle size effects. The particle size as determined from the TEM micrograph is found to be in the range of 300-600 Å in all the powders. Figures 8 and 9 show the magnetization and coercivity as a function of boron concentration. M_s is found to decrease initially with B content, going through a minimum at 15 at% and then increase again for higher B content. The largest change in magnetization was observed in the smallest particles. M_s is expected to decrease with increasing B content and it is usually lower in amorphous samples than in crystalline samples [13]. The large difference between the values of the M_s for the smaller and larger particles may be explained by oxidation effects. Smaller particles have a higher surface to volume ratio than larger particles and as a result smaller particles get more oxidized. When the B content is higher than 20 at%, the particles become amorphous and the particles do not oxidize as much as in the crystalline state and therefore M_s is higher. The coercivity was

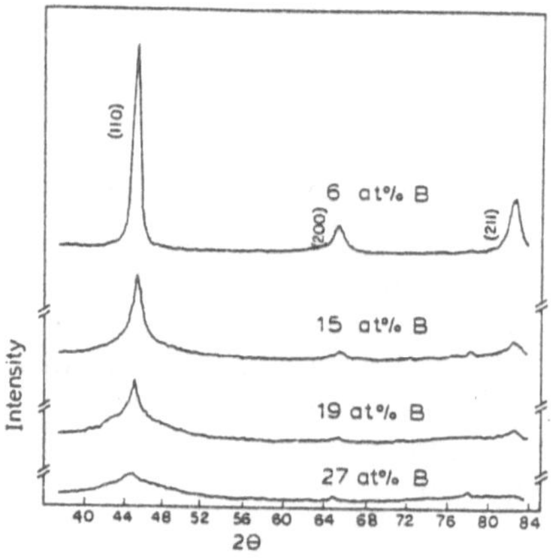

Figure 7. X-ray diffraction pattern for different B content, in Fe-B particles.

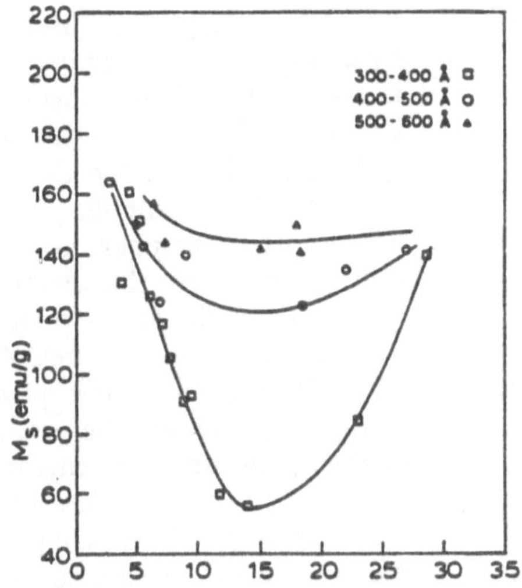

Figure 8. M_s as a function of B content for FeB particles.

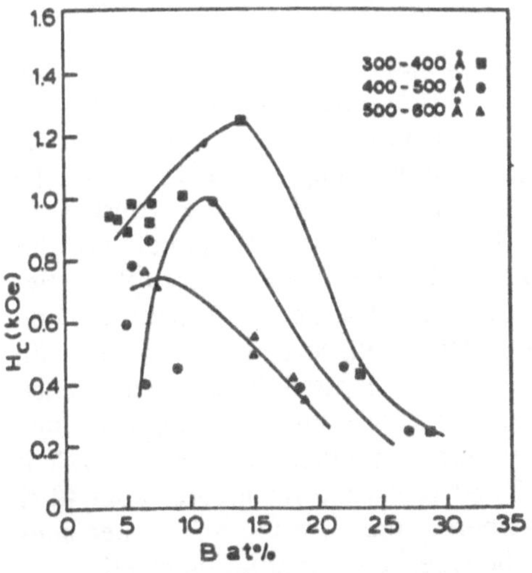

Figure 9. Coercivity as function of B content in FeB particles.

found to increase with decreasing B concentration and go through a maximum at a B content 15 at%. The large scattering of data is due to the uncertainty of size determination and to various degrees of oxidation. The coercivity obtained in fine Fe-B particles is at least two orders of magnitude higher than that reported on metallic glasses. It is worth to note, however, that the highest H_c was obtained in samples with a higher oxide content. This was confirmed by Mössbauer spectroscopy. The decrease of coercivity with decreasing particle size may be due to the multidomain nature of the larger particles.

Typical spectra obtained at 85 K for the amorphous and crystalline samples are shown in Figure 10. The spectrum of the amorphous sample consists clearly of a magnetic hyperfine pattern and a paramagnetic one. The first component with broad lines and an effective field near 273 kG corresponds to amorphous Fe-B. The second component which is a quadrupole doublet at 85 K becomes magnetic at 4.2 K with an average hyperfine field 485 kG. The Mössbauer parameters of this spectral component are typical for small particles of α-Fe_2O_3. The spectrum of the crystalline sample can be analyzed as a superposition of three magnetic hyperfine patterns as shown by the solid line (Fig. 10b). The first component with narrow lines and an effective field 340 kG is of α-Fe. The other two spectral components have broad lines and an average hyperfine field of 490 kG. These magnetic components start to broaden at T > 4.2 K and finally resolve into a doublet at 300 K. The hyperfine parameters of this component are very similar to previously reported parameters of small particles of α-Fe_2O_3 [14]. The values of isomer shifts for Fe^{2+}, $Fe^{2.5+}$ are close to those in bulk Fe_3O_4 and FeO. An indirect experiment supported the presence of

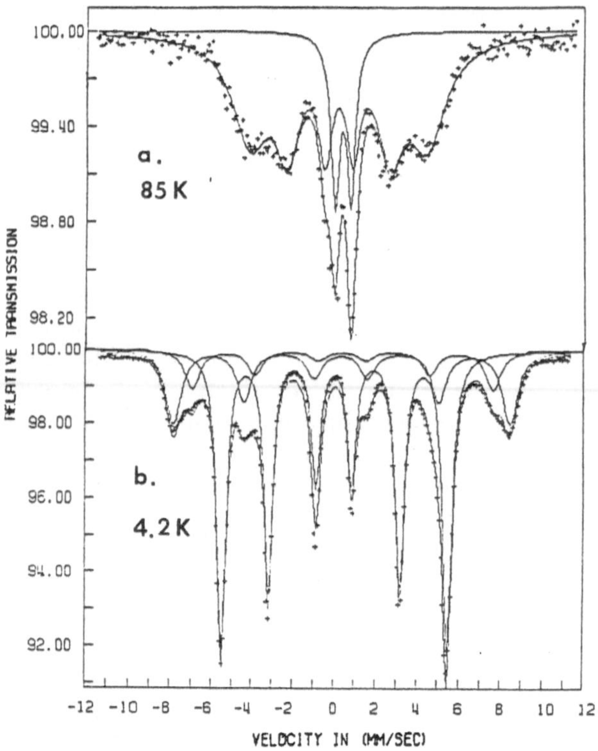

Figure 10. Mössbauer spectra for FeB samples (a) amorphous at 85 K, and (b) crystalline at 4.2 K.

an oxide coating around the Fe-B core. A sample was prepared by drying the as-made powder in a N_2 atmosphere inside a chamber. The sample was then split in two halves; the first was packed in a sample holder outside the chamber and the other one inside the chamber. The two samples were found to have different coercivities, 900 and 750 Oe, with the higher value obtained on the sample packed outside the chamber. Thus it may be speculated that the coercivity of the sample is mainly due to the surface oxide layers which coat the Fe-B core.

4. The "Shell-Core" Model

The structural, magnetic and Mössbauer data indicate a shell/core type of particle morphology. In Fe, Co and Ni particles, the core consists of metallic α-Fe, fcc Co and Ni respectively and the shell of the corresponding oxides (Fe_3O_4/γ-Fe_2O_3, CoO and NiO). In the Fe-B particles the core is an amorphous Fe-B matrix in the amorphous sample (B > 20 at.%) and bcc Fe(B) in the crystalline samples. The shell consists of Fe_3O_4/γ-Fe_2O_3. The shell/core model can explain adequately the magnetic hysteresis behavior of the particles and the size dependence of magnetization. In Fe particles, the Mössbauer data indicates a strong exchange interaction at the core/shell interface. Such exchange interaction at the core/shell interface could enhance the coercivity at 10 K. At temperatures greater than 40 K (crossover temperature in Figure 5) where the oxide shell becomes superparamagnetic, this exchange interaction between the superparamagnetic shell and the hard magnetic Fe core would make the whole particle magnetically soft, resulting in a strong temperature dependence of coercivity in smaller particles where the volume fraction of oxide is very large.

This model was further strengthened by the behavior of Fe/Ag samples where the amount of oxide on the particle surface was minimized [15]. This resulted (as expected) in a strong reduction of temperature dependence of coercivity; furthermore, the crossover among the coercivity vs temperature curves disappeared, as it is shown in Figure 7.

In Fe-B the size of the particles is much larger than the size of the Fe particles (by approximately a factor of 5) and therefore the effect of core/shell interface exchange interaction is not strong enough to affect the magnetic behavior of the particle as a whole. This results in a normal $H_c(T)$ behavior with a small increase in coercivity with decreasing temperature down to the Bloch temperature of the Fe-oxide where a stronger $H_c(T)$ dependence is observed.

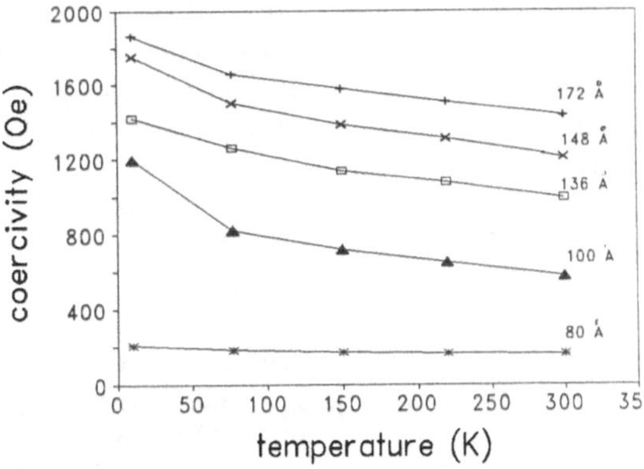

Figure 11. Reduced temperature dependence of coercivity in the case of Fe/Ag samples; the crossover temperature as in the case of passivated Fe samples is not present.

As it was pointed earlier, the magnetization is usually decreased with decreasing particle size, because of dilution effects due to the lower magnetization of the oxides. However, the magnetization values below 90 emu/g can not be explained by the dilution hypothesis. In the past a similar effect was observed in fine Fe-oxide particles, where magnetization values as small as 30 emu/g were observed in γ-Fe_2O_3 [16]. The reduced magnetization values have been attributed to different effects. Berkowitz et al [16] explained the lower M_s by assuming the existence of a non-magnetic layer ("dead" layer) at the surface of the particles. Later Morrish et al [17] and Coey et al [18] claimed that the "dead" layer is due to surface spin canting. Recently this idea has been disputed, however, and Pankhurst [19] claimed that the lower magnetization values are due to non-saturation (of magnetization) effects because of the random distribution of small Fe-oxide particles with enhanced values of magnetocrystalline anisotropy. Research in this area is presently very active and the latest results suggest are that the low M_s values are due to bulk spin-canting in the whole particle [20].

5. Acknowledgments

This work has been supported by NSF CHE-9013930.

6. References

1. Matijevic E. (1989) 'Fine Particles: Science and Technology', MRS Bulletin XIV, 19.
2. Ozaki M. (1989) 'Prepration and Properties of Well-Define Particles', MRS Bulletin XIV, 35.

3. Sharrock M. P. and Bodnar R. E. (1989) 'Magnetic Materials For Recording: An Overview With Special Emphasis On Particles', J. Appl. Phys. 57, 3919.
4. Tasaki A., Takao M. and Tokunaga H. (1974) 'Magnetic Properties of Ferromagnetic Metal alloy Fine Particles prepared by evaporation in inert gases', Jap. J. Appl. Phys. 13, 27.
5. Oppegard A. L., Darnell F. I., Miller H. C. (1961) 'Magnetic properties of single domain iron and iron-cobalt particles prepared by borohydride reduction' J.Appl.Phys. 32, 3184.
6. Kim S. G. and Brock J. R. (1987) 'Growth of ferromagnetic particles from cation reduction by borohydride ions' J. Coll. and Interf. Sci. 116, 431.
7. Linderoth S., Morup S., Koch C. J. W., Wells S., Charles S. W., Van Wonterghem J. and Meagher A. (1988) 'Ultrafine particles of amorphous $Fe_{63}B_{38}$, A study of structural relaxation and crystallization', J. de Physique 49, c-1369.
8. Gangopadhyay S. and Hadjipanayis G. C. 'Magnetic properties of ultrafine particles', Submitted to Phys. Rev. B.
9. Kneller E. F., Luberosky F. E. (1963), 'Particle size dependence of coercivity and remanence of single domain particles', J. Appl. Phys. 34, 656.
10. Cullity B. D. (1972) Introduction to magnetic materials, Addison-Wesely Publishing Company, 347.
11. Papaefthymiou V., Kostikas A., Simopoulos A., Niarchos D., Gangopadhyay S., Hadjipanayis G. C., Sorensen C. M., and Klabunde K. J. (1990), 'Magnetic hysteresis and Mössbauer studies in ultrafine Fe particles', J. Appl. Phys. 67(9), 4487.
12. Haneda K. and Morrish A. H. (1978), 'Mössbauer study of Fe-oxide surface layers formed on small Fe particles', Surf. Sci. 77, 584.
13. Luborsky F. E. and Walter J. L. (1980) 'Preparation and properties of Fe-B-Si-C amorphous alloys', IEEE. Trans. Mag. mag-16, 572.
14. Van Der Kraan A. M.(1973) 'Mössbauer effect studies of surface ions of ultrafine-αFe2O3 particles' Phys. Stat. Sol. (a)1s, 215.
15. Gangopadhyay S., Hadjipanayis G. C., Shah S. I., Sorensen C. M., Klabunde K. J., Papaefthymiou V. and Kostikas A. (1991), 'Effect of oxide layer on the hysteresis behavior of fine Fe particles', J. Appl. Phys. 70.
16. Berkowitz A. E., Schuele W. J. and Flanders P. J. (1968) 'Influence of crystallite size on the magnetic properties of acicular γ-Fe_2O_3 particles', J. Appl. Phys. 39, 1261.
17. Morrish A. H., Haneda K. and Schurer P. J. (1976) 'Surface magnetic structure of small γ-Fe_2O_3 particles', J. Phys. Colloq. 37, C6-301.
18. Coey J. M. D. (1972), 'Noncollinear spin arrangement in ultrafine ferrimagnetic crystallites', Phys. Rev. Lett. 27, 1140.
19. Pankhurst Q. A. and Pollard R. J. (1991) 'Origin of spin-canting anomaly in small ferrimagnetic particles', Phys. Rev. Lett. 67, 248.
20. Parker F. T., Foster M. W., Margulies D. and Berkowitz A. E. (1991) 'Spin canting in γ-Fe_2O_3 particles', J. Appl. Phys. 69, 4505.

SIZE DEPENDENCE OF THE ELECTRONIC AND MAGNETIC PROPERTIES OF ULTRA-SMALL CHROMIUM AND VANADIUM PARTICLES STUDIED USING SYNCHROTRON RADIATION

C. BINNS and C. NORRIS
Dept. of Physics, University of Leicester, Leicester LE1 7RH, UK
H. S. DERBYSHIRE, *Dept. of Physics, Leicester Polytechnic, Leicester, UK*
S. C. BAYLISS, *Dept. of Physics, University of Loughborough, Loughborough, UK*

ABSTRACT. We report a synchrotron radiation study of the electronic and magnetic properties of Cr and V particles in the size range 3nm to 11nm supported on graphite. The average island size of these was determined by a measurement of the wavelength dependence of soft X-ray scattering by the sample. With decreasing particle size there is an increasing disorder of the antiferromagnetic alignment in Cr particles. The appearance of the extra structure in the 3s photoemission spectra of V samples indicates magnetic order localised at the surface of the particles. An increase in the peak asymmetry with decreasing diameter indicates less effective screening of the core hole by the electron gas as it becomes more confined.

1. Introduction

Mesoscopic metal particles in the size range 2nm - 100nm lie between clusters and solid state systems. They are large enough to have the bulk atomic structure but their finite nature modifies the behaviour of the electron gas and produces novel electronic and magnetic properties. The onset of size effects occur in ferromagnetic metals at a dimension of about 100nm with the formation of single domain (SD) particles. The magnetic behaviour of an assembly of such particles depends on the relative magnitude of the thermal energy kT and the anisotropy barrier KV where V is the volume of the system and K is its magnetic anisotropy. When kT>KV the system is an a super-paramagnetic state[1] and an important fundamental question is whether in this regime it is possible to detect increasing magnetic disorder on a local scale, that is between neighbouring spins, with decreasing size. This is also an important question in anti-ferromagnetic systems and can be addressed using spin-polarised photoelectron diffraction (SPPED).

Another size effect in mesoscopic metals is the modification of the collective behaviour of the electron gas. The screening response of the valence electrons to the creation of a core hole in photoionisation can be studied by measuring the lineshapes of core level photoelectrons[2]. Here we report the results of a spin-polarised photoelectron diffraction and photoemision study of Cr and V particles produced by deposition in-situ in ultra-high vacuum onto HOPG graphite.

2. Experimental

The experiment was carried out on beam line 6.1 of the synchrotron radiation source (SRS) at

P. Jena et al. (eds.), Physics and Chemistry of Finite Systems: From Clusters to Crystals, Vol. I, 753–758.

Daresbury laboratory, UK. The Cr and V metal films were deposited from a Knudsen cell and an electron beam evaporation source respectively, in which the charge was surrounded, apart from a 5mm collimation hole by a water cooled shroud. The chamber base pressure was below 2×10^{-11} mbar and was never higher than 3×10^{-10} mbar during deposition. Substrates were cleaved in air and were prepared in UHV by heating to 800°C for 30 minutes.

The average island size in the films was determined by measuring the spectral dependence of the soft X-ray reflectivity[3] of the samples over the usable range of the monochromator (35eV - 200eV). Photoelectron spectra were recorded using a hemispherical analyser employing 16 detection channels. The combined analyser/monochromator resolution was 1.2eV.

3. Results and Discussion

Short-range magnetic order can be studied using the technique of spin-polarised photoelectron diffraction developed by Shirley and co-workers[4]. In its simplest form the experiment looks for an abrupt change in the intensity ratio of the two main 3s core level multiplets as a function of temperature which indicates the loss of short range magnetic order (SRMO). Initially a thick Cr layer was deposited in order to understand changes occurring to 3s core level spectra in the bulk. Figure 1 shows a typical spectrum and the results of a Doniach-Sunjic lineshape[5] fitting routine necessitated by the incomplete resolution of the multiplets. The behaviour of the spin asymmetry, S, given by $S = [R_{HT}/R - 1] \times 100$ where R is the intensity ratio of the multiplets (high spin:low spin) and R_{HT} is the value of R at the high temperature limit, is shown in figure 2 and compared with the results obtained from the anti-ferromagnetic compounds $KMnF_3$[6] and MnO. The step at 680 K is due to the loss of SRMO which compares with the Neel temperature of 308 K.

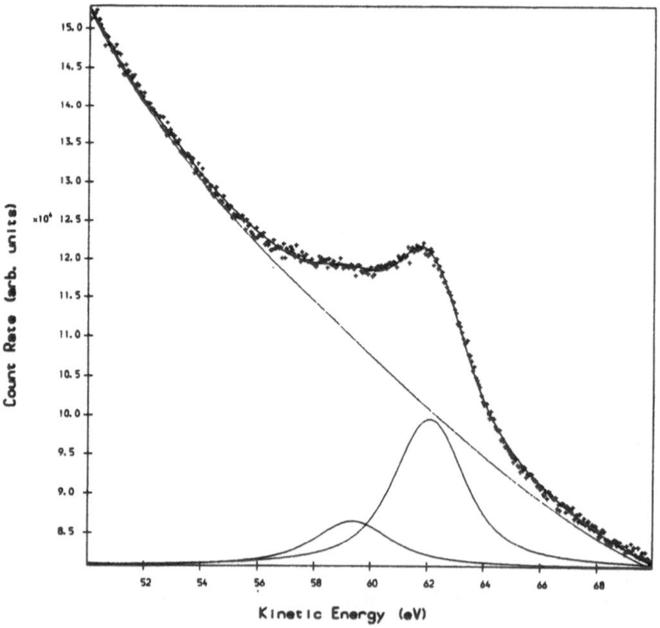

Figure 1. 3s photoemission spectrum (hω=140eV) from a thick Cr film (dots). The fitted Doniach-Sunjic peaks and secondary electron background are also shown.

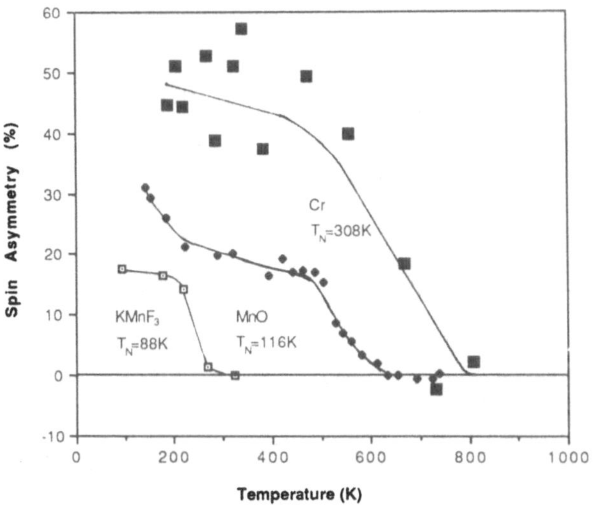

Figure 2. Spin asymmetry vs temperature for two anti-ferromagnetic compounds and Cr. The step in each case is ascribed to the loss of SRMO.

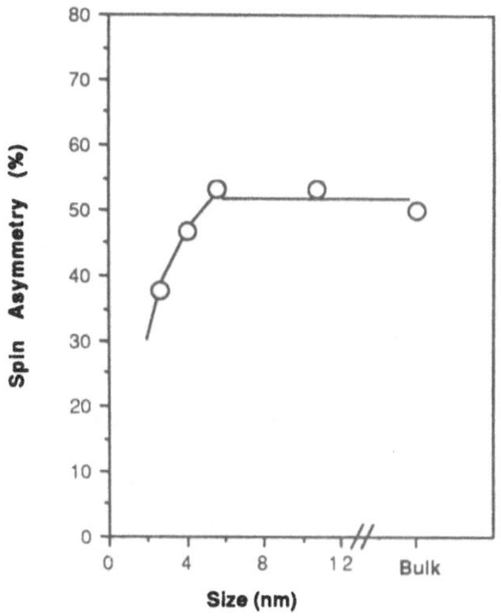

Figure3. Spin asymmetry at room temperature (averaged between 180 K and 450 K) vs Cr particle size.

The islanded films are not stable up to the temperatures required to observe the loss of SRMO so the procedure adopted was to measure the spin asymmetry in the temperature range 180K to 450K to ascertain if there was any trend with temperature. None was found and all the spectra for a given particle size were averaged, to give a value with good statistics around room temperature and the results plotted in figure 3. Below about 6 nm a decrease in the room temperature spin asymmetry towards the disordered value is observed indicating increasing local magnetic disorder with decreasing size. This is a significant result since it confirms that observed size-dependent changes in long-range magnetic behaviour can be related to changes in the degree of magnetic alignment between neighbouring spins.

Figure 4 shows the behaviour of the Cr and V 3p levels as a function of size. We ascribe the smaller peak about 8eV below the main Cr 3p line to a low spin multiplet state. This shows a dramatic decrease in intensity with decreasing particle size. Various studies of transition metal 3p multiplet structures have demonstrated the sensitivity of features to correlation and within a configuration interaction (CI) approach it is possible to distribute a peak intensity over a large number of states. In this case it appears that size dependent changes in the distortion of excited orbitals due, for example, to changes in the screening by the valence electron gas increase the overlap of the hole states with alternative configurations.

Changes in the screening of core holes by the valence electron gas can be monitored more directly by examining photoemission lineshapes[2]. Figure 5 shows the asymmetry (ratio of HWHM measured on the low and high binding energy sides of the peak) of the Cr and V 3p lines as a function of size. The asymmetry which is due to the screening of the core hole by the electron gas can be related directly to the depth of the core hole scattering potential. Increasing asymmetry indicates either an increase in the density of states near the Fermi level or a deepening potential

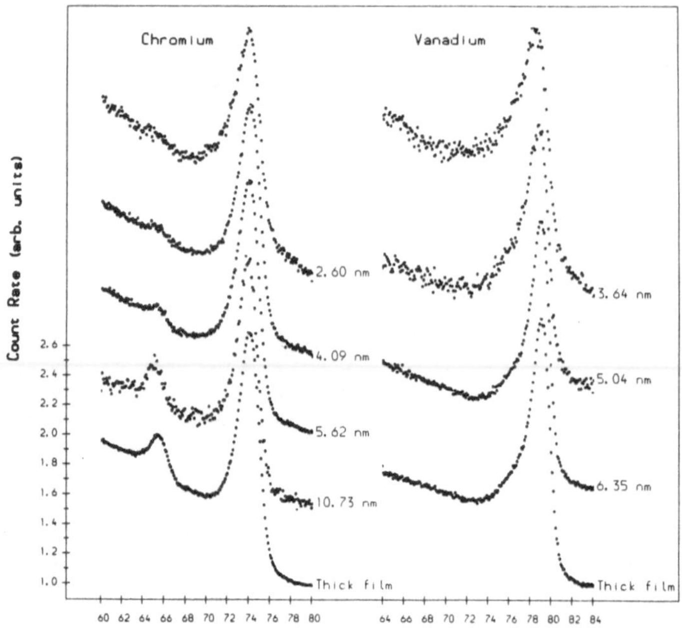

Figure 4. Cr and V 3p photoemission spectra (hω=120eV) as a function of particle size. Note the disappearance of the low spin multiplet in the case of Cr.

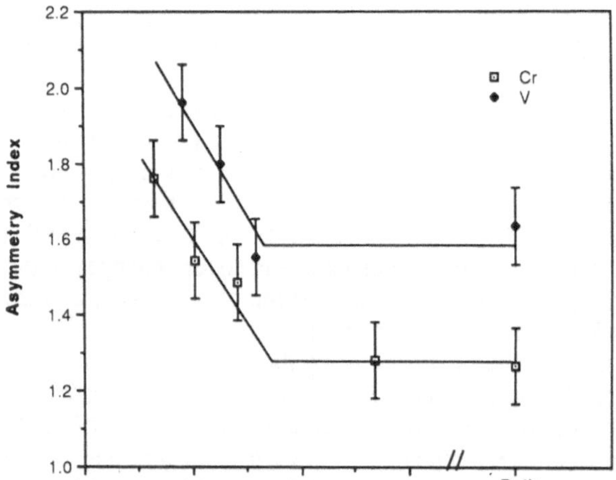

Figure 5. Asymmetry (ratio of HWHM on low and high binding energy sides of the peak) of Cr and V 3p peaks as a function of size.

which in turn indicates a decreasing efficiency of screening by the valence electrons. The Cr and V valence band spectra from the same films as in figure 5 show a narrowing of the Cr 3d band with decreasing size (due to the increased surface/volume ratio) but no significant decrease in the density of states at the Fermi level. Thus the change in asymmetry of the 3p peaks is due to a less efficient response of the electron gas to the core hole created by photoionisation as the particles become smaller. This can be conceptualised as a 'stiffening' of the valence electron gas as it becomes more confined.

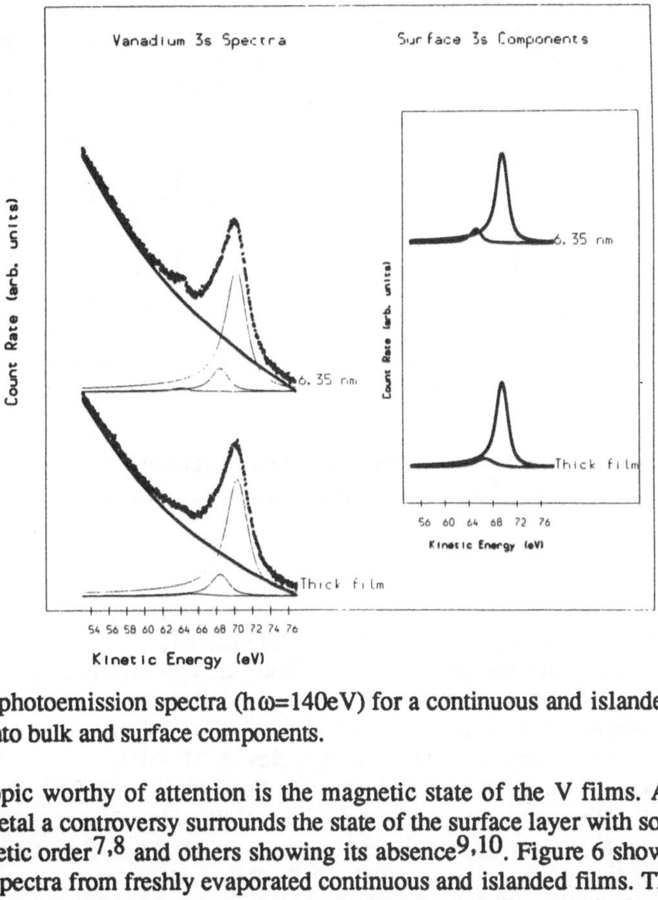

Figure 6. V 3s photoemission spectra (hω=140eV) for a continuous and islanded film showing deconvolution into bulk and surface components.

A final topic worthy of attention is the magnetic state of the V films. Although V is a paramagnetic metal a controversy surrounds the state of the surface layer with some experiments observing magnetic order[7,8] and others showing its absence[9,10]. Figure 6 shows 3s core level photoemission spectra from freshly evaporated continuous and islanded films. The small feature visible 6eV below the main peak is more intense in the islanded films and is thus associated with the surface. We ascribe it to the low spin (^3S) component of a multiplet split 3s spectrum from surface atoms possessing a magnetic moment. The feature is extremely sensitive to contamination and fades rapidly in the ambient (<2×10^{-11} mbar) atmosphere (mainly CO) of the UHV chamber. The high spin (^5S) state is hidden below the main peak and in order to monitor the behaviour of the multiplets with contamination a 3 peak Doniach-Sunjic spectrum was fitted to the data. Figure 7 shows the intensity of the ^3S feature and the position of the ^5S peak as a function of exposure. The disappearance of the ^3S component is accompanied by a shift of the ^5S peak to the centre of gravity of the 2 features which indicates a reduction of the surface moment leading to complete quenching after an exposure of about 1L. This is an upper limit since the chamber ion guage was operating at its X-ray limit and the actual pressure was probably less than quoted. Such an extreme sensitivity to contamination could well resolve the controversy concerning the existence of surface magnetic moments on V.

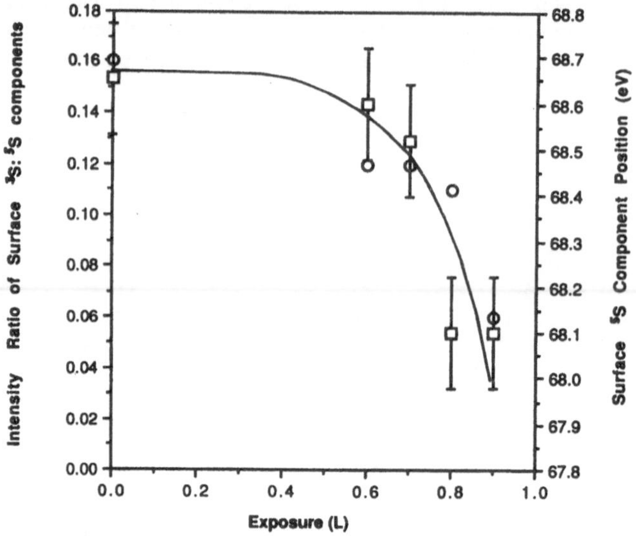

Figure 7. Intensity ratio of the surface 3s peak ^3S and ^5S components as a function of exposure (o) and position of surface 3s peak ^5S component as a function of exposure (□).

References

[1] Laurent , C. and Kay, E. (1989), Z. fur Physik D **12**,465.

[2] Wertheim, G. K. and Citrin, P. H. (1978) ,"Photoemission in Solids I", Cardona, M. and Ley. L . (eds.), Springer-Verlag, Berlin, pp 197 - 236.

[3] Binns, C., Bayliss, S. C. and Derbyshire, H. S. (1991), Z. Phys. D **20**, 391.

[4] Sinkovic, B. and Fadley, C. S. (1985), Phys. Rev. B **31**, 4665.

[5] Doniach, S. and Sunjic, M. (1970), J. Phys. C: Solid St. Phys. **3**, 285.

[6] Sinkovic, B., Hermseiser, B. and Fadley, C. S. (1985), Phys. Rev. Letts. **55**, 1227.

[7] Rau, C., Liu, C., Schmalzbauer, A. and Xing, G. (1986), Phys. Rev. Letts. **57**, 2311.

[8] Moodera, J. S. and Meservey, R. (1989), Phys. Rev. B **40**, 8541.

[9] Stampanoni, M., Vaterlaus, D., Pescia, M., Aeschlimann, F., Durr, W. and Blugel, S. (1988), Phys. Rev. B **37**, 10380.

[10] Fink, R. L., Ballentine, C. A. and Erskine, J. L. (1990), Phys. Rev. B **41**, 10175.

MAGNETIC PROPERTIES OF FREE METAL CLUSTERS

D. C. DOUGLASS, J. P. BUCHER, D. B. HAYNES,
and L. A. BLOOMFIELD
Department of Physics
University of Virginia
Charlottesville, Virginia 22901
United States

ABSTRACT. We report in this paper results of Stern-Gerlach experiments for all the systems we have studied to date. Three distinct behaviors are observed. Cobalt exhibits superparamagnetic behavior, with a magnetic moment per atom of $2.29 \pm 0.08\mu_B$ for N=115. The rare earths gadolinium and terbium exhibit superparamagnetic behavior for certain clusters, while other clusters exhibit behavior consistent with their moment being locked to the crystal lattice. In both cases, the moments are considerably reduced from the bulk values. The third behavior is no measurable deflection at all, as demonstrated by vanadium, palladium, chromium, and aluminum.

1. Introduction

The magnetic properties of free transition metal clusters have been the subject of much recent study[1-4]. These results have not been without controversy, however. The observed magnetic moment of transition metals has been reported to both increase[2] and decrease[4] as a function of temperature. In this paper, we present new results for cobalt clusters as a function of temperature. We continue to observe that the moment decreases with increasing temperature.

Previously we stated that the magnetic moment per atom for cobalt clusters increases with increasing size and applied magnetic field, but decreases with increasing temperature. The temperature behavior is the opposite of that observed by de Heer, *et. al.* We initially reported data for only two different temperatures, room temperature and liquid nitrogen temperature. We have since acquired data at various intermediate temperatures which are in excellent agreement with our previous results.

The magnetic properties of the rare earth metals have been subjected to much less intense study. We have previously reported results for gadolinium[5], and noted two distinct behaviors. Certain size clusters, such as Gd_{22}, Gd_{30}, and Gd_{33}, behave as cobalt clusters do. Their deflection profile is characterized by a relatively narrow deflected beam deflecting uniquely to the strong field side of the magnet. Other sizes, however, exhibit a very broad profile, which, for low fields at least, deflect to both sides of the magnet. This behavior can be explained by the moment being fixed to the lattice of the cluster and rotating with the cluster.

P. Jena et al. (eds.), Physics and Chemistry of Finite Systems: From Clusters to Crystals, Vol. I, 759–765.

We noted that due to isotope effects, we had difficulty differentiating between pure metal clusters and clusters with one or two oxygen atoms attached. Therefore, we have done similar experiments on monoisotopic terbium to determine conclusively if the oxygen was playing a role in the magnetic behavior. We have found that terbium has similar magic numbers compared to gadolinium, displaying the same two types of behavior. We have also found that the addition of one or two oxygen atoms to a cluster seems to have little effect in most cases, with the most dramatic exception being Tb_{22}.

We have also studied several systems that are not ferromagnetic in the bulk[6]. These are of interest for a variety of reasons. Calculations have shown that palladium and vanadium may become ferromagnetic under appropriate conditions[7,8]. Calculations on chromium have predicted large moments for clusters of sufficiently small size[9,10]. Aluminum can be compared to previous results[11]. We were unable to observe deflections of any of these clusters.

2. Experiment

The experiment consists of four basic pieces; three high vacuum chambers and a magnet assembly. The cluster beam is formed in the first chamber by a laser vaporization cluster source (LVCS). In this LVCS, the second harmonic of a Nd:YAG laser (532nm) is focused on the face of a disk sample. The sample vapor jumps into a 0.5cm^3 cavity, filled with helium by a modified General Valve valve. Cluster growth and cooling occur within the cavity. A cluster beam is formed as the vapor exits the source cavity through a conical nozzle. This beam enters the second chamber through a 2mm diameter conical skimmer. The entire source assembly, including the helium valve, can be cooled below 70K.

The second chamber contains the two 0.4mm wide collimating slits and a beam chopper. Together, they form a narrow cluster packet, 0.4mm wide by 2.5mm high. The beam then passes through a 250mm long gradient field magnet. The magnet reproduces one quadrant of a quadrupole field. After the magnet, there is a meter long deflection tube followed by a detection region in the third box. Deflections are measured by scanning a narrow (0.5mm) excimer laser beam across the cluster beam to ionize the clusters. The resultant ions are mass analyzed by time-of-flight mass-spectroscopy.

The beam chopper allows us to measure the velocity of the cluster beam to within a few percent. This velocity is important in calculating the magnetic moments from the measured deflections. The chopper also allows us to measure the time between the firing time of the vaporizing laser and the time the clusters actually leave the source. This residence time τ_{res} is very important. The longer τ_{res}, the closer the vibrational temperature of the clusters T_{vib} is to the source temperature T_{source}. We find for superparamagnetic particles (*e.g.*, cobalt) that the measured moment increases with increasing τ_{res} for relatively short dwell times, but that eventually no further increases in the measured moments can be obtained. At this point, the clusters are leaving the source with T_{vib} equal to T_{source}. The expansion seems to have no significant effect on T_{vib}.

Moments are determined from the deflections and the velocities according to the formula

$$\mu/atom = \frac{dmv^2}{(dB/dz)(DL + L^2/2)},$$

where d is the measured deflection, m is the mass of a single atom of the sample element, v is the cluster velocity, dB/dz is the magnetic gradient, D is the distance from the magnet to the ionization region, and L is the length of the magnet. Thus, the resolution of μ is governed by our ability to measure the deflection d, the cluster beam velocity v, and the field gradient.

3. Cobalt

As we have previously reported[4], cobalt clusters in the size range N=20-200 behave superparamagnetically, as proposed by Khanna[12]. A superparamagnetic cluster has one moment that fluctuates rapidly in direction due to the cluster's vibrational temperature. The clusters deflect uniquely to the strong field side of the gradient magnet, and the observed magnetic moment per atom increases as a function of field and size, but decreases as a function of temperature. The cluster size and field strength are well known, but the temperature is more difficult to establish reliably. It has been a subject of some controversy. To address this point, we have studied one particular cluster, N=115, in extensive detail. We have also modified our experiment to increase our deflection resolution.

To obtain more precise deflection measurements than were previously published, we reduced the collimation slit widths for both the cluster beam and the scanning excimer laser to the values reported here. The previous measurements were performed using wider slits. We also replaced our cylindrical nozzle with a conical nozzle. This nozzle produces a cluster beam that is much narrower than that produced by our pin hole nozzle, which in turn yields a much larger ultimate signal strength. This change has enabled us to reduce our ionizing laser beam width while still retaining a usable signal. The net result is that we are able to resolve deflections that are about half as small as our previous limit.

Our previous work included data taken at both room temperature and liquid nitrogen temperature. To obtain data at intermediate temperatures, we reduced the thermal contact between the liquid nitrogen reservoir and the cluster source, and we attached a heater to the source. The heater was driven by a thermal process controller to maintain a stable source temperature. In this manner we were able to obtain any temperature we wanted above liquid nitrogen temperature.

Table 1. Experimental magnetic moment per atom for Co_{115} at various source temperatures. μ is the calculated true moment given by the Langevin function.

$T_{source}(K)$	Applied B(T)	$\mu_{expt}(\mu_B)$	$\mu(\mu_B)$
307 ± 2	.772	$.366 \pm .014$	$2.39 \pm .05$
246 ± 3	.390	$.209 \pm .024$	$2.27 \pm .14$
190 ± 6	.390	$.289 \pm .025$	$2.35 \pm .14$
142 ± 5	.257	$.224 \pm .008$	$2.20 \pm .08$
96 ± 7	.390	$.507 \pm .013$	$2.24 \pm .10$

As previously mentioned, we find that the measured moment increases as a function of τ_{res} until T_{vib} is in equilibrium with T_{source}. We performed such experiments at five different temperatures, ranging from 96K to 307K. We then calculated the true moments at each of these temperatures from the Langevin function[12]

$$\mu_{eff} = \mu\left[\coth(\frac{N\mu B}{k_b T}) - \frac{k_b T}{N\mu B}\right]$$

where μ is the true moment per atom, N is the number of atoms in the cluster, B is the magnetic field strength, k_b is the Boltzmann constant, and T is the vibrational temperature of the cluster T_{vib}. By assuming $T_{vib} = T_{source}$, we found the true moment to be $2.29 \pm 0.08\mu_B$, where μ_B is the Bohr magneton. The data is reported in Table 1. There is no indication that the true moment is affected by the temperature.

We have also previously reported some interesting magnetic structure near N=55. There appear to be two distinguishable magnetic isomers of the clusters between N=55 and N=66. We have been unable to fully separate the two isomers to date, however. In our original measurements, the resolution of the deflections was not great enough to observe completely separated beamlets. After the modifications discussed above, we were unable to see two isomers. Instead, we see a flat portion in the moment vs. N curve for N=50-55, with an abrupt rise at N=56 to the larger moment. This difference could be attributed to subtle changes in the growth conditions of our source after the modifications.

4. Gadolinium

We have previously reported results for gadolinium. We have observed two different types of magnetic behavior. Some clusters, such as N=22, 30, and 33 appear to be predominantly superparamagnetic. The beam deflection is small, and the profile does not broaden significantly. Other clusters, however, exhibit quite a different behavior. The beam profile is spread broadly, and a significant fraction of the clusters deflect toward the weak field. These clusters most likely have their moments locked on their lattice. Beam profiles can be generated by a computer assuming an isotropic distribution of the moments prior to their entry into the magnetic field, with the only adjustable parameters being the cluster magnetic moment and the rotational temperature. A good estimate of the rotational temperature can be obtained by fitting several profiles obtained with different applied fields and leaving the moment and temperature the same.

Most of the cluster profiles actually appear to be a combination of these two behaviors. The profiles are best fit with a superparamagnetic profile superposed on a locked moment profile. This introduces a third adjustable parameter, but again it should not vary for scans of the same size cluster.

5. Terbium

One question yet to be addressed is the role of oxygen on the magnetic properties. As gadolinium has four nearly equally abundant isotopes, we have difficulty resolving the first oxide from the pure metal cluster for peaks larger than about N=13. Our results indicate that the two observed behaviors for a single cluster are not due to the presence or absence of oxygen atoms, but we are unable to conclusively state this. To settle this issue, we looked at terbium. Terbium is monoisotopic, so our mass resolution is much better. Terbium also has a lower Curie temperature and a larger atomic magnetic moment.

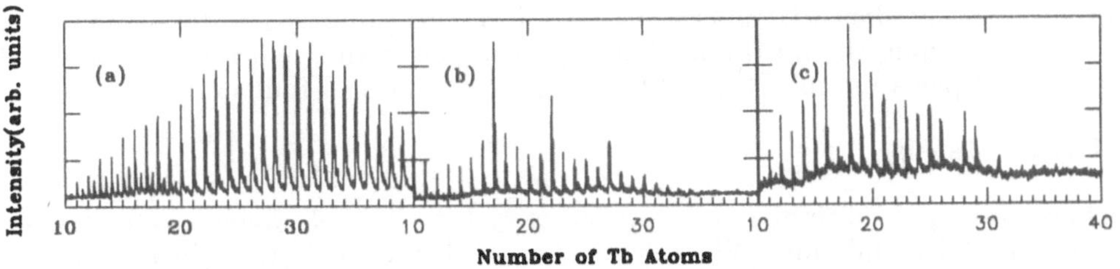

Figure 1. Mass spectra of Tb. (a) is on the beam axis with no field applied. (b) and (c) are with an applied field of .257T and a gradient of 101T/m. (b) is on the beam axis, and (c) is 2.5mm to the strong field side.

The data in Fig. 1 clearly show the two types of behavior we observe. Figure 1(a) is the spectrum on the beam axis with no applied field. No reproducible structure is observed, except a slight superabundance of 13. This is also observed in gadolinium. Figure 1(b) is the spectrum at the same position on axis with an applied field. N=17, 22, 27, and 30 are clearly more abundant than their neighbors. Figure 1(c) is similar to 1(b), only now the ionizing laser is positioned 2.5mm to the strong field side of the magnet. N=17,27, and 30 are clearly depleted relative to their neighbors.

At first glance, Tb_{22} seems to be as abundant as Tb_{21} and Tb_{23} when positioned off axis. However, upon close inspection of the data it is seen that the peak in Fig. 1(c) is actually $Tb_{22}O$. The pure metal cluster signal is very depleted. Complete profiles of Tb_{22} and $Tb_{22}O$ are shown in Fig. 2(d). $Tb_{22}O$ is clearly behaving differently from both Tb_{21} and Tb_{22}. We can attribute such behavior to an intermediate behavior between superparamagnetic moments and completely locked moments. That is, the moment may be somewhat free to move relative to the cluster lattice, but not completely so. It may be trapped in a well or trough of a complicated crystal anisotropy energy surface, and be free to move only in that region. The important point here is that the pure metal cluster Tb_{22} exhibits both behaviors, as does its first oxide. There is a large, narrow superparamagnetic peak, and also a broad locked moment profile.

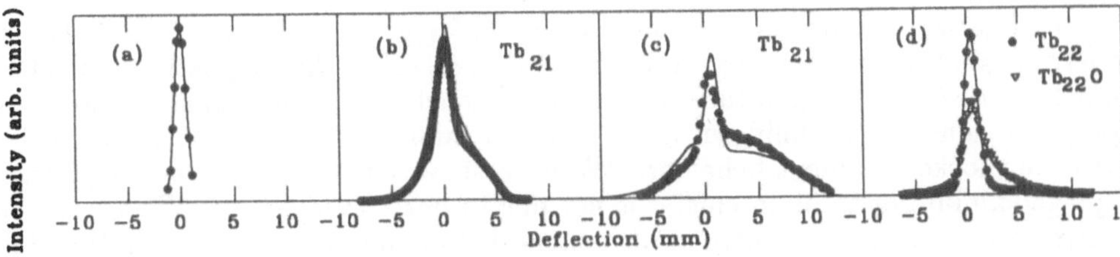

Figure 2. Deflection profiles for Tb_{21} and Tb_{22}. (a)-undeflected profile ($B = 0$, $dB/dz = 0$). (b)-Tb_{21} with $B = .130$T, $dB/dz = 50.6$T/m. Solid line is rigid rotor fit with 21% superparamagnetic peak superposed. $\mu = 4.0\mu_B$, $T_{rot} = 14$K. (c)-Same parameters on fitted curve as (b), but $B = .257$T, $dB/dz = 101.3$T/m. (d) Tb_{22} and $Tb_{22}O$. $B = .257$T, $dB/dz = 101.3$T/m. Solid line is to differentiate the two curves.

The data presented in Fig. 2(a) and 2(b) are deflection profiles at two different fields and gradients. The fitted curve is the theoretical profile of a cluster with its moment fixed on the cluster lattice so that it rotates with the cluster. A superparamagnetic peak of 21% abundance is superposed on top of this profile. The total moment is obtained by finding the maximum deflection and calculating its associated moment. The moment per atom for Tb_{21} is 4.0 μ_B, and the rotational temperature is 14K.

6. Other Metals

We have also reported results on several other systems, namely vanadium, palladium, chromium, and aluminum[6]. We were unable to observe deflections in any of these systems. Upper limits of the measured magnetic moments per atom and the cluster ranges observed are given in Table 2.

These experiments were performed with the source at liquid nitrogen temperature using the longest possible τ_{res}. This should produce clusters with $T_{vib} = T_{source}$. However, as we observed no deflections, we cannot say with certainty that the clusters were in equilibrium with the source. All measurements were performed in a 1T field with a gradient of 310T/m.

Assuming that the transition metals would behave as the ferromagnetic transition metals, we can calculate the maximum possible superparamagnetic moment for these clusters. Assuming that the clusters were in equilibrium with the source, the maximum moments per atom calculated using the Langevin function are $.77\mu_B$ for Cr_9, $.59\mu_B$ for V_9, and $.22\mu_B$ for Pd_{110}.

Table 2. Experimentally measured magnetic moments per atom and the cluster ranges observed.

Element	Moment Per Atom	Size Range	T_{source}(K)
Aluminum	$0.000 \pm .003\mu_B$	$15-48$	88
Chromium	$0.000 \pm .014\mu_B$	$9-31$	86
Palladium	$0.000 \pm .014\mu_B$	$100-120$	86
Vanadium	$0.000 \pm .008\mu_B$	$8-99$	89

7. Conclusion

To summarize, we are able to observe two different types of magnetic behavior for ferromagnetic particles. By systematically measuring the magnetic moment as a function of source temperature, we can state conclusively that cobalt clusters are superparamagnetic. Gadolinium and terbium exhibit both superparamagnetic behavior and locked-moment behavior. We have shown that the dual behavior for a single gadolinium or terbium cluster is not due to unresolved oxides. We are unable to observe any magnetic deflections of clusters that are not ferromagnetic in the bulk.

8. References

1. D. M. Cox, D. J. Trevor, R. L. Whetten, E. A. Rohlfing, and A. Kaldor, Phys. Rev. B **32**, 7290 (1985).
2. W. A. de Heer, P. Milani, and A. Chatelain, Phys. Rev. Lett. **65**, 488 (1990).

3. J. P. Bucher, D. C. Douglass, P. Xia, B. Haynes, and L. A. Bloomfield, Z. Phys. D **19**, 251 (1991).
4. J. P. Bucher, D. C. Douglass, and L. A. Bloomfield, Phys. Rev. Lett. **66**, 3052 (1991).
5. D. C. Douglass, J. P. Bucher, and L. A. Bloomfield (submitted).
6. D. C. Douglass, J. P. Bucher, and L. A. Bloomfield (submitted).
7. H. Chen, N. E. Brener, and J. Callaway, Phys. Rev. B **40** 1443 (1989).
8. V. L. Moruzzi, Phys. Rev. Lett. **57**, 2211 (1986).
9. D. R. Salahub and R. P. Messmer, Surf. Sci. **106**, 415, (1981).
10. G. M. Pastor, J. Dorantes-Dávila, and K. H. Bennemann, Phys. Rev. B **40**, 7642 (1989).
11. D. M. Cox, D. J. Trevor, R. L. Whetten, E. A. Rohlfing, and A. Kaldor, J. Chem. Phys. **84**, 4651 (1986).
12. S. N. Khanna and S. Linderoth, Phys. Rev. Lett. **67**. 742 (1991).

ONSET OF MAGNETISM IN BCC-LIKE Fe_n CLUSTERS

J. DORANTES-DAVILA [1] , H. DREYSSE [2] and G.M. PASTOR [3]

[1] Instituto de Fisica "Manuel Sandoval Vallarta",
Universidad de San Luis Potosi - Mexico.
[2] Laboratoire de Physique du Solide, 54506
Vandoeuvre-les-Nancy, France.
[3] Institüt für Theoretische Physik der Universität zu Köln,
Köln, Germany.

ABSTRACT: The onset of magnetism in bcc-like Fe_n clusters is studied by using a Hubbard-like Hamiltonian in the unrestricted Hartree-Fock approximation. The dependence of the local and average magnetic moments on cluster size, cluster structure and interaction parameter (exchange integral J) is determined. We found sharp first order transitions in the magnetic moments for several values of J. We discuss the size dependence of the transition by comparing cluster results with those of the bulk.

1. Formalism

This conference emphazises that one key question in solid states physics is to understand how and why the physical properties change when the electrons of a single atom become part of a group of several atoms and then delocalize and how bulk-like behavior is reached. This question is particulary interesting for transition metal magnetism. Isolated 3d-atoms are known to bear a localized magnetic moment, whereas the bulk equilibrium structure can be paramagnetic (as V for example) or presenting a magnetic order. In this late case, the magnetism is well described in an itinerant model.

In this short communication, we investigate the magnetic behavior of small Fe clusters, versus the cluster size and the interaction parameters. We shall focus particulary on bcc-like Fe clusters. It is well known that for bulk Fe, the equilibrium structure at normal pressure is the bcc, with a local magnetic moment of 2.2 μ_b . The magnetic moment as a function of the lattice parameter, $\mu = \mu(l)$ is a continuous monotonic curve or , eventually, presents a slight discontinuity for small values of J /1/. The fcc phase presents a completely different behaviour: for the ferromagnetic state, two or three magnetic states can coexist. In this case the curve $\mu = \mu(l)$ shows discontinuities (first order transition) /2/. Within a simple Stoner analysis for the local density of states, it has been shown that such a behaviour is quit

P. Jena et al. (eds.), Physics and Chemistry of Finite Systems: From Clusters to Crystals, Vol. I, 767–772.

common for "3d" transition metal series: the possible values of the magnetic moments are rather limited /1/.

Finally let us mention that antiferromagnetic arrangments must be also considered. For clusters, a supplementary difficult arises, since the sublattices cannot be determined a priori. Nevertheless, we can guess from physical grounds that frustating effects might play an important role /2/.

First principle calculations provide the best realiability and precise on can reach /3/ . Unfortunately, they are limited to small size clusters with high symmetry (usually $n \simeq 15$ /4,5/). Another way is to use a tight binding approach, which has been shown also to give good results for clusters and surfaces /6-8/ . Using a Hubbard like Hamiltonian , Pastor et al /6/ have studied the magnetic properties of various clusters of Cr, Fe and Ni with good agreement with first principles calculations. In this work such approach shall be used for the onset of magnetism as a function of the exchange integral J. Details of the calculation can be found in reference /6/. Let us recall, however, the main points: only "d" orbitals are considered. The interaction Hamiltonian H_I in the unrestricted Hartree-Fock approximation is given by

$$H_I = \sum_{i\sigma} \Delta\varepsilon_{i\sigma} n_{i\sigma} - E_{dc} \quad ; \quad \Delta\varepsilon_{i\sigma} = \sum_{\sigma'} U_{\sigma\sigma'} \Delta\nu_{i\sigma} \tag{1}$$

Here $\Delta\nu_{i\sigma} = \nu_{i\sigma} - \nu_0$ where $\nu_{i\sigma} = < n_{i\sigma} >$ refers to the average occupation and ν_0 to the corresponding average occupation in the paramagnetic solution of the bulk. The intraatomic Coulomb interactions $U_{\sigma\sigma'}$ between d electrons can be written in terms of the exchange and direct Coulomb integrals, $J = U^{\uparrow\downarrow} - U^{\uparrow\uparrow}$ and $U = (U_{\uparrow\downarrow} + U_{\uparrow\uparrow})/2$, respectively. Finally, $E_{dc} = 1/2 \sum U_{\sigma\sigma'} \nu_{i\sigma} \nu_{j\sigma'}$ is the correction due to double counting. Usually, the exchange integral J is choosen to yield the proper magnetic moment of bulk Fe ($\mu_b = 2.21 \mu_B$). Here, however in order to study in a systematic way the apperance of magnetism in small Fe_n clusters, we calculate the local magnetic moments and band energy as a function of J. Note, that this variation gives the behaviour observed by changing the interatomic distance by anisotropic dilatation /1/. In such problem, J or the interparticule distance, is the relevant quantity /1,9/.

The number of "d" electrons ν_i , and the local magnetic moments μ_i at site i given by

$$\mu_i = < n_{i\uparrow} > - < n_{i\downarrow} > \quad ; \quad \nu_i = < \nu_{i\uparrow} > + < n_{i\downarrow} > \tag{2}$$

are determined selfconsistentely by requiring

$$< n_{i\sigma} > = \int_{-\infty}^{\varepsilon_F} \rho_{i\sigma}(\varepsilon) d\varepsilon \tag{3}$$

Here $\rho_{i\sigma}(\varepsilon)$ refers to the local density of states which is calculated using the recursion method . Concerning the accuracy of the continuous fraction expansion of the local Green's function, we compute up to the levels N in order that the results become independent of N. Empirically, we found that N approx 25-30. We assume for the hopping integrals the canonical values Notice that we allow "d" charge transfer between different cluster sites.

2. Results and discussion

In the following we present results for bcc-like Fe_9 and Fe_{15} clusters, in which the geometrical structure is obtained by adding to a central atom(i= 1) the succesive shells of its first and second neighbors respectively in a bcc lattice. We assume for the band filling $N_d = 7$ and take the direct coulomb integral U equal to the band width.

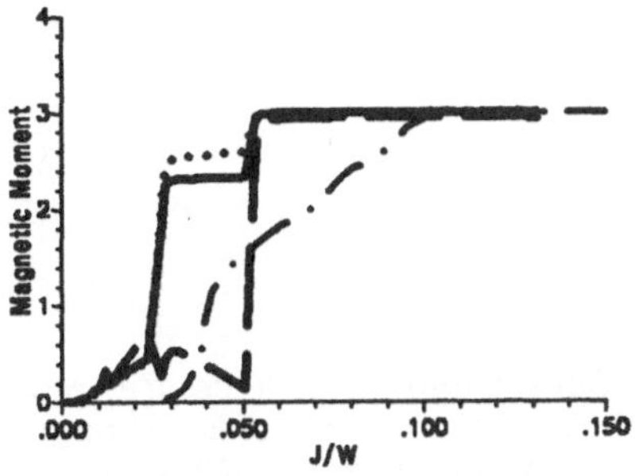

Figure 1: Local magnetic moments $\mu(i)$. Dashed line : $\mu(1)$ (central atom) , dotted line : $\mu(2)$ (surface atom), full line : $\bar{\mu}$, dotted-dashed line : bulk (in units of Bohr magneton) as a function of the integral exchange J in units of the band width, J/W_b for Fe_9 .

We determine the local and average magnetic moments, $\mu(i)$ and $\bar{\mu}$ as a function of the exchange integral in units of the band width W $_b$., Results are shown in figs. 1 and 2 for Fe_9 and Fe_{15} respectively. For comparison we also show results for bulk Fe (solid line). Note that for the considered clusters a sharp first order transition in the magnetic moments occurs for several values of J. Also note that this transition is stronger in the central atom (i=1) than in the surface atoms (i>1). Moreover, $\mu(1)$ shows an oscillating behaviour in contrast to the monotonic decreasing of the surface magnetic moments for decreasing J. This rather strong difference is due to the fact that the perturbation introduced by the cluster surface is symmetric around the central atom and therefore produces very strong changes in the local density of states (DOS). This behavior is more interesting for Fe_{15} since here, the central atom is aligned antiferromagnetically with respect to the surface atoms for some values of J. Also, notice that for the considered clusters, the transition magnetic-non magnetic (mu(i)=0) occurs at values of J smaller when those of the bulk.

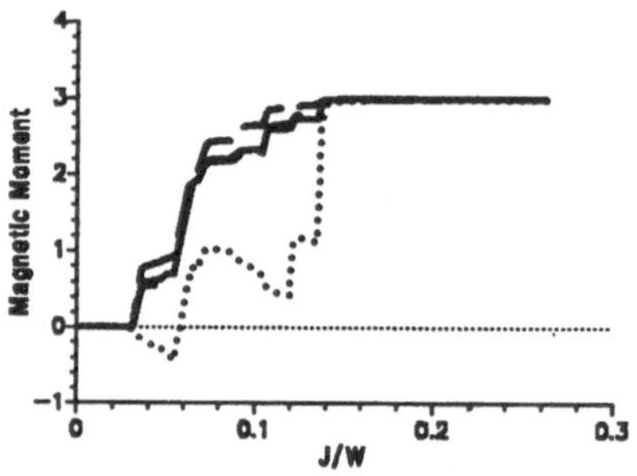

Figure 2: Local magnetic moments $\mu(i)$. Dotted line : $\mu(1)$ (central atom) , dotted-dashed line : $\mu(2)$, dashed line : $\mu(3)$, full line : $\overline{\mu}$ (in units of Bohr magneton) as a function of the integral exchange J in units of the band width, J/W_b for Fe_{15} .

When the first order transitions occur, the selfconsistent calculated DOS show a sharp peak in the Fermi energy. We have also examined the magnetic energy gain, ($\Delta E_{mag} = E_{band}(\mu = 0) - E_{band}(\mu)$ as a function of J/W_b One can notice that, ΔE_{mag} is larger for the smaller systems. This is expected since in this case the local magnetic moments are larger. In the case of Fe_{15}, ΔE_{mag} is smaller than in the case of Fe_9, except in the region where the central atom is aligned antiferromagnetically. From these values we expect that it is necessary to consider larger clusters in order that the magnetic properties converge to the bulk ones.

From figures 1 and 2, we can conclude that, for the bulk equilibrium lattice parameter , all atoms of the Fe_9 cluster bear a satured moment. Moreover these values are obtained for a critical value much smaller than J/W_b in the bulk. So we can conclude that the magnetic moment is higher than in the bulk. For Fe_{15} , the saturation is more subtle: for the external shell, the moments are quasi-satured, but for the central atom, J/W_b is near the transition from $1.15\mu_b$ to $2.80\mu_b$. Moreover the average magnetic moment is, for a large range of values around J/W_b (bulk), much larger than in the bulk. These results are in agreement with the experimental results of Cox et /10/, which observed a magnetic moment bigger than in the bulk for Fe_n clusters ($n \leq 17$). A similar effect has been observed recently by Bucher et /11/, for Co_{20-200} clusters. Study of bigger clusters in now under investigation.

3. Conclusions

In summary we have studied bcc-like Fe_9 and Fe_{15} magnetism clusters by determining the local and average magnetic moments as a function of the exchange integral J. We found that for Fe_9 this transition starts to take place at smaller values of J than those of Fe_{15} . On the other hand the considered cluster still have local magnetic moments different from zero at values of J where the bulk is non magnetic. We found interesting behaviour of magnetic moment of the central atom for both clusters. Moreover, for Fe_{15} an antiferromagnetic aligment of the central atom with respect to the surface atoms occurs for some values of J. This behaviour was not found for Fe_9 . In order to have definitive conclusions about the structural dependence of the magnetic properties is necessary to consider fcc-like clusters. Note that in calculating the cluster size dependence of the magnetic properties and cohesive energy we neglect for simplicity sp electrons contribution. However, as discussed previously, for the magnetic properties of Fe_n seems this effect not to contribute much /6/.

Acknowledgments: This work has been partially supported by a CONACyT (Mexico)- CNRS (France) agreement and DGICSA (Mexico).

REFERENCES

/1/ Stoeffler D. and Dreysse H. (1991) Solid State Comm. 79, 645.

/2/ Christensen N.E., Gunnarsson O., Jepsen O. and Andersen O.K. (1988), J. de Phys. (Paris) C8, 17.

/3/ Physics and Chemistry of Small Clusters , edited by Jena P., Rao B.K. and Khanna S.N., NATO ASI Series B: Physics, Vol. 158, (1987) .

/4/ See, for instance, " Small Particles and Inorganic Clusters", Proceedings of the 4th International Meeting on Small Particles and Inorganic Clusters, Aix-en-Provence, July 1988, edited by Chapon C., Gillet M.F., and Henry C.R., Z. Phys. D 12 (1989) .

/5/ in " Comparison of ab-initio Quantum chemistry with experimental for small molecules, the State of Art" (ed. Bartlett R.J., D. Reidel Publ. Comp., Dordrecht, Holland, 1985).

/6/ Pastor G.M., Dorantes-Davila J. and Bennemann K.H. (1989), Phys. Rev. B 40, 7642 .

/7/ Victora R.H., Falicov L.M. and Ishida S. (1984), Phys. Rev. B 30, 3896
Victora R.H. and Falicov L.M. (1985) Phys. Rev. B 31, 7335 .

/8/ Dreysse H., Mokrani A., Bouarab S. and Demangeat C. (1991), Surf. Sci. 251-252 41.

/9/ Mokrani A., Demangeat C. and Dreysse H. (1990) , Phys. Rev. B 42, 10319.
Liu F., Khanna S.N. and Jena P. (1991) Phys. Rev. B 43, 8179.

/10/ Cox D.M., Trevor D.J., Whetten R.L., Rohlfing E.A. and Kaldor A. (1985), Phys. Rev. B 32, 7290.

/11/ Bucher J.P., Douglass D.G. and Bloomfield L.A. (1991) Phys. Rev. Lett. 66, 3052.

THEORY FOR THE MAGNETIZATION OF AN ENSEMBLE OF TRANSITION METAL CLUSTERS

P.J. JENSEN, S. MUKHERJEE, and K.H. BENNEMANN
Freie Universität Berlin
Institute for Theoretical Physics
Arnimallee 14
W-1000 Berlin 33, Germany

ABSTRACT. Using a previously developed model the magnetization of an ensemble of small transition metal clusters resulting from an external magnetic field is studied for two different systems. We discuss the magnetization of a Fe_n-cluster beam as measured by a Stern-Gerlach deflection experiment, and also the magnetization of a cluster ensemble suspended in a liquid. In addition, the dependence of the Curie temperature on the cluster size is studied.

1. Introduction

The magnetization of a beam of small neutral Fe_n- and Co_n-clusters ($20 \lesssim n \lesssim 250$) [1,2] resulting from an inhomogeneous magnetic field has been determined by Stern-Gerlach deflection. In contrast to experiments on single atoms, the metal clusters were found to be deflected only in one direction by the applied inhomogeneous magnetic field. Furthermore, the magnetization of the cluster beam was much lower than expected assuming bulk magnetic moments for the cluster atoms. The cluster magnetization was found to increase with applied magnetic field and cluster size. Also, the magnetization depends on the cluster temperature, which has been determined mainly by the residence time τ_{res} of the clusters inside their source. Due to the generation process, the clusters are expected to have higher temperatures. Only for sufficient long τ_{res} the clusters can reach thermal equilibrium with the surrounding medium.

Theoretically, it has been proposed that the transition metal clusters behave largely as superparamagnetic particles [2,3,4]. In this case the cluster beam is viewed as an ensemble of giant independent magnetic moments, which reach statistical equilibrium within the measuring time [5]. Therefore, the cluster beam magnetization should be interpreted as resulting from an ensemble average of particles with random orientation of the various cluster spins. This will lead to a small cluster beam magnetization even for atomic magnetic moments having bulk like value. In addition, the magnetization depends strongly on whether the cluster temperature T is above or below the so called "blocking temperature" T_{bl} estimated to be $T_{bl} = n \cdot K_2 / \ln(\tau_{exp}/\tau_o)$, K_2 being the anisotropy energy per atom, τ_{exp} the time of measurement and τ_o the gyromagnetic precession time. For temperatures below T_{bl} the cluster magnetization will mainly be tied to the cluster geometry by the lattice anisotropy. For $T \gg T_{bl}$ (and for $T < T_c$) superparamagnetic behavior is expected. In this case the lattice anisotropy has only little effect on the cluster beam magnetization, whereas in the former case the consideration of the lattice anisotropy is very important. Note that T_{bl} will be the higher the larger the anisotropy energy K_2, the larger the cluster size, the lower the cluster temperature, and the shorter the characteristic measuring time τ_{exp}. The lattice anisotropy can be much larger at surfaces than in bulk [6].

P. Jena et al. (eds.), Physics and Chemistry of Finite Systems: From Clusters to Crystals, Vol. I, 773–778.

The asymmetric Stern-Gerlach deflection profiles can be explained in principle by a transfer of electronical angular momentum to the cluster rotation or other degrees of freedom, allowing for a depopulation of the high energy magnetic levels. We have proposed this previously [3]. Note that the total angular momentum of the isolated clusters is by no means conserved.

We have applied our theory previously to Co_n-clusters with the assumption that they behave approximately as superparamagnets. The calculation simulates the experimental results by assuming the bulk value for the Co- atomic magnetic moment and the same temperature for all three cluster sizes investigated [3]. It is of interest to remark that stronger magnetic anisotropy, as expected for rare-earth cluster beams, will yield even a smaller cluster beam magnetization. Furthermore, then $\langle \mu_z(B, T, K_2) \rangle$ will exhibit in general a different behavior as in the case of transition metals. In particular note that we expect a broarder deflection profile due to a variation of easy axes from cluster to cluster. In the following we treat in a similar way the Fe_n-clusters.

2. Fe_n - Clusters

The Fe_n-clusters are assumed to be superparamagnetic. The magnetic moment at Fe surfaces can be enhanced ($\mu_s \sim 3\mu_B$) [7]. For steps or edges expected to be present on

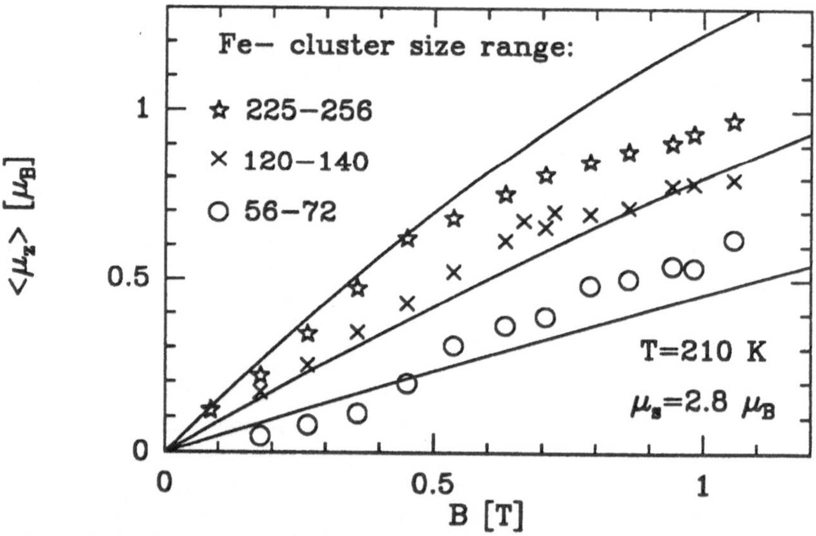

Fig.1: Cluster magnetization $\langle \mu_z \rangle$ vs. applied magnetic field B. Results of our model calculation for clusters containing 64, 130 and 240 atoms, respectively, are compared with experimental results [1] for Fe_n-clusters, marked by $\star, \times, \circ$. For the cluster temperature we use T = 210 K for the three cluster sizes investigated. The magnetic moment is assumed to be the bulk value $\mu_b = 2.2\ \mu_B$ for the core Fe-atoms, whereas we choose $\mu_s = 2.8\ \mu_B$ per atom for the surface shell. The cluster size n is assumed to be the average of the experimental size range as indicated in the figure.

clusters the Fe magnetic moment might be still larger [8]. Thus, we decompose the Fe-cluster in a core of n_b Fe-atoms having almost bulk-like environment $\mu_b = 2.2\mu_B$, and a surface shell containing n_s atoms with a reduced coordination number and an enhanced surface magnetic moment μ_s. The numbers n_b and n_s depend on the cluster shape and size. The average magnetic moment per cluster atom is then given by

$$\bar{\mu} = \frac{n_b\mu_b + n_s\mu_s}{n_b + n_s}. \tag{1}$$

Furthermore, one expects a similar cluster temperature for the different cluster sizes if nearly identical experimental conditions exist. Indeed, a similar temperature was derived for Co_n-clusters [2]. Thus, we use the same temperature for the three cluster sizes studied. In Fig.1 we compare results of our theory with experimentally observed results for Fe_n clusters assuming the same cluster temperature $T = 210$ K and a surface magnetic moment $\mu_s = 2.8\mu_B$ for all cluster sizes. As noted in [2], this high cluster temperature compared to the source temperature is plausible because of the short residence time τ_{res} of the cluster inside the source. For the total number $n = n_b + n_s$ of atoms in the clusters we took the average of each experimentally indicated size range. The number n_s is evaluated by calculating the volume fraction of a one- atom shell of a spherical cluster. We have used the following values for n, n_b, and n_s: (a) $n = 64$: $n_b = 13$, $n_s = 41$; (b) $n = 130$: $n_b = 55$, $n_s = 75$; (c) $n = 240$: $n_b = 139$, $n_s = 101$. The lattice anisotropy was neglected in this calculation. Note the discrepancy between theory and experiment for the larger clusters. This may indicate the necessity for a more refined theory (deviation from superparamagnetism, anisotropy effects, and effects due to cluster rotation, for example).

3. Fe_n - Particles Suspended in a Liquid

As a second application of our model we compare the magnetization of a system of magnetic fine particles in a matrix of an organic solvent [9]. The particles contain amorphous Fe with a diameter of 3.2 nm corresponding to ca. 1200 Fe-atoms per particle. The total magnetic moment per particle is $\mu_{tot} = 2650\ \mu_B$. Because of the low particle density and the long measuring time the clusters should behave superparamagnetically. The magnetization curve measured at $T = 100$ K shows a deviation from the familiar Langevin-behavior, s. Fig.2. For the calculation we assume an uniaxial lattice anisotropy,

$$E_{ani} = -K_2 \cos^2\theta - K_4 \cos^4\theta, \tag{2}$$

θ being the angle between the easy axis and the magnetization direction, and a random orientation of the easy axes in the system of particles. The best fit to the experimentally observed results is obtained by taking for both K_2 and K_4 negative values. This means that these clusters merely prefer an easy plane magnetization. The anisotropy energies are given in units of the magnetic energy $\mu_{tot}B$ at $B = 1$ T. In Fig.2 we present results for the cluster magnetization calculated from our model for (a) $K_2 = 0.0$, $K_4 = -2.75$, and (b) $K_2 = -0.8$, $K_4 = -1.0$. For comparison, the curve of an easy axis magnetization is depicted assuming $K_2 = 1.0$, $K_4 = 0.0$. The anisotropy energies are of the order 1 - 4 K per Fe atom, which is much higher compared to the bulk value (~ 0.03 K), but agrees with calculations for a freestanding (100)-Fe-monolayer [6].

The measurements on magnetic particles suspended in liquid permits a much more precise determination of the cluster temperature. Also a wider range of the measuring time can be obtained with an oscillating magnetic field. However, the magnetic moments and the lattice anisotropy are not well known, because these quantities are affected by the environment. The particle size has not been determined as exactly as in the beam experiment.

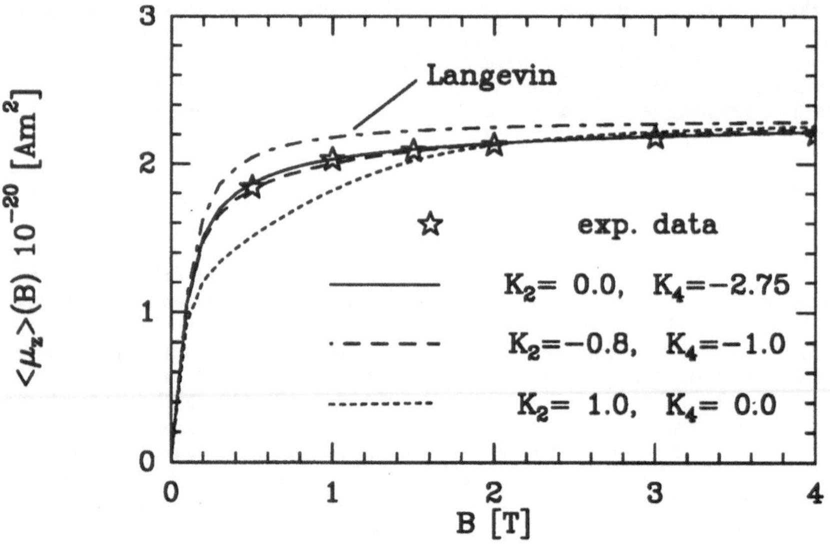

Fig.2: Comparison of the calculated magnetization of Fe_n-clusters suspended in liquid with experimental data [9]. Different curves correspond to different set of anisotropy constants K_2 and K_4. For comparison the Langevin-function for the Fe_n-particles is also shown.

4. The Curie Temperature of Clusters

In the above calculations the clusters are treated as superparamagnetic particles and therefore behave as an ensemble of giant single magnetic moments. This may be approximately true for the cases of larger Fe- and Co-clusters, because the Curie temperature $T_C(n)$ of these clusters are expected to be much larger than the cluster temperature T. For Ni- and rare-earth- clusters one expects $T_C(n) \sim T$. Then the consideration of the finite magnetic ordering is important. Hence, it is of interest to calculate the size dependence of the Curie temperature $T_C(n)$. We use a simple mean field theory. Only clusters with complete atomic shells are considered. The i-th shell magnetization $m_i(T, D)$ of a cluster with D shells and $n(D)$ atoms is obtained by solving a system of D equations selfconsistently. For Ising 1/2-spins these equations are given by

$$m_i = \tanh\left(\beta \mathcal{J} \mu_i \sum_{j=1}^{D} z_{ij} \mu_j m_j\right). \tag{3}$$

Here, μ_i is the magnetic moment of the i-th shell, z_{ij} the number of nearest neighbors between shells i and j, and $\beta = (k_B T)^{-1}$, k_B Boltzmann's constant. The exchange coupling constant $\mathcal{J}$ is assumed to be the same for all spin pairs. For $T \lesssim T_C(n)$ Eq.(3) can be linearized leading to a $D \times D$ matrix. In Fig. 3 we present our results for $T_C(n)$ as a function of the total number n of cluster atoms for a cubooctahedral and an icosahedral cluster geometry. Although the overall behavior of $T_C(n)$ is similar for these two cases, some differences are clearly visible. Note that a cluster containing about 50 atoms possesses already a Curie temperature $T_C(n) \sim \frac{3}{4}T_C(\text{bulk})$.

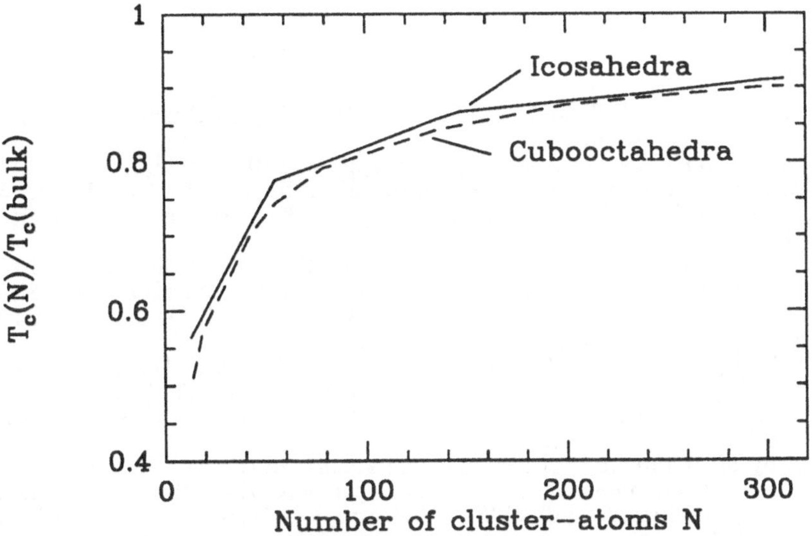

Fig.3: Calculated Curie temperature $T_C(n)$ as a function of the number n of cluster atoms for icosahedral and cubooctrahedral clusters.

5. Conclusion

In this study we presented two applications of our theory [3] for the magnetization of a cluster ensemble resulting from an external magnetic field. In both applications the clusters are in the superparamagnetic region ($T > T_{bl}$) and possess a large Curie temperature T_C compared to the cluster temperature T. Thus, our calculations need to be extended: First, $T \sim T_C$ should be taken into account in the calculation (as has already been done by us [3] for the case $T < T_{bl}$). Second, a crossover between the two limiting cases, $T > T_{bl}$ and $T < T_{bl}$, should be observable by varying the cluster temperature or the measuring time. For $T \sim T_{bl}$ the clusters no longer behave as an ensemble of superparamagnetic particles, because at this temperature the magnetization begins to be tied to the cluster geometry via the lattice anisotropy. In this case the system is no longer in thermal equilibrium and should be calculated in a different manner. Furthermore, experimental results seem to indicate deviation from Langevin behavior and a tendency towards earlier saturation of $\langle\mu_z(B)\rangle$, which becomes more pronounced for lower temperatures. One expects such behavior from cluster rotation, if $\omega_{rot} \approx \omega_o$, ω_{rot} the cluster rotation frequency and ω_o the gyromagnetic precession frequency. These problems are currently investigated and numerical calculations are in progress.

We thank M. Hanson for sending her results to us prior to publication. This work has been partly supported by the Deutsche Forschungsgemeinschaft, Sonderforschungsbereich 337.

References

[1] de Heer, W.A., Milani, P., and Châtelain, A. (1990) 'Spin Relaxation in Small Free Iron Clusters', Phys. Rev. Lett. 65, 488.
[2] Bucher, J.P., Douglass, D.C., Xia, P., Haynes, B., and Bloomfield, L.A. (1991) 'Magnetic Deflection of Neutral Metal Clusters in a Beam', Z. Phys. D 19, 251; Bucher, J.P., Douglass, D.C., and Bloomfield, L.A. (1991) 'Magnetic Properties on Free Cobalt Clusters', Phys. Rev. Lett. 66, 3052.
[3] Khanna, S.N. and Linderoth, S. (1991) 'Magnetic Behavior of Clusters of Ferromagnetic Transition Metals', Phys. Rev. Lett. 67, 742; Jensen, P.J., Mukherjee, S., and Bennemann, K.H. (1991) 'Theory for Spin Relaxation in Small Magnetic Metal Clusters', Z.Phys. D 21, 349.
[4] Merikoski, J., Timonen, J., Manninen, M., and Jena, P. (1991) 'Ferromagnetism in Small Clusters', Phys. Rev. Lett. 66, 938. Note that it was necessary in this work to assume a cluster temperature T larger than the Curie temperature T_C to obtain agreement with experiment. However, $T < T_C$ is the case.
[5] Studies by other groups [2,3] also assumed superparamagnetism. They use values for the atomic magnetic moment and cluster temperature which differ somewhat from ours.
[6] Gay, J.P. and Richter, R., (1986) 'Spin Anisotropy of Ferromagnetic Films', Phys. Rev. Lett. 56, 2728.
[7] Blügel, S., Drittler, B., Zeller, R., and Dederichs, P.H. (1989) 'Magnetic Properties of 3*d* Transition Metal Monolayers on Metal Substrates', Appl. Phys. A49, 547.
[8] Vega, A., et al. (1991) 'Stepped Fe(001) Surface Magnetism', ICM 91, Edinburgh.
[9] Hanson, M. (1991), private communication.

DIAMAGNETIC SUSCEPTIBILITY AND CONDUCTIVITY OF COLUMNAR STRUCTURES

M. KUMAN
Department of Electrical Engineering
University of Tennessee
Knoxville, TN 37996

ABSTRACT. Expressions are derived for free electron diamagnetic susceptibility and energy gap of columnar structures. As an illustration it is shown that the derived expressions are in agreement with the experimentally measured electrical and magnetic properties of metal-halogenic columnar structures and ciano-metal columnar structures, as well as their doped derivatives.

1. INTRODUCTION

All ligand-bridged complexes of the type

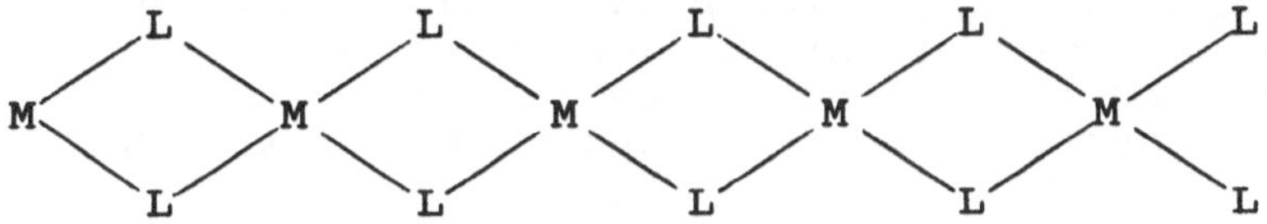

are named columnar structures.

The compound I

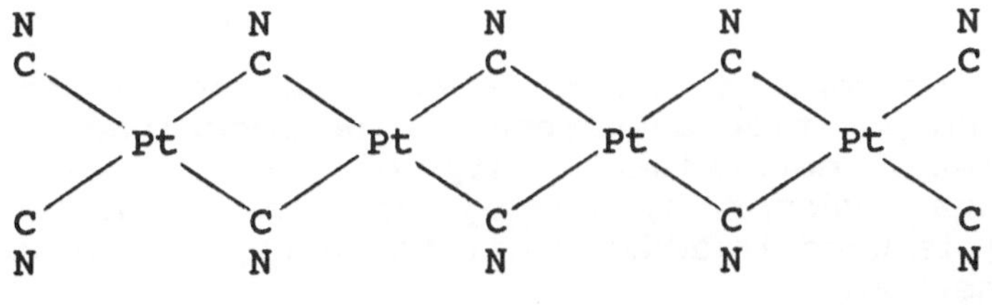

formed by segregated stacking from $K_2(Pt(CN)_4).3H_2O$ is an example of columnar solid. It has a closed shell electronic structure with a pronounced diamagnetic

P. Jena et al. (eds.), Physics and Chemistry of Finite Systems: From Clusters to Crystals, Vol. I, 779–784.

susceptibility. It has semiconducting properties ($\sigma = 5x10^{-1}(om.cm)^{-1}$) [2].

The compounds II

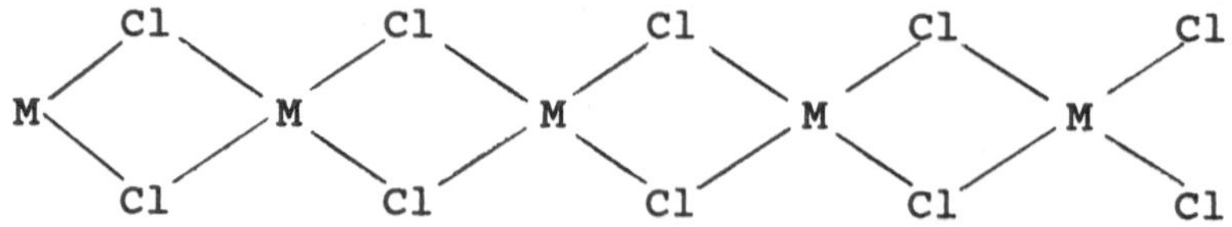

(M = Au, Pt, Pd) are prototypes of halogenic columnar structures with closed shells of electrons and much smaller diamagnetic susceptibility than the compound I [1]. They are insulators.

2. MATHEMATICAL APPROACH

The aim of this paper is to show that a combination of a few simplification techniques like the symmetry group theory [3] and London's theory [4] can lead to relatively simple analytical expressions for diamagnetic susceptibility and energy gap (that determine the type of conductivity) of columnar structures.

Columnar structures are planar [1], [5]. Hence, one can consider the set of almost free electrons, responsible for both conductivity and free electron diamagnetic susceptibility, separately from the localized electrons.

The delocalized free electrons will be subject of this paper. The way the free electron diamagnetic susceptibility is separated from the measured total diamagnetic susceptibility is described by the author in [6].

The free electrons of the whole columnar solid chain and their cooperative phenomena can be described by a characteristic determinant only of rank 3, if a translational vector k, analog of the Bloch wave vector [3] is used in addition to the group theory and London's method.

If by analogy with the London's theory, one ascribes formally to each of the bonds metal-ligand ML one quarter of the susceptibility of the ring MLML the characteristic determinant will have the form (since it is symmetric only one half of it is represented):

$$\begin{vmatrix} d - E & ..... & ..... \\ \exp(iw/4)+\exp(-ik+iw/4); & -d-E & ..... \\ \exp(iw/4)+\exp(-ik-iw/4); & 0 & -d-E \end{vmatrix} = 0$$

where $d = (a_1 - a_2)/b$ is a parameter of alternation (a_1 and a_2 are expressed through the electronegativities of the alternating atoms x_1 and x_2), $w = (e/\hbar c)SH$ (e - electric charge of electron, h - Plank's constant ($\hbar = h/2\pi$), c - light velocity, H - magnetic field, S - area of the ring LMLM.

One has to be aware of the fact that the so introduced parameter of alternation d and the determinant are shifting the origin of the energy scale. In our case this shift is not important, since both diamagnetic susceptibility and type of conductivity depend only on energy differences.

After some transformations the characteristic determinant can be written in the form:

$$\begin{vmatrix} d - E & 2\cos(w/4 - k/2) & 2\cos(w/4 + k/2) \\ 2\cos(w/4-k/2) & -d - E & 0 \\ 2\cos(w/4+k/2) & 0 & -d - E \end{vmatrix} = 0$$

From this it follows:

$$E^2 = d^2 + 4(1 + \cos(w/2).\cos k)$$

or

$$E_{\pm} = \pm (d^2 + 4(1 + \cos(w/2).\cos k))1/2$$

The energy levels E_- will be occupied, the levels E_+ - free. The energy gap is:

$$\Delta E = 2(d^2 + 4(1 + \cos(w/2).\cos k))1/2$$

Beside dependence on the magnetic field H through the parameter w, the energy gap is proportional to the parameter of alternation d expressed through the difference in the electronegativities of the alternating atoms. In other words the energy gap is bigger and the compound is more insulating when the alternating atoms have affinities to electrons that are more different.

In the given example of columnar structures type II the electronegativity differences are:

for Cl and Au complex x(Cl) - x(Au)=3.0 - 2.4=0.6
for Cl and Pt complex x(Cl) - x(Pt)=3.0 - 2.2=0.8
for Cl and Pd complex x(Cl) - x(Pd)=3.0 - 2.2=0.8

From the formula for the energy gap it follows that:

1/ The columnar structures type II are insulators, because the large electronegativity differences metal-ligand (0.6 - 0.8) lead to a large energy gap.

2/ The columnar structures type I are semiconductors because the small electronegativity differences ligand-metal lead to a small energy gap ($x(CN) - x(Pt) < 2.5 - 2.2 = 0.3$ ($x(CN) < 2.5$ because the nitrogen atom with electronegativity 3.0 will reduce the electronegativity of CN, that is normally 2.5).

To receive an expression for free electron diamagnetic susceptibility calculation it is necessary to differentiate twice the expression for the energy with respect to the magnetic field intensity H and take the result with negative sign at H = 0.

The result is:

$$X=(eS/2\hbar c)^2 \int_0^{\pi} (2\cos^2(k/2) - 1)/(d'^2+ 2\cos^2(k/2))^{1/2} dk$$

The integral can be expressed through a hypergeometric function of two variables F(k',r) and a second order elliptic integral E(k',r) in the following way:

$$X = (eS/4\hbar c)^{1/2}(d'^2+2)^{1/2}[E(\pi/2,k') - (d'^2+1)/(d'^2+2)F(\pi/2,k')],$$

where $d'= d/2$; $k'^2= 2/(d'^2+2) = 4/(d^2+4)$.

As one can see the diamagnetic susceptibility depends strongly on the parameter of alternation d, i.e. on the electronegativity difference of the alternating atoms. This dependence is not only in the coefficients. The functions E(k',r) and F(k',r) depend on it through k'.

Numerical calculations are presented in Table 1.

TABLE 1. Free electron diamagnetic susceptibility X and free electron diamagnetic susceptibility per ring X' for columnar structures.

d	k'	F	E	-X	X'
0.1	0.9756	4.3854	1.0049	3.3695	2.3811
0.3	0.9780	2.9786	1.0538	1.3049	0.9222
0.6	0.9174	2.3624	1.1500	0.2373	0.1641
0.8	0.8621	2.1432	1.2158	0.1905	0.1298

One can see that the more different the alternating atoms are in electronegativities, the lower is the free electron diamagnetic susceptibility. This seams natural or something one can expect - large difference in electronegativities of alternating atoms will be like a barrier for the almost free moving electrons. The larger is this barrier, the slower is the movement of the electrons in a magnetic field, i.e. the larger is the electronegativity difference of the alternating atoms, the lower is the diamagnetic susceptibility.

The same mathematical approach was used by the same author [7] for calculation of magnetic susceptibilities of boron-nitride and graphite bands. The results were in good agreement with experiments and calculations of other authors, using different mathematical technique [8].

As another illustration let us now take doped columnar structures. Cianoplatinides (as compound I) with common formula $Y_2Pt(CN)_4X_{0.3} \cdot 3H_2O$, i.e. doped with X=Cl, Br, I and with Y=2K, 2Li, 2Rb, Ba, Sr, were found to have metallic conductivity at room temperature [9]. Hence, the doping has shifted their conductivity from semiconducting to conducting. These structures can be synthesized by partial oxidative halogen doping of cianoplatinides (structure I) or mixture of halogeno- and cianoplatinides (structures I and II)([1],p.191) .

According to our results if a dopant has electronegativity, that minimizes the existing metal-ligand electronegativity difference, it will shift the conductivity toward metallic. In this particular case the used dopants have electronegativities: x(Cl)=3.16, x(Br)=2.96, and x(I)=2.66. As it was already mentioned x(CN) < 2.5. The availability of a dopant with electronegativity higher than that of the CN group, will pull more electrons out of CN, and makes its electronegativity even lower, closer to x(Pt)=2.2. Hence, the dopant makes the material conducting through decrease of the existing metal-ligand electronegativity difference, that liberates the movement of the electrons. In other words the dopant lowers the parameter of alternation d and in this way decreases the energy gap and increases the diamagnetic susceptibility.

3. CONCLUSION

This approach can be used as a leading idea in attempts to synthesize new columnar structures with metallic conductivity.

References

Keller H.J. (Ed.), a:(1976) Chemistry and Physics of One Dimensional Metals, ps. 191, 391, Plenum Press, New York; b:(1978) Low Dimensional Cooperative Phenomena, p.287, Plenum Press, New York.

Ferraro J. and Williams J. (1987) Introduction to Synthetic Electronic Conductors, p. 156, New York.

Landau L.D. and Lifschitz E.M. (1958) Quantum Mechanics, Pergamon Press, London - Paris.

London L. (1937) J. Phys. Rad. 8, 397.

Pauling L. (1960) Nature of the Chemical Bond, New York.

Kuman(ova) M. (1971) J. Str. Chimii (Russ.) 12, 556; ibid. (1973) 14, 364.

Kuman(ova) M. (1970), Vestnik LGU (Russ.) 4, 158.

Chausse J.-P. and Hoarau J. (1969) J. Chim. Phys. 66, 1062.

SUPERPARAMAGNETISM IN ISING CLUSTERS

J.MERIKOSKI, M.MANNINEN and J.TIMONEN
University of Jyväskylä, Department of Physics
P.O.Box 35
SF–40351 Jyväskylä
Finland

ABSTRACT. Recent experiments on small ferromagnetic clusters have inspired introduction of a number of seemingly quite different theoretical models. We shall argue that all these models show superparamagnetic behaviour above the blocking temperature but below the Curie temperature. In particular, we shall show that Ising clusters display superparamagnetism and introduce a simple correction to the usual tanh behaviour of magnetisation which has to be included for very small clusters. We also discuss the dependence of magnetisation on coordination number.

1. Introduction

Interest in magnetism of ultrafine (monodomain) particles has increased considerably over the last year or so, mainly because advancement in cluster research has made it possible to make controlled experiments on the magnetic behaviour of small transition metal clusters. Previous theoretical considerations had been mainly directed to calculating the zero temperature magnetic moments of very small metal clusters[1]. The results of these calculations indicated an increased magnetic moment when the size of the cluster was decreased[1]. This result can be qualitatively understood as arising from the decreased band width as the average coordination number is decreased in small clusters.

It was therefore rather a suprise to find that the first experiments on small iron clusters[2] produced values for the magnetic moment per atom which were much smaller than the 2.2 Bohr magnetons observed in bulk ferromagnetic iron. Magnetisation of small iron clusters was also observed[2] to increase linearly with the applied magnetic field, and to inrease with the cluster size. The most disturbing of the results reported in [2] was, however, the claimed *increase* with temperature of the magnetic moment. This kind of obscure behaviour led us to suspect[3] that the temperature of the clusters may be much higher than expected.

If the temperature of the clusters is rather high, one would then expect that temperature induced spin disorder and finite size effects might explain the experimental observations. To this end we proposed[3] a qualitative high–temperature model based on Ising spins. Bulk ferromagnetic iron has a bcc lattice structure, but small clusters seem to prefer more closed packed structures[1], and we therefore chose to put the Ising spins on an fcc structure.

P. Jena et al. (eds.), Physics and Chemistry of Finite Systems: From Clusters to Crystals, Vol. I, 785–791.

Very much as expected, the model produced the increase of magnetisation with the cluster size and the applied magnetic field. It now seems that, even though the model is rather qualitative, the slope of the magnetisation curve as a function of the applied field is definitely smaller than what is found in experiments, which are also available for cobolt clusters[4].

The experiments on cobolt clusters[4] display magnetic behaviour which is very similar to that of iron clusters. It was possible to measure in these experiments the recidence time in the source chamber of the clusters. It was therefore possible to measure magnetisation as a function of residence time, and it was not found to saturate until for residence times longer than 1.6ms. Interpretation of this result is obvious: clusters are not thermalised for short residence times. Cluster temperatures for short but not too short residence times are higher than the source temperature, but not as much as we suspected[3] based on the iron experiments.

It was now evident that a different explanation had to found for the behaviour of magnetisation as a function of cluster size and applied field. The model proposed in [5] draws on the interpretation of the previous NMR results[6] for suspended iron clusters. These show superparamagnetic behaviour: magnetisation of the clusters rotate so that they appear in an applied magnetic field as giant paramagnetic spins. Magnetisation is given in this case simply by the Langevin function[7]. It seems to fit the observed behaviour of iron and cobolt clusters rather well[4, 5].

In experiments clusters are seen[2, 4] to deflect only to the direction of increasing field strength. Therefore their spin angular momentum cannot be conserved, and there must be a coupling to the orbital angular momentum of the cluster. An obvious mechanism of mediating this coupling is through spin anisotropy, usually of an easy axis type. This is the starting point of [8], which also assumes saturated magnetisation and describes clusters by one spin variable only. The results reported in [8] do not seem to conform with experimental findings, but if one looks more carefully into the parameter values used in that work, it is obvious that temperatures which are used are below the blocking temperature. At blocking temperature thermal energy is equal to the anisotropy energy. It is also obvious from the model of Ref.[8] that, for temperatures larger than the blocking temperature, it shows superparamagnetic behaviour.

The latter feature of the model is in fact confirmed by the results reported in [9], where instead of one spin variable, all the individual magnetig momenta of clusters are included. Otherwise the results of [8] and [9] cannot be compared very easily. It has also been shown in [10] that isotropic clusters of Heisenberg spins display superparamagnetic behaviour in the classical limit. As we shall show below, clusters of Ising spins behave in a similar way.

We can therefore conclude that all proposed models agree in that, above the blocking temperature, small ferromagnetic clusters show superparamagnetic behaviour. It would be very interesting if cluster experiments could be done at extremely low temperatures. In this way one could see the effects of anisotropy, and find if the scaling behaviour reported in [8] is indeed satisfied for very small clusters.

2. Ising superparamagnet

As discussed above, the model of Ising spins which we proposed in [3], was considered there only for temperatures close to the Curie temperature. It is of some interest to find if, at low temperatures, it produces the superparamagnetic behaviour now believed to be displayed by ferromagnetic transition metal clusters. This means that we have to carefully keep track of the reversals of the total spins of the clusters. As it turns out, in Monte Carlo calculations, even several million Monte Carlo steps per spin is not enough to produce statistically reliable averages. Fortunately Ising clusters have only two 'degrees of freedom' per spin, and the cluster Hamiltonian can be diagonalised exactly for rather big clusters. In this way we have determined the *exact magnetisation* of Ising clusters for up to 27 spins. We have also determined the magnetisation for a number of different lattice structures in order to find its dependence on the coordination number.

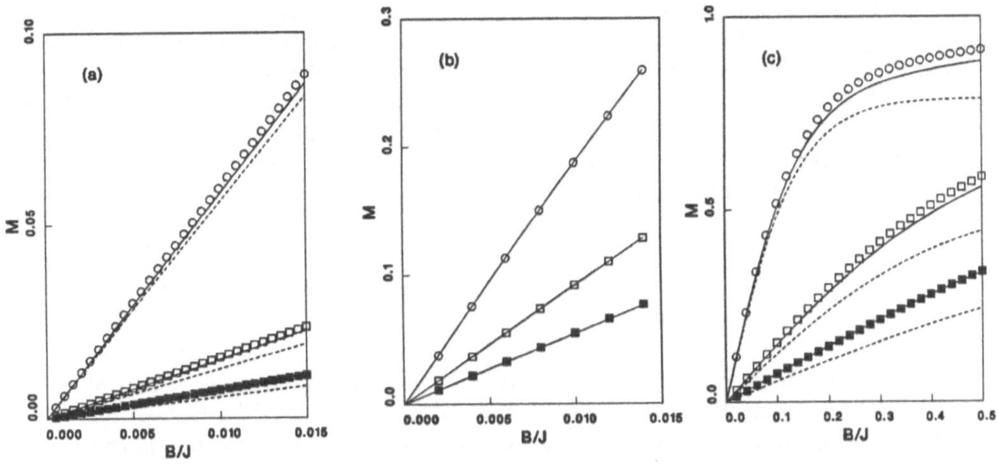

Fig.1 Magnetisation of a bcc cluster with 9 spins (a and c) and of an fcc cluster with 19 spins (b). Circles denote the result for T = J, open squares for T = 2J and filled squares for T = 3J.

In Fig.1 we show the exact magnetisation as a function of the applied field for a bcc clusters with 9 spins and for an fcc cluster with 19 spins. Temperature and magnetic field are both expressed in units of the Ising exchange energy J,

$$H = -J \sum_{\langle i,j \rangle} S_i S_j - B \sum_i S_i. \tag{1}$$

If one uses that value for J which fits[8] the Curie temperature of the Heisenberg model with that of the bulk bcc iron, J$\simeq$11.6 meV, then the temperature scale is J$\simeq$100 K and the scale of the magnetic field is J$\simeq$100 T (= 10^3 kG). One should not in fact use this value here (it was used in [4]), because then the slope of the low field magnetisation of the Ising model is three times too big. The difference is that between $tanh(x) \sim x$ and $coth(x) - \frac{1}{x} \sim \frac{x}{3}$. The latter of these functions is the Langevin function and the former is the magnetisation of an Ising superparamagnet: if an Ising spin cluster has a *total* magnetic moment μ_0 (in units where $g\mu_B = 1$), then its magnetisation per spin in an applied field B is

$$M = \frac{\mu_0}{N} tanh(\frac{B\mu_0}{kT}), \tag{2}$$

where k is the Boltzmann constant and N is the number of spins. Superparamagnetic magnetisation (2) is shown in Fig.1 by a dashed line. In Fig.1b the dashed line overlaps with the solid line to be explained below.

It is clear from Fig.1 that the Ising spin cluster indeed behaves like a superparamagnet for temperatures well below the Curie temperature. It is also evident that for very small clusters single spin excitations make a noticeable deviation from the ideal superparamagnet. We can however take the single spin deviations into account by the following very simple way. It is clear that the *effective* fields felt by up and down *total* spins are slightly different and, therefore, the effective magnetic moments of the two cases are also different. Denote these two effective magnetic moments by $\mu_0 \pm \delta\mu$, respectively. It is then easy to show that the magnetisation per spin in this case is

$$M = M_0 + \frac{\delta\mu}{N}, \tag{3}$$

where M_0 is given by eqn.(2). In order to satisfy the proper boundary conditions at $B \to 0$ and $B \to \infty$, $\delta\mu(B)$ must satisfy $\delta\mu(0) = 0$ and $\delta\mu(\infty) = N - \mu_0$. The form of $\delta\mu(B)$ can be calculated in many ways (we have calculated it exactly in 1D), for our purposes it is enough to use a simple mean field argument which gives

$$\delta\mu(B) = (N - \mu_0) tanh(\chi\frac{B}{kT}), \tag{4}$$

where $\chi = \chi(T)$ is the magnetic susceptibility. The solid curves in Fig.1 are found from (3) with (4) for $\chi = 1$. To show how well this simple theory applies to small clusters even for $\chi = 1$, we have included Fig.1c which shows magnetisation also for larger values of the applied field. For clusters with $N \gtrsim 15$ the correction term (4) is negligible, especially for clusters with a large coordination number.

It is now clear that magnetisation of bigger clusters can be determined from eqn.(2) whenever the absolute value of the total magnetic moment $\mu_0 = \mu_0(T)$ is known. This quantity is in fact very easy to calculate by the Monte Carlo method for example. In this way we have determined the magnetisation of a few clusters shown in Fig.2. For comparison we also show in Fig.2 the corresponding results when the saturated (T = 0) value for μ_0 is used in (2). Even at room temperature the error from using $\mu_0 = N$ in (2) seems to be noticeable. A similar analysis can be done for the Heisenberg model. In this case magnetisation is given by the Langevin function which replaces the tanh in (2). For small clusters there is also a correction term in analogy with $\delta\mu$ in (4).

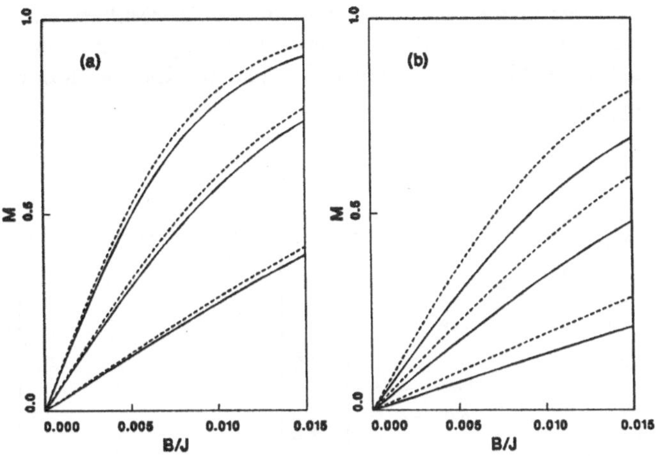

Fig.2 Magnetisation (solid lines) of fcc clusters with 57, 137 and 229 spins (from down upwards). Dashed lines denote the corresponding values from eqn.(2) with $\mu_0 = N$. T = 2J in (a) and 3J in (b).

3. Ising versus Heisenberg clusters

It is often so much easier to deal with Ising clusters than with Heisenberg clusters, that it is worth while to consider, how closely the qualitative behaviours of the two cases are similar. As we noted above, there is a factor of three difference in the slope of the magnetisation against applied field curve. To further probe the similarities, we have calculated the exact magnetisation of 1, 2 and 3 dimensional configurations of 1–8 spins for both cases. The results are summarised in Fig.3 where we show magnetisation as a function of the average coordination number. It is evident that the figures are almost identical apart from the scales, where there is again a difference by a factor of 2–3.

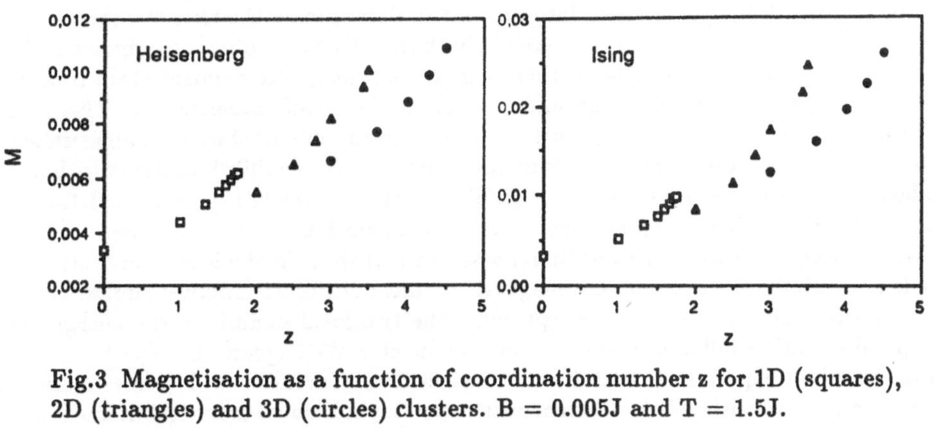

Fig.3 Magnetisation as a function of coordination number z for 1D (squares), 2D (triangles) and 3D (circles) clusters. B = 0.005J and T = 1.5J.

Encouraged by the striking similarity of the Ising and Heisenberg clusters evidenced by Fig.3, we have calculated exactly the dependence of magnetisation on the coordination number for a number of bigger Ising clusters. The results are shown in Fig.4 for two different temperatures.

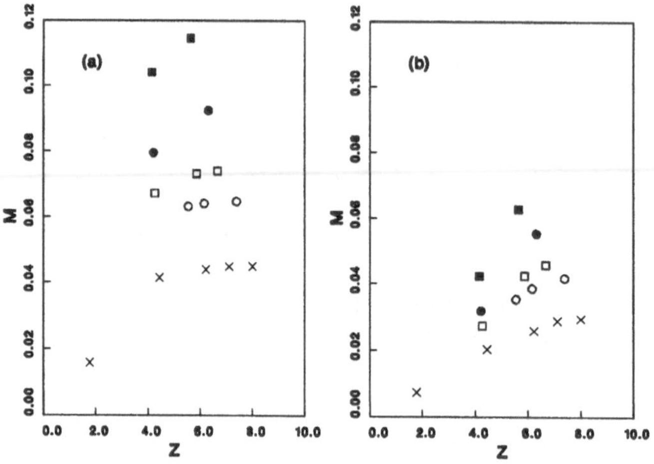

Fig.4 Magnetisation as a function of coordination number z for clusters with 9 (crosses), 13 (open circles), 15 (open squares), 19 (filled circles) and 27 (filled squares) spins. B = 0.02J, T = 2J in (a) and T = 3J in (b).

Similar results should hold for the corresponding Heisenberg clusters. Figure 4 gives an estimate for the possible fine structure in the cluster magnetisation arising from structural isomers. The magnetic moments of the individual atoms may also depend on the lattice structure[11] which has to be taken into account in determining the true fine structure.

Reproducible fine structure has been seen in both iron[2] and cobolt[4] experiments. There was a very interesting isomeric transition from a small magnetic moment state to a large magnetic moment state observed around N = 55 in the cobolt experiments. This isomer effect was tentatively attached to a structural change from a cuboctahedral to an icosahedral cluster structure. We think that there is no such change in the equilibrium structure because icosahedral structure seems to be more stable for all $N \lesssim 10000$[12]. We shall therefore conclude this paper by offering another tentative explanation for the transition.

It can be concluded from the results reported in [11] that, in the icosahedral structure, there is a possibility for two different magnetic moment states with almost similar energies and with essentially the same lattice spacing. The two local minima in the energy curve correspond to rather different total magnetic moments. We suggest that the two isomers seen in the cobolt experiments both have an icosahedral structure but they have a different electronic structure with rather different values for the individual magnetic moments.

References

1. Pastor, G. M., Dorantes–Davila, J. and Bennemann, K. H. (1989) Phys. Rev. B 40, 7642.
2. de Heer, Walt A., Milani, Paolo and Châtelain, A. (1990) Phys. Rev. Lett. 65, 488–91.
3. Merikoski, J., Timonen, J., Manninen M. and Jena P. (1991) Phys. Rev. Lett. 66, 938–941.
4. Bucher, J. P., Douglass, D. C. and Bloomfield, L. A. (1991) Phys. Rev. Lett. 66, 3052–55.
5. Linderoth, S. and Khanna, S.N. (1991) Phys. Rev. Lett. 67, 742–45.
6. Xiao, G., Liou, S. H., Levy, A., Taylor, J. N. and Chien, C. L. (1986) Phys. Rev. B 34, 7573.
7. Kittel, Charles (1971) Introduction to Solid State Physics, John Wiley, New York.
8. Ballone, P., Milani, Paolo and de Heer, W. A. (1991), to be published.
9. Jensen, P. J., Mukherjee, S. and Bennemann, K.H. (1991), to appear in Z. Phys. D.
10. Bucher, J. P. and Bloomfield, L. A. (1991), to be published.
11. Dunlap, Brett I. (1990) Phys. Rev. A 41, 5691–94.
12. Cleveland, C. L. and Landman, U. (1991) J. Chem. Phys. 94, 7376; Martin, T. P., Bergmann, T., Götlich, H. and Lange, T. (1991) Chem. Phys. Lett. 176, 343; Valkealahti, S. and Manninen, M., to be published.

MAGNETIC NANOPARTICLES IN A METALLO-ORGANIC POLYMER: MAGNETIC PROPERTIES AND STRUCTURAL AND CHEMICAL CHARACTERIZATION

F. PALACIO, C. CASTRO, J. REYES, G. STURGEON* AND F.J. LAZARO
Instituto de Ciencia de Materiales de Aragón, CSIC-Universidad de Zaragoza. E-50009 Zaragoza (Spain).
J. GONZALEZ-CALBET
Depto. de Química Inorgánica. Universidad Complutense. E-28040 Madrid (Spain).

ABSTRACT. The metal-organic polymer $[\{Fe(C_{13}H_{17}N_3)_2\}SO_4 \cdot 6H_2O]_n$ is synthetized from the reaction of $FeSO_4 \cdot 7H_2O$ with the polymeric Schiff base $(C_{13}H_{17}N_3)_n$. On the basis of its anomalous magnetic properties, this material has been recently characterized as a superparamagnet with a blocking temperature above room temperature. In this communication we show by means of electron microscopy techniques that the amorphous polymeric material contains small particles of iron oxides which are, indeed, responsible for the magnetic properties. The average size of the particles is in the range of 100 to 300Å. The chemical process through which the particles are formed is also investigated. A more complete account of the magnetic properties of this material is also given. The presence of α-Fe_2O_3 in these preparations is confirmed; the data also suggest the presence of α-FeO(OH) rather than the γ-Fe_2O_3 previously reported.

1. Introduction

In the course of extensive research on the magnetic properties of metal-organic polymers, the anomalous magnetic behavior exhibited by the Schiff-base polymeric material $[\{Fe(C_{13}H_{17}N_3)_2\}SO_4 \cdot 6H_2O]_n$, drew our attention [1, 2]. Indeed, as we briefly review below, the magnetic behavior of this polymeric material is consistent with that of superparamagnetism, its origin being due to the presence of small particles of nanometric size produced in the synthetic process. So far, important questions, such as those concerning the structural characterization of the particles and the understanding of the synthetic process through which these particles have been formed, have not been considered. In particular, the solving of the last problem is of prime importance for the developing of a more general method to synthesize metal oxide nanoparticles embedded in polymeric matrices.

The aim of this paper is to contribute to the solution of both problems. The structural characterization of the small particles is carried out by transmision electron microscopy (TEM), electron diffraction and high resolution x-ray powder diffraction. The study of the chemical process through which the particles are formed is made indirectly by

P. Jena et al. (eds.), Physics and Chemistry of Finite Systems: From Clusters to Crystals, Vol. I, 793–798.

modifying the reaction components in such a way that the formation of the particles is inhibited.

2. Experimental features

The material was prepared following synthetic methods already described in the literature [3] and it is the same material for which magnetic properties have been reported recently [2]. The synthesis was repeated several times modifying the temperature of the reaction in the polycondensation step. This does not affect qualitatively the magnetic properties of the resulting material. It is worth remarking that two repetitive units of the organic polymer are required to coordinate each Fe^{2+} ion. This gives natural amorphicity to the resulting metal-organic polymeric material.

Since Mössbauer spectra indicate the presence of a certain proportion of Fe^{3+} [1,4,5], the synthesis of a polymeric material containing Fe^{3+} as the only metallic ion was attempted with unsuccessful results.

Electron diffraction and microscopy were carried out by means of a JEOL 2000FX electron microscope fitted with a double tilting goniometer stage (± 45°). Samples were prepared following two different methods. In one method, samples were crushed in an agate mortar, ultrasonically dispersed in n-butanol and then mounted on carbon-coated microgrids. In the other procedure, a polymerized block of epoxy resin containing the sample was trimmed with a diamond knife on a Reichert-Jung Ultracut E ultramicrotome to obtain thin sections which were deposited onto copper grids. Samples prepared following the first method tended to give better micrographs and were easier to scan than those prepared according to the second method.

X-ray powder diffraction data ($5° < 2\theta < 100°$) were collected in a Siemens D501 diffractometer using graphite-monochromated CuK_α radiation. The step scanning technique was employed with step size 0.04° and counting time 22 s/step.

3. Magnetic properties

In this section we summarize the magnetic behavior observed in this polymeric metal-organic material. The interested reader can obtain a more detailed discussion in previously published articles [1,2, 4].

The Mössbauer spectrum [1,4,5] at zero dc magnetic field can be explained as the superposition of two spectra, A and B. Spectrum A represents six peaks consists with hyperfine splitting due to an internal magnetic field while spectrum B consists of two peaks of quadrupole splitting. The isomer shift δ, quadrupole splitting ΔE_q and internal field H_i values indicate that spectrum A corresponds to the presence of high-spin ($S = 5/2$) Fe (III) in the samples and spectrum B to low-spin ($S = 0$) Fe (II). The fraction [Fe(III)/total iron] in the samples, calculated from the ratio of the spectral areas, is 0.20±0.09 which agrees within the error bars with the value of 0.3 estimated from X-ray photoelectron spectroscopy [6].

The temperature dependence of the ac susceptibility curve presents a broad maximum at about 250 K and then decreases monotonically down to 20 K, where it remains temperature independent. Two interesting features are worth remarking in this curve. The first is the anomalous high values of χ'. At 250 K χ' = 4.5 emu/mol, a value about 100 times higher than the expected for a typical paramagnet with S = 5/2 and g = 2. The second is the presence of a rather high out-of-phase component of the susceptibility indicative of the presence of net magnetic moments in the material. The χ'' curve shows a maximum around 160 K and then slowly drops to almost zero at 4.2 K. The frequency dependence of χ suggests relaxation effects consistent with superparamagnetic behavior with a blocking temperature around room temperature.

The magnetization *vs*. temperature as measured in the presence of a constant external magnetic field as low as 25 Oe show strong differences between field-cooling and zero-field-cooling curves which are commonly observed in systems of small magnetic particles below the blocking temperature.

The magnetization results as a function of the external field at a constant temperature of 4.2 K present a first irreversible (hysteretical) region for fields up to about 1500 Oe and a reversible one for fields higher than 1500 Oe. The hysteretical cycle exhibits a rather large remanence (300 emu.Oe/mole (430 A/m)) and a high coercive field (420 Oe (320 A/cm)). It is well known that an assembly of ferromagnetic particles shows no magnetic hysteresis above its blocking temperature T_B, while it can be present below T_B. According to the size, shape and anisotropy of the particles a wide range of hysteresis cycles can be observed. The cycle observed in this material is in qualitative agreement with the Stoner-Wohlfarth model for a random assembly of prolate ellipsoidal particles [7]. The reversible region of the magnetization curve does not show saturation even at 50 kOe. This may be due to the existence of high anisotropies or to the presence of additional antiferromagnetic particles.

4. Structural characterization

Transmission electron microscopy (TEM) micrograph is shown in figure 1. The picture evidences the existence in the sample of electrodense particles impregnating an electronically lighter material. Qualitative analysis of selected regions indicate that there is a non negligible concentration of iron in the material, however, within the electrodense particles iron concentration is higher than in the rest. Two kinds of electrodense particles can be observed. Some are acicular with diameter of about 100-150 Å and length around 700 Å-1500 Å while other are polyhedric with a size of about 100 Å and 300 Å.

Electron diffraction experiments were carried out in order to verify the degree of crystallinity and, if possible, to identify the structure of the samples. Although the greater part of each sample seems to be amorphous, electron diffraction patterns, such as that shown in figure 2, which are characteristic of a polycrystalline material can be recorded in some areas. Diffraction rings can be indexed on the basis of the hexagonal unit cell (a = 5.035Å, c = 13.717Å) corresponding to α-Fe_2O_3.

In addition, bigger crystals (>2000 Å) were occasionally observed. Their diffraction pattern, depicted in figure 3, can be indexed along the $[\bar{1}\,1\,2]$ zone axis of α-Fe_2O_3. The corresponding electron micrograph (figure 4) shows the (110) crystallographic planes of a well ordered crystal.

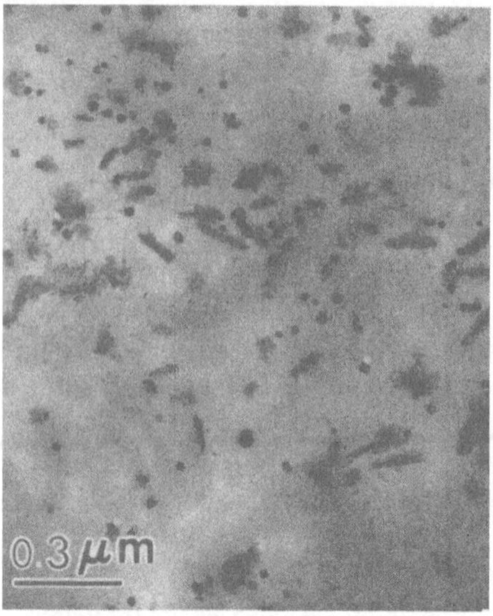

Figure 1. TEM micrograph showing acicular and polyhedric small particles.

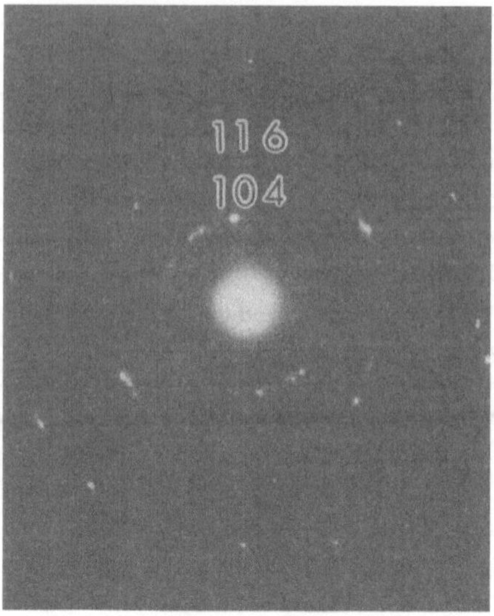

Figure 2. ED pattern of a polycrystalline material indexed on the basis of the α-Fe_2O_3 unit cell.

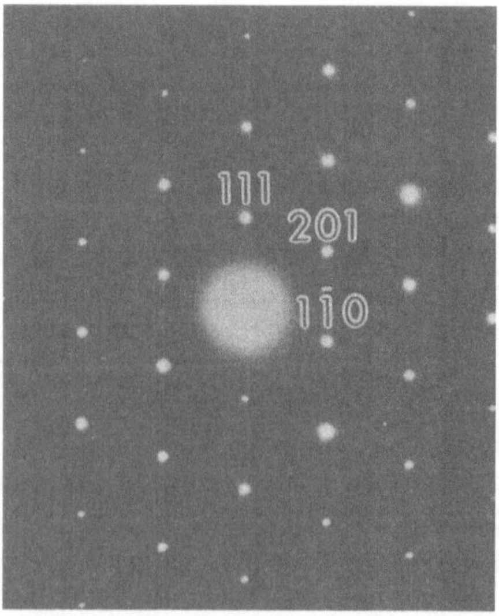

Figure 3. ED pattern of α-Fe_2O_3 along [$\bar{1}$ 1 2].

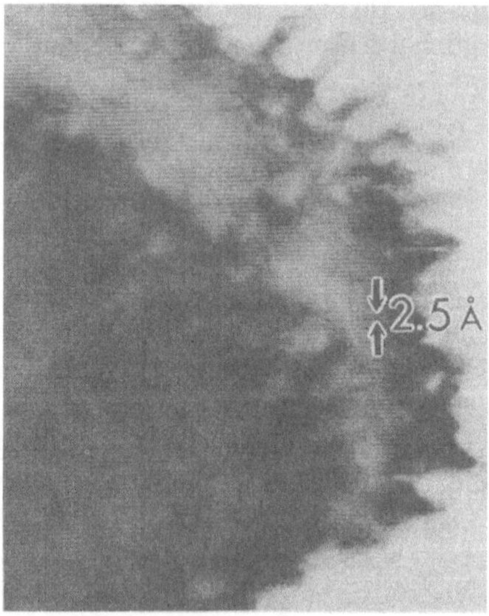

Figure 4. Corresponding micrograph showing the d_{110} spacing.

Hematite (α-Fe_2O_3) is a weak ferromagnet below $T_N = 950$ K, however, at $T_M = 263$ K it exhibits the so called Morin transition becoming a true antiferromagnet below this temperature. Although this might exclude the existence of net magnetic moments at temperatures less than T_M, studies on small hematite particles indicate a decrease of T_M while decreasing the particles size; and for particles sizes of around 200 Å or below the Morin transition disappears [8]. This fact results in the existence of a net magnetic moment in the whole temperature range.

High resolution X-ray diffraction patterns show radial distribution functions characteristic of amorphous materials. However, some small peaks that might be indicative of the presence of crystalline specimens in the material can also be observed [2]. The positions of these peaks are consistent with the presence of α-FeO(OH) (goethite) in the material. This result is in good agreement with TEM results, the goethite particles corresponding to the ellipsoid elongated ones shown in figure 1.

Our results are in good qualitative agreement with other observations previously reported [5] in the sense that small iron oxide particles are responsible for the anomalous magnetic behavior of this polymeric metal-organic material. There is, however, disagreement with respect to the composition of the particles concerned. The existence of γ-Fe_2O_3 has been reported on the basis of Mössbauer and x-ray diffraction measurements. Our results show, unambiguously, the presence of α-Fe_2O_3 in the material and indicate that particles of α-FeO(OH) may also be present. Although it is difficult to differentiate α-FeO(OH) from γ-Fe_2O_3, goethite is thermally stable at the temperatures used in the synthesis and experiments here reported while γ-Fe_2O_3 is metastable. We conclude that α-FeO(OH) is more thermodynamically favoured than γ-Fe_2O_3 and, therefore, more likely to be present in the material.

5. Chemical process of particle formation

It is not simple to make a direct determination of the chemical process involved in the formation of the iron-oxide particles since it involves side reactions that occur parallel to the synthesis of the metal-organic polymer itself. However, it is reasonable to assume that, under the coordinative reaction conditions, a certain amount of $Fe(OH)_2$ is also formed. Then, oxidation with residual oxygen contained in the solvent produces α-FeO(OH) that, during the process of drying the material, yields α-Fe_2O_3:

$$Fe^{2+} + 2OH^- \longrightarrow Fe(OH)_2$$

$$Fe(OH)_2 + O_2 \longrightarrow FeO(OH) + H_2O$$

$$2FeO(OH) \longrightarrow \alpha\text{-}Fe_2O_3 + H_2O$$

In order to verify the suitability of the above process, indirect checks have been made. When the reaction is carried out in the presence of $Na_2S_2O_4$, which is known as a reduction agent [9], an homogeneous polymeric material is formed showing no magnetic anomalies.

Another indirect verification of the above process can be made by chelating the Fe^{3+} formed in the process. A good chelating agent for Fe^{3+} and not for Fe^{2+} is EDTA [10]. When this organic ligand is included in the reaction the yield is an homogeneous, particle-free, polymer.

6. Conclusions

The results presented here confirm the presence of small particles of nanometric size of α–Fe_2O_3 and indicate the likely existence of α–FeO(OH) in the metal-organic polymer $[\{Fe(C_{13}H_{17}N_3)_2\}SO_4 \cdot 6H_2O]_n$. Both particles are responsible of the anomalous magnetic behavior of the material, particularly of its superparamagnetic properties.

References

* On leave from the Department of Chemistry. University of Nebraska. Lincoln, NE 68588-0304.

1. F. Palacio, M.C. Morón, J. Garín, J. Reyes and J. Fontcuberta, Mol. Cryst. Liq. Cryst., **176**, 415 (1989).
2. F.J. Lázaro, M.C. Morón, J. Reyes, J. Garín and F. Palacio, Solid State Comm. (in press).
3. F. Lions and K.V. Martin, J. Am. Chem. Soc., **79**, 2733 (1957).
4. T. Sugano, M. Kinoshita, I. Shirotani and K. Ohno, Solid State Comm., **45**, 99 (1983).
5. H. Nishide, N. Yoshioka, E. Tsuchida and H. Inoue, J. Polymer Sci. A, **27**, 497 (1989).
6. T. Sugano, M. Nomura, K. Agawa, P.L. Kieng, T. Ohta and M. Kinoshita, Bull. Chem. Soc. Japan, **59**, 2615 (1986).
7. E.C. Stoner and E.P. Wohlfarth, Phil. Trans. Roy. Soc., **A240**, 599 (1948).
8. N. Yamamoto, J. Phys. Soc. Japan, **24**, 25 (1968).
9. N.N. Greenwood; "*Chemistry of the Elements*", Ed. Pergamont Press (1986).
10. I.H. Koltoff, P.J. Elving; "*Treatise on Analytical Chemistry*" Part II, Ed. Interscience Publishers (1961).

MAGNETISM IN SMALL ANTIFERROMAGNETIC CLUSTERS

B.V.REDDY AND S.N.KHANNA
Physics Department
Virginia Commonwealth University
1020 W. Main Street, Box 2000
Richmond, Va. 23284-2000

ABSTRACT. The effect of size and geometry on the magnetic behavior of systems with antiferromagnetic interactions has been studied. Our studies based on clusters containing between 13 and 561 atoms model them as Ising spins located at the atomic sites, interacting via a nearest neighbor antiferromagnetic coupling. The magnetic moments of these Ising clusters at various temperatures and magnetic fields have been calculated using Monte Carlo simulations. We have studied clusters having icosahedric and cuboctahedric geometries and different sizes. For icosahedric clusters, the calculated behavior as a function of magnetic field and temperature indicates that the clusters have free spins and behave as frustrated paramagnets. On the other hand, cuboctahedric clusters show well defined ground states indicating that the finite size quenches the frustration inherent in the FCC lattice. The theoretical predictions will be compared with recent measurements.

I. Introduction

Free atomic clusters generated in supersonic beams have emerged as a new class of materials where one can study the progression of properties as one goes from bulk to restricted sizes, reduced dimensions and new geometries. The current interest has focussed on the evolution of the magnetic behavior.[1-3] Various groups have recently investigated the magnetization in small Fe_n and Co_n clusters containing between 13 to several hundred atoms. It has been found that the clusters of these ferromagnetic elements have much lower saturation magnetization than the bulk metals and the magnetization per atom increases with applied field and particle size. These observations have been recently explained in terms of a superparamagnetic model.[4] Since the general behavior of ferromagnetic clusters seems to be understood, the interest is growing towards antiferromagnetic materials[5]. It is well known that certain crystalline structures such as face centered cubic (FCC) or triangular lattice show special features under antiferromagnetic arrangement[6]. In these structures two atoms which are nearest neighbors to another

P. Jena et al. (eds.), Physics and Chemistry of Finite Systems: From Clusters to Crystals, Vol. I, 799–805.

atom may themselves be nearest neighbors. It is therefore not possible to minimize all interaction energies and the lattice is frustrated. The bulk situation however changes as one goes to finite sizes. Here, the existence of low coordinated surface sites may reduce the frustration and lead to a ground state. We show it explicitly for the case of 13 atom cuboctahedron. Small clusters also often exhibit geometrical arrangements not seen in the bulk phase. For example small clusters of several transition metals are icosahedric[7]. It is then important to investigate how the finite size as well as the new geometries absent in the bulk phase accommodate the antiferromagnetic ordering.

In this paper we investigate the magnetic behavior of clusters having atoms interacting via antiferromagnetic coupling. We consider clusters having truncated cuboctahedric (FCC) and the icosahedric shapes. For the magnetic behavior[8], the clusters are modelled as Ising spins with nearest neighbor antiferromagnetic coupling. The magnetization of the clusters as a function of size and magnetic field is calculated using the Monte Carlo simulations[8]. Our studies at low temperatures and magnetic fields show that the spins in cuboctahedric arrangement are locked and the cluster moment does not change appreciably with applied field or temperature. On the other hand the icosahedric clusters contain spins frustrated by the underlying lattice and free to orient. The cluster, therefore, behaves as a simple paramagnet with a small moment per atom which increases with field and decreases with increasing temperature.

In section II we give details of the calculations and results. Section III contains the main conclusions.

II. Details of Calculations and Results

We model the cluster as a collection of localized spins $S_i = \pm 1$ located at the atomic sites labelled by the site index i. The spins interact via a nearest neighbor exchange coupling J and the cluster is described by the hamiltonian

$$H = J\sum_{i<j} S_i \cdot S_j - B\sum_{i=1}^{N} S_i$$

The summation in the first term is carried over the nearest neighbors and B is the magnetic field. To calculate the average magnetization of the cluster as a function of magnetic field and the temperature, we carried out the standard Monte carlo simulations. Basically we start with a given configuration S of spins and generate a sequence of configurations as follows. A spin is selected and the change in energy ΔE associated with flipping this spin is calculated. The new configuration S' with spin flipped is then accepted with probability P given by

$$P = \begin{cases} 1 \text{ if } \Delta E \leq 0 \\ \exp(-\Delta E/K_B T), \end{cases}$$

otherwise. Such an importance sampling[8] should provide a fair estimate of the equilibrium properties. In all simulations, we have carried out 200,000 monte carlo steps (MCS) per spin. The average magnetization was calculated by averaging the cluster moments calculated after every 5N MCS per spin where N is the number of atoms in the cluster. The relaxation times in small clusters can be long and the starting configuration may affect the calculated M_{eff}. To minimize these effects the initial 250,000 MCS per spin were ignored in each calculation. To further minimize these effects we carried out calculations starting with ferromagnetic, random and high temperature high field configurations as starting configurations for each size. The calculated M_{eff} was then averaged. As mentioned before we have studied clusters having 13, 55, 147, 309 and 561 atoms arranged in icosahedric and cuboctahedric arrangements. The above sizes correspond to clusters having a complete shell of atoms. For icosahedric clusters we found that the fluctuations from the average value were minimal (less than 5%). For 13, 55 and 147 atom cuboctahedric clusters, the fluctuations were less than (30%). For 561 atom cuboctahedric cluster, we did observe larger fluctuations (up to 50%) at certain points. As we will show, these fluctuations do not alter the conclusions of this work. The effective magnetic moment M_{eff} (in units of moment per atom) per atom for these clusters was calculated as a function of the magnetic field B (in units of $J\mu_B^{-1}$) and the temperature T (in units of $J K^{-1}{}_B$).

In Fig 1 we show M_{eff} as a function of magnetic field for 13,147 and 561 atom icosahedric and cuboctahedric clusters at T = 0.2. One notices a striking difference in the behavior of small clusters (13 and 147 atoms) with different geometries. M_{eff} in icosahedric clusters increases as the magnetic field is increased reaching a saturation value at large fields. This behavior is reminiscent of a paramagnetic material and shows that icosahedric clusters behave as paramagnets. Using the limiting M_{eff}, one can calculate the effective number of spins free to respond to the magnetic field and one finds 3 and 13 spins for 13, and 55 atom clusters respectively. The proportion of the free spins does not continuously increase as the cluster size is increased. We find that up to 147 atom clusters the limiting M_{eff} is nearly the same as for 13 or 55 atoms. However for clusters containing more than 147 atoms, the number of free spins does not increase with cluster size and a 561 atom cluster only has a limiting M_{eff} of about 0.09. The existence of the free spins and the reduction in the number of free spins per cluster atom for sizes beyond 147 atoms is intimately linked to the cluster topology. Consider a 13-atom icosahedric cluster shown in Fig. 2(a). It can be considered as consisting of two pentagons (B and D) and three atoms A, C, and E. Any attempt to minimize energy of the spins in pentagons B and D leaves the spins A, C, E free to rotate in either direction. It is this frustration arising due to the underlying topology that generates free like spins and gives the cluster a paramagnetic behavior. As the cluster size is increased, the icosahedric topology generates successive shells containing atoms along the edges of the triangles and in the faces of the triangles. For 55 and 147 atoms, the additional layers generate triangular plaquattes and the frustration increases with cluster size. For 309 atom clusters, the surface faces begin to have

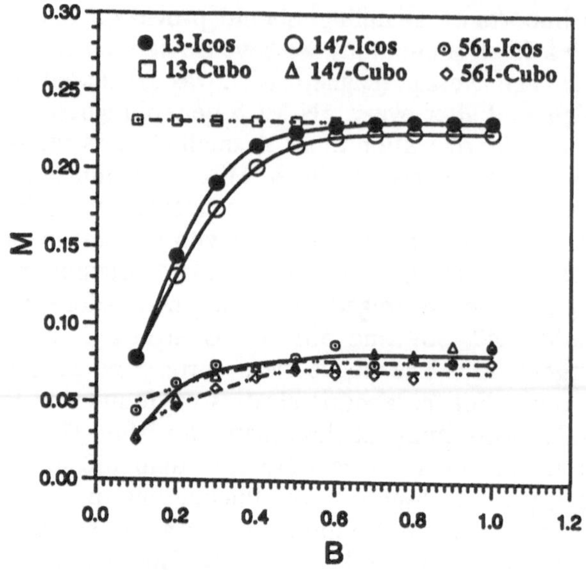

Fig.1. M_{eff} as a function of Magnetic field B (in units of J/μ_B) for icosahedric and cuboctahedric clusters of various sizes at a temperature of T=0.2.

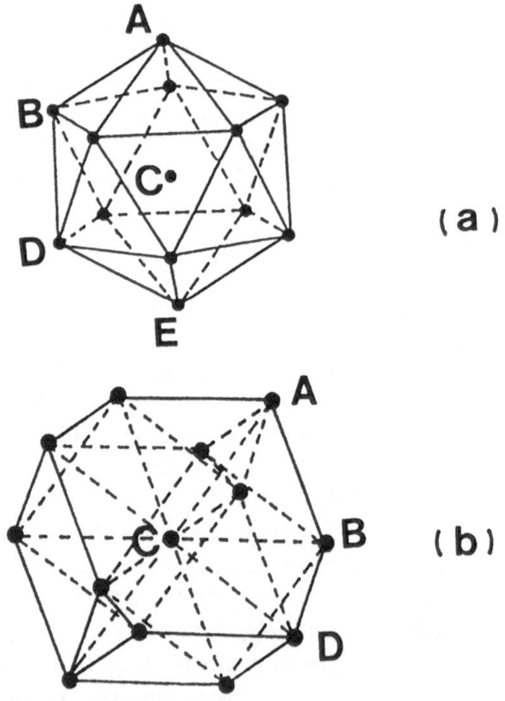

Fig.2. The geometrical arrangement of atoms in an icosahedric (a) and cuboctahedric (b) 13 atom cluster.

atoms with most neighbors forming rectangles instead of triangles. The frustration is thus diminished leading to a decrease in M_{eff}. As opposed to icosahedric clusters, the cuboctahedric clusters show an entirely different behavior. For the 13-atom cuboctahedric cluster, M_{eff} remains at a constant value as the field is increased indicating that the cluster has non-zero spin in the ground state but the spins are not free to rotate as was the case for icosahedric clusters. As the cluster size is increased, the same behavior remains true. To show how the finite size quenches the frustration inherent in the FCC lattice and to get an idea about the magnitude of M_{eff}, consider a 13 atom cuboctahedron shown in Fig. 2(b). It consists of a top square of atoms (A), middle square (B) with a central site (C) and the bottom square (D). Squares A and D can be individually satisfied with a choice of up and down spins. However, irrespective of the choice made in A and D, the central site C and the square D are free to orient. One can now choose a spin on C which requires all D spins to be oriented opposite to C. The cluster has no free spins and the frustration has been quenched by the finite size. Further, the new moment is 3 which yields an M_{eff} of 0.23 as calculated numerically. Similar reasoning holds for higher sizes but M_{eff} decrease as the size is increased.

The existence of free and blocked spins is also visible in the behavior of clusters as a function of temperature. In Fig. 3 we show M_{eff} for 13, 147 and 561 atom

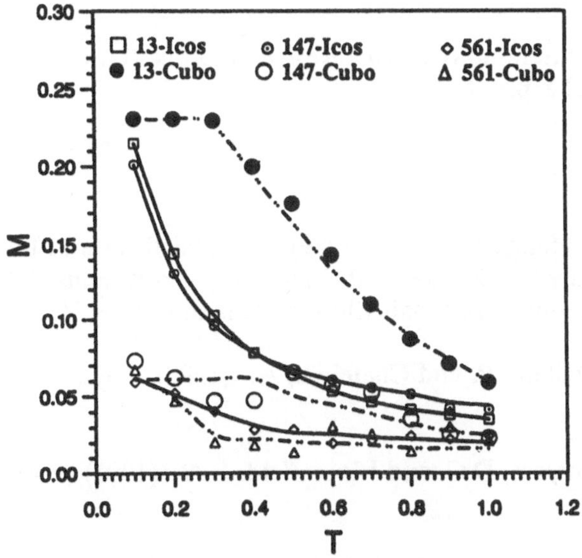

Fig.3. M_{eff} as a function of temperature for icosahedric and cuboctahedric clusters of various sizes in a magnetic field of 0.2.

icosahedric and cuboctahedric clusters. Once again notice that icosahedric clusters behave as paramagnetic systems while cuboctahedric clusters show constant M_{eff} at small T. As T is increased, the loosely bound spins in cuboctahedric clusters become disordered and the magnetization decreases. This is particularly visible for 13-atom cuboctahedric cluster.

III. Conclusions

To conclude, we have investigated the effect of geometry on the behavior of antiferromagnetic clusters. As mentioned earlier, experiments[5] are already underway to determine the behavior of clusters of antiferromagnetic materials such as Cr. In order to make contact with experiments one has to know J and the moment per atom. Both these quantities are not known for small clusters. However even if one takes J ~ 1meV[9] (value for Fe) and T ~ 100K, the observed M_{eff} for fields of even 1 T will be less than 0.01μ for small icosahedric or cuboctahedric clusters, where μ is the moment per atom. This moment is much smaller than the corresponding moments of $1\mu_B$ for Fe or Co clusters. In a recent experiment, Douglas et al.[5] tried to measure M_{eff} for Cr_n clusters. While they could not measure M_{eff} they concluded that M_{eff} if any was much less than $0.01\mu_B$, which is consistent with our findings. However, an actual measurement of M_{eff} requires experiments with better resolution which are underway.

IV. Acknowledgements

The authors are grateful to the Army Research Office (Grant No. DAAL-03-89-K-0015) for financial support.

V. References

1. Gillet, M.F. and Henry, C.R. eds. (1989) Small Particles and Inorganic Clusters, (Z. Phys. D 12); Jena, P. Rao, B.K., and Khanna, S.N. (1987) Physics and Chemistry of Small Clusters, (Plenum, New York).

2. de Herr, W.A., Milani, P. and Chatelain, A. (1990) Phys. Rev. Lett. 65, 488.

3. Bucher, J.P., Douglass, D.C. and Bloomfield, L.A. (1991) Phys. Rev. Lett. 66, 3052.

4. Khanna, S.N. and Linderoth, S. (1991) Phys. Rev. Lett 67, 742.

5. Douglas, D.C., Bucher, J.P., Haynes, D.B. and Bloomfield, L.A. (1991) (Submitted to Richmond Conference).

6. Sherrington, D., (1989) in "Disordered Solids" edited by Baldossare DiBartolo (Plenum Press, N.Y.).

7. Moraweck, B. and Renouprez, A.J. (1981) Surf. Sci. 106, 35.

8. Binder, K and Heermann, D.W., eds. (1988) Monte Carlo Simulations in Statistical Physics, (Springer-Verlag).

9. Kittel, C., ed. (1971) Introduction to Solid State Theory, p. 532 (Wiley, New York).

SIZE EFFECTS IN ORBITAL MAGNETISM

J. M. VAN RUITENBEEK AND D. A. VAN LEEUWEN
Kamerlingh Onnes Laboratorium
P.O. box 9506
NL 2300 RA Leiden
The Netherlands

ABSTRACT. A brief review of magnetic susceptibility measurements on small metallic particles is presented. Many features are poorly understood. Here, the contribution of the orbital motion of the electrons is investigated. Size effects in the orbital magnetic susceptibility studied by considering some simple model systems. The atom-like properties at low temperatures gradually evolve towards a diamagnetism close to the bulk Landau value at high temperatures, with only small size dependent deviations. For $k_B T$ smaller than the energy level spacing, the susceptibility of an ensemble average of different particles is expected to show orbital paramagnetism. This mechanism is proposed as an explanation for the observations of Kimura and Bandow [1]. Also, the implications for other experiments are discussed.

1. Magnetic Suseptibility in Small Particles.

The study of the properties of small metallic particles generally concentrates on the question how the properties of an atom, evolve into the properties of a bulk metal, as more and more of these atoms are connected together. Here we discuss one such property, namely the magnetic suceptibility. The magnetic susceptibility naturally devides into a spin- and an orbital contribution.

The spin of the electrons in an atom gives rise to a magnetic response only if there is one or more non-paired spin. In that case, the degeneracy of spin-up and spin-down states is lifted in a magnetic field and the temperature dependent occupation of the split set of levels gives rise to Curie type paramagnetism. In a metal there is a continuum of states for each spin direction, and these are shifted with respect to each other by the application of a magnetic field. The spin states that are shifted to lower energy increase in population and the resulting magnetic response is known as Pauli spin paramagnetism. A small metallic particle is expected to show a transition between these two behaviors, depending on the ratio of the thermal energy, $k_B T$, and the energy level spacing, ΔE. Small particles at low temperatures are then devided into those with an even or an odd number of electrons, and accordingly expected to show no magnetism or Curie paramagnetism, repectively. At high temperatures, higher levels above the ground state are excited and a gradual transition to the bulk-like Pauli susceptibility is expected. The shape of this transition depends on the energy level distribution, several models of which have been discussed in the literature [2].

P. Jena et al. (eds.), Physics and Chemistry of Finite Systems: From Clusters to Crystals, Vol. I, 807–812.

It proves to be fairly difficult to observe this transition in small metallic particles directly from the susceptibility, for a review see Halperin [3]. The problem comes from the fact that the signal from the odd-electron particles dominates at low temperatures and the resulting Curie-like behavior is hard to distinguish from impurity contributions. The most convincing evidence, in our view, is in the experiment by Kimura and Bandow [1] on magnesium particles, which all have an even number of electrons. They find a decease in the susceptibility at low temperatures which cannot be explained by impurities and agrees with the expected behavior for even electron systems. In addition they observe a nearly temperature independent paramagnetism that increases for smaller particles. The latter effect could not be explained and is further discussed below.

A geat variety of large metal cluster compounds has been developed in recent years, and these would seem to be the ideal model systems for the study of size effects [4]. Metal cluster compounds are molecules produced by chemical means, which have a core of metal atoms, stabilized by a shell of organic ligands. Each cluster in the sample has the same structure and a clear distinction between even and odd clusters is expected. However, most metal cluster compounds investigated to date show Curie type paramagnetism corresponding to a fraction of a spin per cluster. In table 1 we show a compilation of results of magnetic susceptibility measurements on various metal cluster compounds. An interpretation of these results was given by Drake *et al.*[5] in terms of intra-cluster electron-electron and spin- orbit coupling. Further evidence is for this model is needed to test its validity.

Table 1. The susceptibility of most metal cluster compounds shows a Curie paramagnetic contribution. For a selection of cluster compounds the effective moment is calculated from the measured Curie constants. A moment of $\mu = 1.73$ corresponds to one free spin per cluster and by comparison the effective number, n, of free spins per cluster is derived.

Cluster compound	*no. electrons*	$\mu_{eff}(\mu_B)$	*n*	*References*
$[Pt_6(CO)_{12}](Bu_4N)_2$	even	0.38	0.05	[6]
$[Pt_{19}(CO)_{22}](Bu_4N)_4$	even	0.49	0.08	[6]
$[Pt_{38}(CO)_{44}H_2](PPN)_2$	even	1.1	0.4	[7]
$[Ni_{12}(CO)_{21}H_2](PPh_3CH_2Ph)_2$	even	1.2	0.5	[7]
$Os_7(CO)_{21}$	even	0	0	[5]
$Os_{10}C(CO)_{24}I_2$	even	0.40	0.05	[5]
$Os_{20}HgC_2(CO)_{48}H_2$	even	2.02	1.36	[5]
$Ru_3(CO)_{12}$	even	0	0	[8]
$Ru_4(CO)_{13}H_2$	even	0.72	0.17	[8]
$Au_{55}(PPh_3)_{12}Cl_6$	odd	1.0	0.33	[9]
$Pt_{55}(AsBu_3)_{12}Cl_{20}$	even	1.0	0.33	[9]
$Rh_{55}(PBu_3)_{12}Cl_{20}$	odd	0.8	0.21	[9]
$Pt_{309}Phen^*_{36}O_{30}$	even	1.5	0.72	[9]
$Pd_{561}Phen_{36}O_{200}$	even	≤ 0.4	≤ 0.05	[10]

The other contribution to the magnetic susceptibility is due to the orbital motion of the electrons, and has again very different properties in an atom or in a bulk metal. For a bulk metal in a magnetic field the electrons circle on cyclotron orbits resulting in Landau diamagnetism and de Haas-van Alphen oscilations. The latter are ignored here because they are readily smeared out by temperature or impurities. In an atom, on the other hand, we have Larmor- (or Langevin-) diamagnetism, which scales as the square of the extend of the electron wavefunction. In addition the magnetic quantum numbers are split by a term

$\mu_B \mathcal{L}_z B$ in the Hamiltonian (Zeeman splitting), producing Curie paramagnetism. When the atom is bound to other atoms, the spherical symmetry is broken and the orbital moment is 'quenched', $\langle \mathcal{L}_z \rangle = 0$. In second order perturbation, the Zeeman term produces a temperature independent contribution to the susceptibility, known as van Vleck paramagnetism. The latter contribution was held responsible for an increasing paramagnetism as a function of cluster size observed in a series of Osmium clusters [11],[5]. In the end, however, this paramagnetism should give way to the bulk Landau diamagnetism. The rest of this paper is concerned with the question of how the transition between bulk and atomic orbital magnetism takes place.

2. Model Calculations of Size Effects in Orbital Magnetism.

The van Vleck paramagnetism, that is increasing with cluster size, as described above, is not the only orbital contribution. Also the Larmor diamagnetism, which scales with the square of the radius of the cluster, increases with cluster size. In order to investigate which of the two dominates, and why, simple model calculations have been performed; for details see refs.[12] and [13]. As a first model for a small metallic particle a rectangular box is considered, having infinite potential walls, containing a number of free and independent electrons. This is, of course, a considerable simplification, but it illustrates some important aspects of the magnetic properties of small systems. The effect of a magnetic field is treated in second order perturbation and the resulting susceptibility is plotted in fig.1, as a function of the number of electrons in the box. In this case the box is taken to be nearly cubic, $L_x = L_z, L_y = 1.005 L_x$, and the field is along the z direction. The small difference in the x and y dimensions is introduced in order to avoid degenerate levels.

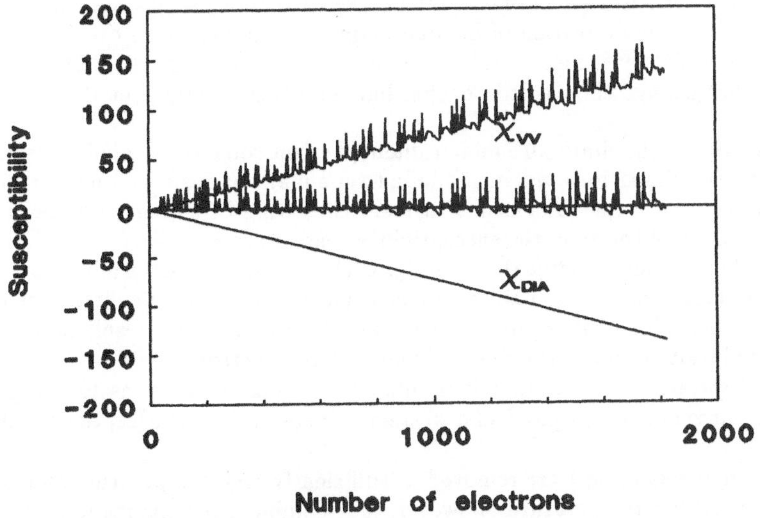

Fig. 1 The orbital susceptibility for a rectangular box at $T = 0$ as a function of the number of particles in the box. The dia- and paramagnetic contributions approximately cancel. Here, the ratio of sizes along the x- and y-axes is $L_x/L_y = \sqrt{1.01}$ and $L_z = L_x$. The units for χ are $\mu_0 e^2/m_e L_1$.

The diamagnetic contribution depends almost linearly on the number of electrons in agreement with the fact that the extent of the wavefunction is about the same for all electrons. The paramagnetic curve is more erratic as a result of the variations in energy level spacing close to the highest occupied level. It is obvious from fig.1 that the two contributions nearly cancel and the total susceptibility fluctuates around zero. In fact, the separate para- and diamagnetic contributions have no physical relevance, since their value depends on the choice of gauge for the vector potential [12]. The sum of the two is gauge independent and is also the quantity that is actually measured. From these results we may conclude that neither a large paramagnetism, nor a large diamagnetism is expected for increasing cluster size. An alternative interpretation for the results on the Osmium series of clusters [5],[11] was given in an earlier paper [12]. It is not possible to judge from simple arguments whether a cluster will be paramagnetic or diamagnetic, the result depending sensitively on details of the energy level structure. This statement holds for small particles as well as large ones, since we have not yet specified how large the dimensions L_x, L_y, L_z are. Then, how do we arive at Landau diamagnetism for the bulk limit?

This transition to the bulk Landau result has been treated in a recent paper [13] and can be summarized as follows: When the temperature is increased the $T = 0$ erratic behavior (fig.1) smears out as soon as $k_B T > \Delta E$. The resulting high temperature value corresponds to Landau's result *independent of the size of the system.* The size of the particle only determines the energy level spacing and, thus, the temperature at which the transition occurs. That this result may be very general was demonstrated by a similar calculation for a spherical box. The spherical potential well is different in the sense that we have a zeeman splitting of the magnetic quantum numbers which produces the Curie $1/T$ divergence of the susceptibility. However, again the paramagnetism is nearly canceled by the diamagnetic terms and Landau diamagnetism (with only a minor size correction) survives. This result is quite astonashing, since the derivation of Landau diamagnetism hinges on the formation of quantized cyclotron orbits by the electons in a metal. For the spherical potential well the magnetic field does not change the electron orbits, but only their energy, yet the Landau result emerges.

In order to investigate the limits of Landau diamagnetism some other model systems were invesigated. In two dimensions, again a Landau type diamagnetism is found at high temperatures. However, for a one-dimensional system consisting of a ring with the field perpendicular to the plane of the ring, the susceptibility tends to zero for $k_B T \gg \Delta E$. The diamagnetic term, which is now exactly the same for each electron and determined by the radius of the ring, is balanced by the paramagnetism due to the Zeeman-splitting of the doubly degenerate levels. This is illustrated in figs.2 and 3. Possibly, this result may help us to develop a more intuitive understanding of Landau diamagnetism.

Above, we have stressed the fact that it is difficult to make predictions for the low temperature orbital magnetism of a particular system. There are a few exceptions which we discuss next.

The paramagnetic peaks in fig.1 are removed at sufficiently high temperature and give way to diamagnetism. At low temperatures, however, they dominate the diamagnetic values so much that an ensemble average of particles of different size (in the model corresponding to different numbers of electrons in the box) produces a net orbital paramagnetism. It has been argued [13] that this effect explains the enhanced paramagnetism in small particles observed by Kimura and Bandow [1] that was mentioned above.

A second exception applies to particles with a well defined high symmetry. As an example, consider the benzene molecule. It is well-known that the diamagnetism of the ring for the field perpendicular to the plane of the ring is accurately described by assuming a current due to free, perfectly conducting electrons around the ring. This is the same as taking just the diamagnetic term in the hamiltonian (for a symmetric gauge with the origin in the center of the ring) and ignoring any effects of the $\mu_B \mathcal{L}_z B$ paramagnetic term. Ignoring this term is reasonable by the following arguments: Let us first picture the ring as having a perfect cylindrical symmetry. The wavefunctions of the six delocalized π-electrons of benzene are then given by

$$\Psi_k(\phi) = (2\pi)^{-1/2} e^{ik\phi}, \quad k = \ldots -2, -1, 0, 1, 2, \ldots,$$

where ϕ is the angular coordinate around the ring. The energy levels are given by $E_k = \hbar^2 k^2 / 2mR^2$, with R the radius of the ring. The degeneracy of the levels k and $-k$ is lifted by the magnetic field and produces paramagnetism if the degenerate set of levels is partially occupied. For benzene the $k = 0$ and $k = \pm 1$ are occupied, and the first excited levels $k = \pm 2$ are high enough that they have negligible occupation at room temperature. For this closed shell configuration only the diamagnetic term in the hamiltonian produces a net effect. Introducing a lattice potential of the carbon atoms around the ring modifies the wavefunction and the energy spacings, but does not lift the degeneracy. As long as the lattice potential is sufficiently weak the argument of closed shell diamagnetism remains valid. At the same time it is clear that rings with 4 or 8 carbons should be paramagnetic and a 10-ring may again be diamagnetic. This is indeed observed experimentally[14].

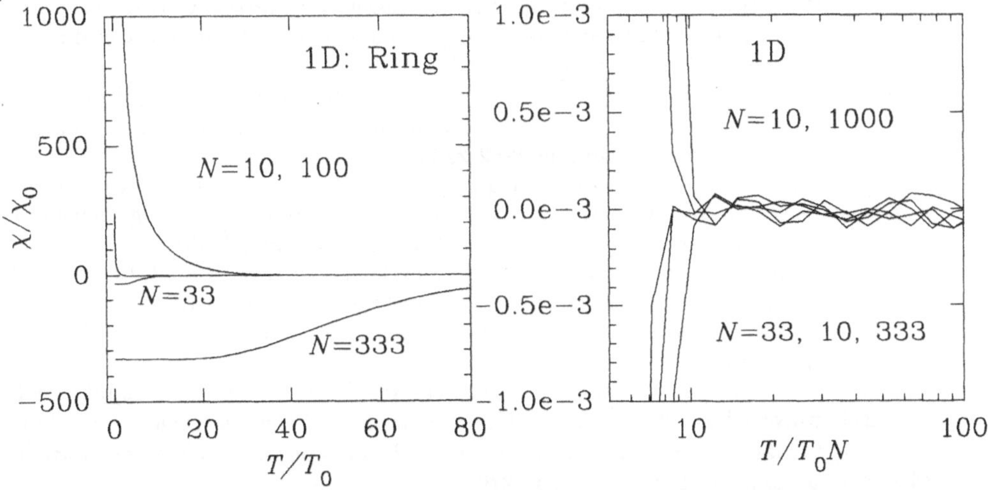

Fig. 2 The orbital susceptibility as a function of T for a one- dimensional ring, occupied by N electrons; the number N is indicated for each curve. The expanded scale on the right ($-1 \cdot 10^{-3}$ to $1 \cdot 10^{-3}$) shows that the susceptibility vanishes at high temperatures to within the accuracy of the calculation. The temperature in the right plot is scaled by N in order to get the curves in one graph. The units for χ and T are $\mu_0 e^2 a / 4m_e$ and $\hbar^2 / 2m_e a^2 k_B$, where a is the radius of the ring.

For the fashionable C_{60} molecule speculations of large orbital diamagnetism in analogy to benzene were soon proven wrong. In our approach one could model the π-wavefunctions of C_{60} as spherical harmonics $Y_{lm}(\theta, \phi)$. Filling the levels with 60 electrons we find that the $l = 5$ level is partially filled. Thus we expect paramagnetism, rather than diamagnetism. Diamagnetism would only result for filled shells, e.g. 50 or 72 electrons in this example, however, the carbon potentials probably modify the wavefunctions too far from free- electron like behavior, eliminating diamagnetism also for that case.

This work is part of the research program of the Leiden Materials Research Center (WFMO) and the Stichting voor Fundamenteel Onderzoek der Materie (FOM) and was made possible by financial support from the Nederlandse organisatie voor Wetenschappelijk Onderzoek (NWO).

References.

1 Kimura, K. and Bandow, S. (1987) 'Paramagnetic enhancement in the magnetic susceptibility of ultrafine magnesium particles', Phys. Rev. Lett. 58, 1359-1362.

2 Kubo, R. (1962) 'Electronic properties of metallic fine particles.I.', J. Phys. Soc. Japan 17 975-986.; Denton, R., Mühlschlegel, B., Scalapino, D. J. (1973) 'Thermodynamic properties of electrons in small metal particles', Phys. Rev. B 7, 3589-3607.

3 Halperin, W. P. (1986) 'Quantum size effects in metal particles', Rev. Mod. Pys. 58, 533-606.

4 de Jongh, L.J., de Aguiar, J.A.O., Brom, H.B., Longoni, G., van Ruitenbeek, J.M., Schmid, G., Smit, H.H.A., van Staveren, M.P.J., Thiel, R.C. (1989) 'Physical properties of high-nuclearity metal cluster compounds' Z. Phys. D 12 445-450.

5 Drake, S.R., Edwards, P.P., Johnson, B.F.G., Lewis, J., Marseglia, E.A., Obertelli, S.D., Pyper, N.C. (1987) 'The emergence of paramagnetism in transition-metal carbonyl clusters', Chem. Phys. Lett. 139 336-344.

6 Teo, B.K., DiSalvo, F.J., Waszczak, J.V., Longoni, G., Ceriotti, A. (1986) 'Weak Curie-type paramagnetism of high-nuclearity platinum carbonyl clusters: anomalous magnetic behavior', Inorg. Chem. 25 2262-2265.

7 Pronk, B.J., Brom, H.B., de Jongh, L.J., Longoni, G., Ceriotti, A. (1986) 'Physical properties of metal cluster compounds I: magnetic measurements on high-nuclearity nickel and platinum carbonyl clusters', Solid State Comm. 59 349-354.

8 de Aguiar, J.A.O., Mees, A., Darriet, J., de Jongh, L.J., Drake, S.R., Edwards, P.P., Johnson, B.F.G., Lewis, J. (1988) 'Physical properties of metal cluster compounds V: magnetic measurements on low- nuclearity ruthenium carbonyl clusters', Solid State Comm. 66 913- 916.

9 van Leeuwen, D.A., van Ruitenbeek, J.M., Schmid, G., de Jongh, L.J., unpublished.

10 van Ruitenbeek, J.M., Jurgens, M.J.G.M., Schmid, G., van Leeuwen, D.A., Zandbergen, H.W., and de Jongh, L.J. (1991) 'Metallic susceptibility in a giant molecule: $Pd_{561}Phen_{36}O_{200}$.', Z. Phys. D 19 267-270.

11 Johnson, D.C., Benfield, R.E., Edwards, P.P., Nelson, W.J.H., Vargas, D. (1985) 'Study of magnetism in osmium cluster compounds as molecular models for small metallic particles', Nature 314 231- 235.

12 van Ruitenbeek, J.M. (1991) 'Orbital magnetism in finite size systems.' Z. Phys. D 19 247-250.

13 van Ruitenbeek, J.M., van Leeuwen, D.A. (1991) 'Model calculation of size effects in orbital magnetism' Phys. Rev. Lett. 67 640-643.

14 Carey, F.A., Sundberg, R.J., 'Advanced organic chemistry', Plenum Press, New York.

SIZE AND SHAPE DEPENDENCE OF SMALL SPIN LATTICES DESCRIBED BY AN ISING HAMILTONIAN.

E.E. VOGEL, S. CONTRERAS and J. CARTES
Departamento de Física, Universidad de La Frontera
Casilla 54-D, Temuco, Chile

J. VILLEGAS
Departamento de Física, Universidad de Concepción
Casilla 3-C, Concepción, Chile.

ABSTRACT. Two-dimensional square spin lattices, with Ising Hamiltonian up to nearest-neighbor interactions are considered. The total number of spins (N) defines the size while the shape corresponds to the way of arranging the N spins in a particular rectangular array. Bonds can be either ferromagnetic ($-J$) or antiferromagnetic ($+J$), N of each, randomly distributed throughout the lattice. The number of possible distributions of bonds increases rapidly with size. Therefore, a large number (500) of bond distributions (samples) for each size and shape is prepared in order to allow a statistical analysis. For each sample we solve the exact problem looking for ground state properties. It is found that all properties depend strongly on size. Some of them depend also on shape (i.e. ground state energy per bond). In spite of the size limitation due to computer facilities, namely N$\leq$ 40, it is possible to observe that saturation is clearly present and it is consistent with analytic studies already reported for the thermodynamic limit.

1. Introduction

There has been a great amount of work on systems possessing a mixture of ferromagnetic and antiferromagnetic bonds described by an Ising Hamiltonian hoping they can describe spin glasses. [1]

In spite of the fact that in the thermodynamic limit such a simple model does not produce a spin glass, several connected problems have made substantial progress based on a similar description. One example is the development of neural networks.[2] Another example is the mathematical description of frustration by means of topology.[3] In the present article we want to go back to the original problem, attempting exact solutions for small systems and discussing the extrapolations towards the thermodynamic limit.

We consider finite square lattices, consisting of N spins distributed in rectangular arrays, therefore prime values of N are lef out. Nearest neighbors interact ferromagnetically or antiferromagnetically with equal magnitude for the strength of either interaction. The abundances of the two kind of bonds can be varied in general. However the most interesting case from the point of view of frustration occurs for equal number of ferromagnetic and antiferromagnetic bonds,[4] which will be a common property for all systems to be studied below.

P. Jena et al. (eds.), Physics and Chemistry of Finite Systems: From Clusters to Crystals, Vol. I, 813–818.

We shall concentrate here on the ground state properties of small square Ising lattices. Due to space limitations we have left out of the present article other results and developments for square lattices as well as the complete analysis for triangular lattices. In particular, we would like to stress that besides computer calculations there was a parallel analisis based on topological techniques that show advantages in dealing with frustration.

In Section II we perform a presentation of a few basic theoretical definitions. In Section III we report the results for several ground state properties of small square lattices. In Section IV we discuss the results and compare them with both analityc results valid for the thermodynamic limit and approximate solutions found for large systems.

2. Theory

Let subindexes (i, j, etc.) indicate the site at the lattice. Then the z-component of the spin at site i will be denoted by S_i . The exchange interaction of this spin and the one at site j is given by the Ising term $J_{ij}S_iS_j$. We will restrict our calcultions to the case in which sites i *and* j are nearest neighbors. By adding the contributions of all such pairs we can write the total Ising Hamiltonian of the sistem in the form:

$$H = \sum_{i<j} J_{ij} S_i S_j \tag{1}$$

A state is defined as an array of signs, corresponding to the orientations of the spins. It is obvious that the Hamiltonian above is invariant with respect to an inversion of all spins in the system. This means that any state is degenerate with another one in which all signs are reversed. This trivial degeneracy will be left out of the present analysis. In particular we will talk about singlets when a level is shared by just one state and its reversed associate.

The main variable for most of the analysis below is the size of the system, given by the total number of spins N which we will vary between 4 and 40. The"shape" is defined as a particular rectangular array associated to a given size. Thus, for the systems of size 24, there are three different shapes 2×12, 3×8 and 4×6.

For a system of given size and shape there is a fairly large number of combinations in which both types of exchange interactions or "bonds" can be distributed. Each way of distributing N positive bonds and N negative bonds will be referred to as "sample" and will be considered as fixed or frozen during the calculations. Physical properties of samples of a given size and shape can be quite different among them. Since we are interested in the tendency of the systems towards the thermodynamic limit the particular deviations due to particular samples are of no inteterest here. In order to avoid such deviations we will report statistical results based on 500 samples for given size and shape. The reason for choosing this particular number will be discussed later on.

Two important usual simplifications to this problem have been considered. First, we assume equal strength for ferromagnetic and antiferromagnetic interactions; in arbitrary energy units J=−1 for ferromagnetic interaction while J=+1 for antiferromagnetic interaction. Second, we assume periodic boundary conditions to deal with spins and bonds at the borders. Actually, we have also performed studies using antiperiodic boundary conditions for the same collection of 500 samples obtaining the same statistical results in spite of strong variations of the results for particular samples.

In the present paper we will report the following results : a) Ground state energy per bond ϵ_g ; b) Degeneracy of the ground level W ; c) Site correlation parameter q. ; and d) Stability of the statistical results considering up to 500 samples.

3. Results and Discussion

We have performed exact calculations by means of computer programs that flip the spins, going through all of the lower energy portion of the Hilbert space using an algorithm that minimizes the computer time. In spite of this, CPU time grows rapidly as the systems get larger which is a practical limitation of all exact calculations. This is the reason to stop at N=40 at the present stage, which is enough to draw the conclusions given below.

The ground state energy per bond, ϵ_g, is given in Fig. 1 as a function of N. The dependence on shape is enhanced by using different symbols to represent the dimension with lower number of spins. Thus all systems having two spins on one dimension (2×2, 2×3,...etc.) are represented by plus signs; all systems having 3 spin on the shorter side (3×3, 3×4, ...etc.) are represented by triangles, and so on. Several comments are in order. First, the shape has an important role on ϵ_g; systems with two spins on one dimension saturate far from the Monte Carlo lower limit –0.70. [4,5,6,7] Second, very small systems, namely $N < 25$, depart significantly from such a lower bound value. Third, for $N \geq 25$ a saturation value of about –0.68 is apparent which is in reasonable agreement with asymptot at –0.70. Fourth, the ideal lower bound limit of –0.75[4] is not reached by small samples.

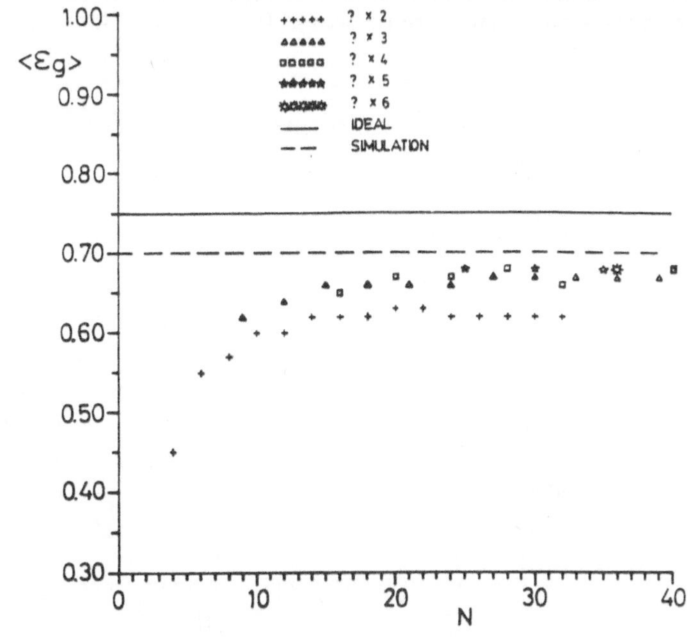

Fig.1. Ground state energy per bond ϵ_g as a function of the size N of the lattice. (Average over 500 randomly generated samples).

The degeneracy of the ground level W is a measure of the number of different points in Hilbert space capable of freezing physical properties of the system in different low temperature phases. In Fig. 2 we present the dependence of W with respect to the size and shape of the lattices. The size matter the most: as the systems grow W increases rapidly. The dependence on shape is rather surprising: systems where the shorter side has an even number of spins are more degenerate than systems where the shorter side has and odd number of spins. This fact is so clear in Fig. 2 that systems of the kind $2\times?$ show the larger degeneracy, just opposite to systems $3\times?$ that show the lower degeneracy. Systems of the kind $4\times?$ stay close to those $2\times?$, while slightly over the kind $3\times?$ we find the systems $5\times?$ and 6×6 lays in between.

The site correlation parameter q can be defined in several diferent ways. [1] We adopt the traditional manner:

$$q = \frac{\sum_{i=j}^{N} < S_i^{\alpha} S_j^{\beta} >}{N} \quad , \tag{2}$$

where <> corresponds to the average over all the possible states α and β degenerate in the ground level in a way that each pair is counted just once. Obviously for a singlet $q = 1.0$ which is characteristic of ferromagnetic, antiferromagnetic and ideal spin-glass behavior. For the systems under present consideration q maximizes at about 0.76 decreasing slowly as N grows. The family of systems $2\times?$ separates out from the general trend, showing less site correlation and a quicker descent. Due to the increase in the computer time required to calculate q as the systems grow, we report results up to $N = 30$ only.

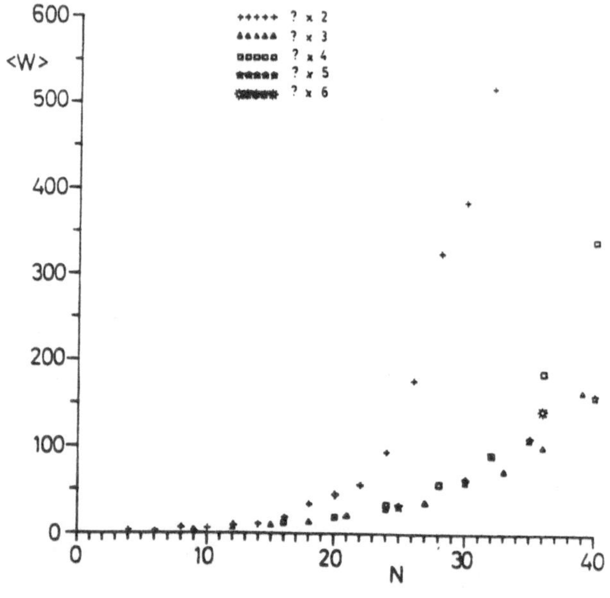

Fig. 2. Degeneracy of the ground level W as a function of the the size of the lattice. (Average over 500 randomly generated samples).

Let us finally examine the stability of the statistical results.In Fig. 4 we present the progression of two quantities as the number of samples increases from 1 to 500 for arrays 4×10. On the ordinates at the left-hand side axis we can read the percentage of singlets S_P while on the right-hand side axis we report the average degeneracy W of the ground level. Both magnitudes stabilize over 300. Other physical magnitudes oscillate less than S_P or W. We conclude that a statistic over 500 samples is enough to define stable results.

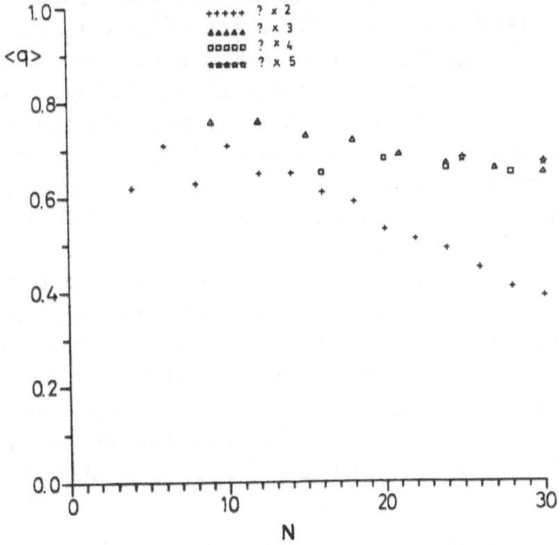

Fig. 3. Site correlation parameter q as a function of the the size of the lattice. (Average over 500 randomly generated samples).

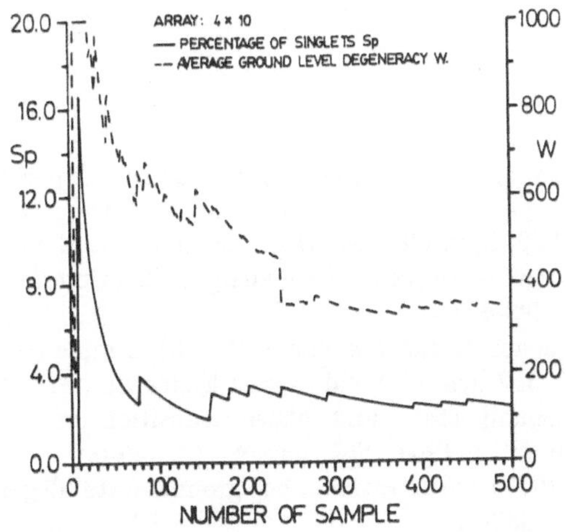

Fig.4. Stability of the number of singlets S_P and degeneracy W for the array 4×10 with respect to the increase of the number of samples up to a maximum of 500.

4. Conclusions

The size of small lattices clearly determines their properties. It can be risky to draw general conclusions towards the thermodynamic limit based on results valid for very small samples. Asymmetric shapes (such as strips) also matter significantly even for fairly large systems. A different asymptotic behavior is observed in accordance to the minimum number of spins on a particular side of the lattice as can be observed from Fig. 1 and Fig. 3.

However there is a tendency to stability of the different results as the number of spins approaches 40. By looking at the most symmetric arrays we can observe that the energy per bond tends to a clear saturation to a value of −0.68 in Fig. 1, which is consistent with the value on −0.70 obtained by Monte Carlo simulations in large systems.

The degeneracy of the ground level increases with size as can be expected from the sole increase of the total Hilbert space. A more elaborate approximation considering the remnant entropy of the system (not developed here) shows that the thermodynamics of the system is still well defined as N increases.

The site correlation parameter q presents a maximum for N around 9 to 12 decreasing slowly towards larger lattices. It is not clear whether an asymptotic behavior will be reached for N large enough. The CPU time involved in such calculations lead us to stop at N=30. Very recently we have defined a more sensitive site memory parameter p which is also simpler to calculate. Initial results for p show a sharp decrease as a function of N which is consistent with the expected behavior in the thermodynamic limit. This and other developments related to triangular lattices will be soon submitted for publication.

Acknowledgements. The authors are grateful to FONDECYT (Contract 90-0040), Dirección de Investigación y Desarrollo de la Universidad de La Frontera and Dirección de Investigación de la Universidad de Concepción for partial support.

References.

[1] Binder, K. and Young, A.P. (1986) 'Spin glasses: Experimental facts, theoretical concepts, and open questions', Reviews of Modern Physics **58**, 801-976.

[2] Morgenstern I. (1986)'Spin-Glasses, Optimization and Neural Networks', in J.L van Hemmen and I. Morgenstern (eds.), Heidelberg Colloquium on Glassy Physics, Springer Verlag, Berlin, pp. 399-427.

[3] Derrida B., Vannimenus J. and Pomeau Y. (1978) 'Simple frustrated systems: Chains, strips and squares', J. Phys. C: Solid State Physics **11** , 4749-65.

[4] Ono,I. (1976) 'Ground state and phase transition in the quenched mixture of ferro-antiferromagnets', J. Phys. Soc. Japan **41** , 345-6.

[5] Kirkpatrick, S. (1977) 'Frustration and ground-state degeneracy in spin glasses', Phys. Rev. B **16** , 4630-41.

[6] Wang, J.-S and Swendsen, R.H. (1988) 'Low-temperature properties of the $\pm J$ Ising spin glass in two dimensions' Phys. Rev. B **38** , 4840-4.

[7] Vogel E.E., Vargas, P. and Pérez, G.(1989) 'Magnetic phases in Ising square lattices with mixed bonds', Phys. Rev. B **40**, 4369-74.